NUTRITION *for* FOODSERVICE *and* CULINARY PROFESSIONALS

NUTRITION *for* FOODSERVICE *and* CULINARY PROFESSIONALS

KAREN EICH DRUMMOND | **LISA M. BREFERE**

Ed.D., R.D., L.D.N., F.A.D.A., F.M.P.

C.E.C., A.A.C.

8th EDITION

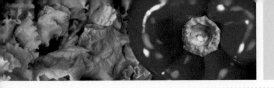

CONTENTS

PART 2 · **BALANCED COOKING AND MENUS**

PART 3

APPLIED NUTRITION

Nutrition for Foodservice and Culinary Professionals, Eighth Edition is written for students in culinary programs, as well as those in hotel, restaurant, and onsite management programs. Practicing culinary and management professionals will find it useful as well. As with previous editions, this is meant to be a practical how-to book tailored to the needs of students and professionals.

Nutrition is constantly in the news, with reports on overweight Americans, vegetarian eating, school lunch, and many more topics being streamed on the Internet 24/7 and reported by the media as well. Hectic lifestyles force many to eat out or get take-out meals at least several times a week. The typical American purchases a meal, whether breakfast, lunch, dinner, or snack, from a foodservice operation about three to five times a week. As a foodservice professional, you have a responsibility to your clients to understand contemporary cooking techniques that are balanced, moderate in rich ingredients, well prepared, and, of course, great tasting. You have a captive audience of people who depend on the chef, cooks, and foodservice employees to prepare nutritious food for them with the limits and balance they require to maintain their current lifestyles.

This book is written to help you use nutritional principles to evaluate and modify menus and recipes, as well as to respond knowledgeably to customers' questions and needs. As in the previous editions, co-author Lisa Brefere, C.E.C., A.A.C., lends her firsthand experiences applying nutrition to selecting, cooking, and menuing healthy foods in restaurants and foodservices. After all, we eat foods, not nutrients!

WHAT'S NEW FOR THE EIGHTH EDITION

Many important changes and additions have been made to *Nutrition for Foodservice and Culinary Professionals* to make this text even more accessible and practical. Among the most significant changes are the following:

- The *Eighth Edition* includes a discussion of the *Dietary Guidelines for Americans, 2010,* and ChooseMyPlate.gov graphic and key content.

- *WileyPLUS*, Wiley's online teaching and learning environment, is now available for this textbook. *WileyPLUS* is a research-based online environment for effective teaching and learning. *WileyPLUS* builds students' confidence because it takes the guesswork out of studying by providing a clear roadmap and many opportunities to practice and apply concepts.

- More photos, charts, and recipes are used to effectively convey nutrition concepts and applications in a visual manner.

- A new section, called **Culinary Focus**, is included in Chapters 3 to 7. Culinary Focus examines each food group from the perspective of a chef, from picking your ingredients to putting a new dish on the menu.

- The chapters on balanced cooking and menus (Part Two) have been thoroughly updated and expanded with more examples and tips. Recipe makeovers, which were previously

1. In lactose intolerance, lactase is deficient so lactose (milk sugar) is not split into its components in the small intestines. Instead, it travels to the colon where it attracts water and causes bloating and diarrhea. In addition, intestinal bacteria ferment lactose and produce gas. Symptoms usually occur with 30 minutes to 2 hours and clear up within 2 to 5 hours.
2. Lactose intolerance is an inherited problem and is especially prevalent among Asians, Native Americans, African Americans, and Latinos.
3. Treatment for lactose intolerance includes a diet limited in lactose (present in dairy and added to some other foods), use of lactose-free milk and milk products and/or lactase, and consuming small servings of dairy with a meal and/or dairy products lower in lactose (such as hard cheeses and yogurt) as tolerated.

Summary Designed to help students focus on the important concepts within each chapter, a summary is given after each section within a chapter.

Chef's Tips Chef's Tips provide an experienced chef's advice on all aspects of cooking, including which foods go together, how to use foods' natural colors to create an attractive dish, and how to use culinary techniques to create healthy and delicious dishes.

VELOUTÉ SAUCE

CHEF'S NOTES

Velouté sauce is a classic mother sauce, used in a variety of preparations as a base for sauces, soups, and stews. It is made with a flavorful stock, usually chicken and fish, and thickened with a traditional butter and flour, cooked blond roux. The flavor and mouthfeel of this sauce is critical to a variety of popular dishes, including fricassee, a la king sauces, fish and oyster stew, as well as soups such as Billy Bi soup (Cream of Mussels), cream of chicken, mushroom, asparagus, and broccoli soup.

This alternative sauce can be the foundation of a lighter cooking style. A good, defatted stock with a lighter thickening option can be an accepted alternative to this classic sauce with extremely favorable results using high-quality fresh ingredients. This base sauce can be used in a variety of cooking applications, in sauce making (such as seven onion sauce, marsala, horseradish, and curry) or in soup preparation (such as corn and fish chowders, shrimp and lobster bisque, or cream of broccoli and spinach). Selections using this balanced sauce, such as pot pies; winter vegetable stews; veal and wild mushroom; pork with chilies, new potatoes, onions, and peppers; snapper or shellfish stews create interesting choices while still maintaining balance.

CULINARY FOCUS: GRAINS AND LEGUMES

Most Americans need to get more whole grains and legumes such as beans into their diet. Both whole grains and legumes are very nutritious, contain little fat, and are filling due to their fiber content. This section will help you highlight dishes with grains and legumes on the menu.

Culinary Focus Culinary Focus, which replaces "Food Facts," examines various food groups from the perspective of a chef. Organized into Product, Preparation, and Menuing and Presentation, Culinary Focus is full of tips for you to use to produce tasty and healthy menu items. You will find this feature in Chapters 3-7.

HOT TOPIC
ALTERNATIVE SWEETENERS

Hot Topics Hot Topics promote critical thinking and discussion forums on current issues related to nutrition, including functional foods and phytochemicals, gluten-free cooking, and sustainable foods. Every chapter contains a Hot Topic.

The introduction of diet soda in the 1950s sparked the widespread use of alternative sweeteners, substitutes for sugar that provide no, or almost no, kcalories. If you drink diet soda, look at the food label and see which alternative sweeteners are present. The following alternative sweeteners are approved for use by the Food and Drug Administration: saccharin, aspartame, acesulfame potassium, sucralose, and neotame. The only one that contains kcalories is aspartame—but because so little is used, the kcalories are close to 0. Besides offering virtually no kcalories, alternative sweeteners are beneficial because they do not cause tooth decay or force insulin levels to rise as do added sugars such as high-fructose corn syrup.

Because they are considered food additives, the FDA requires that they be tested for safety before going on the market. The FDA uses the concept

Approved Alternative Sweeteners

SACCHARIN

Saccharin, discovered in 1879, has been consumed by Americans for more than 100 years. Its use in foods increased slowly until the two World Wars, when its use increased dramatically due to sugar shortages. Saccharin is about 200 to 700 times sweeter than sucrose and is excreted unchanged directly into the urine. It is approved for use at specific maximum amounts in foods and beverages and as a tabletop sweetener. Known as Sweet'N Low or Sweet Twin, it is sold in liquid, tablet, packet, and bulk form. Because saccharin leaves some consumers with an aftertaste, it is frequently combined with other alternative sweeteners, such as aspartame.

ASPARTAME

Aspartame is approximately 160 to 220 times sweeter than sucrose and has an acceptable flavor with no bitter aftertaste. It is marketed under the brand names NutraSweet and Equal. Aspartame is approved as a general-purpose sweetener and is found in diet sodas, cocoa mixes, pudding and gelatin mixes, fruit spreads and toppings, and other foods. If you drink diet soft drinks, chances are, they are sweetened with aspartame. Fountain-made diet soft drinks are more commonly sweetened with a blend of aspartame and saccharin, because saccharin helps provide increased stability.

Aspartame breaks down during prolonged heating and starts to lose its sweetness. For stovetop cooking, it is best to add aspartame at the end of cooking or after removing the food from the heat. For baking, it is best to use aspartame with regular sweeteners such as brown sugar in specially sug-

Check-Out Quiz At the end of each chapter, a Check-Out Quiz allows students to check their comprehension of the chapter's concepts. Answers to these quizzes are found in Appendix C.

Nutrition Web Explorer This feature encourages students to visit specific websites in order to learn more about a wide variety of topics. Students are usually asked to complete a specific, written assignment for each website.

In the Kitchen In the Kitchen is a new feature found in all chapters that brings the content of the text directly into the kitchen. Students are first asked to prepare menu items, such as low kcalorie foods in the chapter on weight management. Then students complete a brief written assignment based on the topic and recipes. The recipes and assignment sheets are in the Instructor's Manual.

Glossary All key terms and definitions are listed in the glossary, easily found in the back of the book.

Appendices A very useful reference for readers, the appendices include a variety of useful information, including serving sizes for MyPlate food groups and body mass index charts. Additional appendix materials on the nutritive value of foods, food patterns, Dietary Reference Intakes, sample menus, and growth charts for children and adolescents can also be found on the Book Companion Website: www.wiley.com/college/drummond.

MEDIA AND SUPPLEMENTS

WileyPLUS, Wiley's online teaching and learning environment, integrates the entire digital textbook with the most effective instructor and student resources to accommodate every learning style.

With *WileyPLUS*:

- Students achieve concept mastery in a rich, structured environment that's available 24/7.

- Instructors personalize and manage their courses more effectively with assessment, assignments, grade tracking, and more.

WileyPLUS can complement the current textbook or replace the printed text altogether.

For Students

Different learning styles, different levels of proficiency, different levels of preparation—each student is unique. *WileyPLUS* empowers each of them to take advantage of individual strengths:

- Students receive **anytime/anywhere access** to resources that address their demonstrated needs, and they get immediate feedback and remediation when needed.

- Integrated multimedia resources—including **videos, animations, worksheets, interactive activities such as Create-a-Plate and Revise-a-Recipe**, and much more—offer multiple study paths to fit each student's learning preferences and encourage more active learning.

- Students can take advantage of many opportunities for self-assessment linked to relevant portions of the text. They can take control of their own learning and practice until they master the material.

For Instructors

WileyPLUS empowers instructors with tools and resources that make teaching even more effective:

- Instructors can customize their classroom presentations with a wealth of resources and functionality, from PowerPoint slides to a database of rich visuals. They can even add their own materials to their *WileyPLUS* courses.

- Instructors can identify students who are falling behind and intervene accordingly, without waiting to see them during office hours.

- Instructors can take advantage of the program's pre-built assignments and extensive question bank, therefore simplifying and automating such tasks as student performance assessment, creating assignments, scoring student work, keeping grades, and more.

Create-a-Plate and Revise-a-Recipe

Create-a-Plate interactive exercises help students create their own virtual plate of food and also provide those guidelines and hints on how to make a healthy meal. Students can build a plate by selecting a menu and seeing real-time nutritional analysis based on their selections. As modifications are made, nutritional information adjusts, helping to create a balanced meal. In addition to creating balanced meals, students can use this exercise to create meals with specific guidelines, such as low in kcalories or high in fiber or protein.

Through the *Revise-a-Recipe* interactive exercises, students are provided recipes to choose from, and then immediately can see how the recipe's nutritional values change as ingredients are adjusted or substituted based on the goals of the activity.

SUPPLEMENTARY MATERIALS

A *Study Guide* (ISBN: 978-1-118-50721-6) for students is available to help reinforce nutrition concepts and allow students to make nutrition applications.

A *Companion Website* (www.wiley.com/college/drummond) provides links to both the Student and Instructor websites. The **Student Website** includes self-tests as well as a number of interactive activities that are designed to reinforce key concepts from each chapter. The Student Website also includes supplementary recipes.

From the **Instructor Website**, instructors can download the *Instructor's Manual* as well as PowerPoint slides, *Study Guide* solutions, test bank questions and answers, and student worksheets and activites for each chapter. A selection of bonus recipes is also available here.

An *Instructor's Manual* (ISBN: 978-1-118-50696-7) that includes class outlines, student worksheets, key terms and definitions, In the Kitchen recipes and activities, and test questions and answers is available. Please contact your Wiley representative for a copy or visit the companion website to download a copy.

ACKNOWLEDGMENTS

We are grateful for the help of all the educators who have contributed to this and previous editions through their constructive comments.

Chuck Becker, Pueblo Community College, CO

Marian Benz, Milwaukee Area Technical College, WI

Benjamin Black, Trident Technical College, SC

Alex Bladowski, North Georgia Technical College, Currahee Campus

Cynthia Chandler, Sullivan University, KY

Nicole Dowsett, The Art Institute of Charlotte, NC

William J. Easter, Des Moines Area Community College, IA

Collen Engle, Sullivan University, KY

Jo Anne Garvey, Metropolitan Community College, NE

Dona Greenwood, Florida International University

Keith E. Gardiner, Guilford Technical Community College, NC

Julienne M. Guyette, Atlantic Culinary/McIntosh College, NH

Chef Herve Le Biavant, California Culinary Academy

Marjorie Livingston, The Culinary Institute of America, NY

Debra Macchia, College of DuPage, IL

Jacinda Martin, Niagara College, ON, Canada

Aminta Martinez Hermosilla, Le Cordon Bleu College of Culinary Arts—Las Vegas, NV

Kevin Monti, Western Culinary Institute, OR

Renee Reagan-Moreno, California Culinary Academy

Mary L. Rhiner, Kirkwood Community College, IA

Richard Roberts, Wake Technical Community College, NC

Vickie S. Schwartz, Drexel University, PA

Joan Vogt, Kendall College, IL

Donna Wamsley, Hocking Technical College, OH

Wes Wilkinson, Algonquin College, ON, Canada

Jane Ziegler, Cedar Crest College, PA

WileyPLUS

WileyPLUS is a research-based online environment for effective teaching and learning.

WileyPLUS builds students' confidence because it takes the guesswork out of studying by providing students with a clear roadmap:

- **what to do**
- **how to do it**
- **if they did it right**

It offers interactive resources along with a complete digital textbook that help students learn more. With *WileyPLUS*, students take more initiative so you'll have greater impact on their achievement in the classroom and beyond.

Now available for

Bb
Blackboard

WileyPLUS

ALL THE HELP, RESOURCES, AND PERSONAL SUPPORT YOU AND YOUR STUDENTS NEED!

www.wileyplus.com/resources

1st DAY OF CLASS ...AND BEYOND!

2-Minute Tutorials and all of the resources you and your students need to get started

WileyPLUS

Student Partner Program

Student support from an experienced student user

Wiley Faculty Network

Collaborate with your colleagues, find a mentor, attend virtual and live events, and view resources
www.WhereFacultyConnect.com

WileyPLUS

Quick Start

Pre-loaded, ready-to-use assignments and presentations created by subject matter experts

Technical Support 24/7
FAQs, online chat, and phone support
www.wileyplus.com/support

Your *WileyPLUS* Account Manager, providing personal training and support

PART

1

FUNDAMENTALS OF NUTRITION & FOODS

INTRODUCTION TO NUTRITION

CHAPTER 1

3

- Explain what nutrition is and why it should be important to you on a personal level and as a culinary/foodservice professional.

- Identify three food groups we don't eat enough of and two food groups we eat too much of.

- Define flavor and explain how it involves all five senses.

- Discuss five factors that influence what you eat.

- Define kilocalories, identify the three factors that influence the number of kcalories you use every day, and explain the effect of the following on basal metabolic rate: gender, age, exercise, and growth.

- Name the six classes of nutrients and their characteristics.

- Give two examples of foods that are nutrient dense and two that are empty kcalorie foods and explain why you chose these foods.

- Describe four characteristics of a nutritious diet.

- Identify a given food as a whole food, processed food, enriched or fortified foods, and/or organic food.

- Explain what is meant by Recommended Dietary Allowance, Adequate Intake, and Tolerable Upper Intake Level of a nutrient.

- Explain how food is digested and absorbed in the gastrointestinal tract.

- To run a sustainable facility, list five steps chefs are taking in the kitchen and five steps managers are taking in the dining room and production areas.

NUTRITION AND YOU

Americans are fascinated with food: choosing foods, reading newspaper articles on food, finding recipes online, preparing and cooking foods, checking out new restaurants, and, of course, eating foods. Why are we so interested in food? Of course, eating is fun, enjoyable, and satisfying, especially when we are eating with other people whose company we like. You probably know that eating the right foods and eating the right amounts of foods are part of a healthy lifestyle. A healthy lifestyle also involves maintaining a healthy weight, being physically active, getting enough sleep, and not smoking.

Nutrition A science that studies nutrients and other substances in foods and in the body and the way those nutrients relate to health and disease. Nutrition also explores why you choose particular foods and the type of diet you eat.

Nutrients The nourishing substances in food that provide energy and promote the growth and maintenance of your body.

Diet The food and beverages you normally eat and drink.

Nutrition is a science that studies nutrients (such as protein or vitamin C) found in foods and the body. Nutrition is important because what you eat can affect your health. Almost daily you are bombarded with news reports that something in the food you eat, perhaps a nutrient such as sugar, may not be good for you—that it may indeed cause or complicate conditions such as diabetes or heart disease. Nutrition researchers look closely at the relationships between nutrients and disease, as well as the processes by which you choose what to eat and the balance of foods and nutrients in your diet.

Nutrients are the nourishing substances in food that give you energy, allow your body to grow, and keep you feeling healthy. They help regulate many processes that go in your body, such as the beating of your heart or the digesting of food in your stomach. Examples of nutrients include carbohydrates, fats, protein, water, and vitamins. In summary, nutrition is a science that studies nutrients and other substances in foods, and how they affect the body, especially in terms of health and disease. Nutrition also explores why you choose the foods you do and the type of **diet** you eat.

Diet is a word that has several meanings. Anyone who has tried to lose weight has no doubt been on a diet. In this sense, diet means a weight-reducing diet and is often thought of in a negative way. But a more general definition of diet is the foods and beverages you normally eat and drink every day. Of course, your normal diet may change, such as when you started college and had new places to eat.

Your lifestyle choices, including diet, strongly influence whether you might get heart disease, cancer, and stroke—the three biggest killers in the United States. Your genetics and environment can also put you at greater risk for disease. Table 1-1 takes a look at diet-related diseases and some of their risk factors. A risk factor is anything that affects your chance of get-

TABLE 1-1	A Look at Diet-related Diseases
Cardiovascular disease	• 37 percent of Americans have cardiovascular disease. • Major risk factors include high levels of blood cholesterol, overweight and obesity, high blood pressure, physical inactivity, type 2 diabetes, and smoking.
High blood pressure (hypertension)	• 34 percent of American adults have high blood pressure. • 36 percent of American adults have blood pressure numbers that are higher than normal, but not yet in the high blood pressure range. • Dietary factors that increase blood pressure include excessive sodium intake, overweight and obesity, and excess alcohol consumption. • High blood pressure is a risk factor for heart disease, stroke, heart failure, and kidney disease.
Cancer	• Almost one in two men and women will be diagnosed with cancer during their lifetime. • Dietary factors are associated with risk of some types of cancer, such as breast (post-menopausal), colon, kidney, mouth, and esophagus.
Diabetes	• Almost 11 percent of Americans ages 20 years and older have diabetes. The vast majority of cases are type 2 diabetes, which is heavily influenced by diet and physical activity. • About 35 percent of American adults have pre-diabetes. Pre-diabetes means that blood glucose levels are higher than normal, but not high enough to be called diabetes.

Source: US Department of Agriculture and US Department of Health and Human Services. *Dietary Guidelines for Americans, 2010,* 7th ed. (Washington, DC: US Government Printing Office, December 2010).

ting a disease. For example, if you get little to no physical activity (such as walking), you are at a higher risk of getting heart disease.

Poor diet and physical inactivity are the most important factors contributing to an epidemic of overweight and obesity in this country. The most recent data indicate that 73 percent of men and 64 percent of women are overweight or obese, and 32 percent of children and adolescents ages 2 to 19 years are overweight or obese.

If you are overweight or obese, you have an increased risk of many health problems, such as type 2 diabetes and heart disease. These increased health risks are not limited to adults. More children and adolescents are being diagnosed with type 2 diabetes and high blood pressure than in the past. Preventing obesity in childhood is an important way to combat and reverse the obesity epidemic, because overweight children are more likely to be overweight or obese as adults than normal-weight children.

Eating healthy can help you maintain a healthy body weight as well as reduce your risk of heart disease, high blood pressure, diabetes, and several types of cancer. As described in the *Dietary Guidelines for Americans, 2010,* eating healthy means focusing on foods such as the following:

- Vegetables
- Fruits
- Beans and peas
- Whole grains (such as whole wheat bread or oatmeal)
- Fat-free or low-fat milk and milk products
- Lean meats and poultry
- Seafood
- Nuts and seeds

Healthy eating also includes eating minimal solid fats (such as butter), sugars, or sodium (found in salt). Figure 1-1 shows how typical American diets compare to the *Dietary Guidelines for Americans.* Unfortunately, we eat too few whole grains, fruits, vegetables, seafood, and dairy, and we eat too many foods high in solid fat and sugar (such as cookies and most sweets), as well as too many foods high in sodium.

Now that you know that good nutrition, as part of a healthy lifestyle, is beneficial for your long-term health, we can look at why good nutrition is important for foodservice and

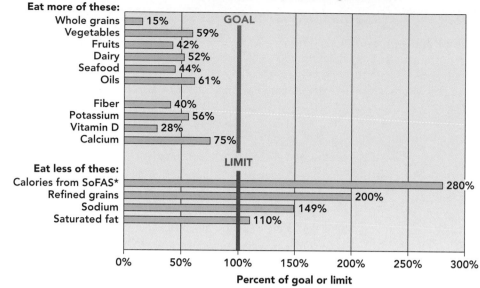

FIGURE 1-1 How American diets compare to recommendations.

Source: US Department of Agriculture and US Department of Health and Human Services. *Dietary Guidelines for Americans, 2010,* 7th ed. (Washington, DC: US Government Printing Office, December 2010).

Usual intake as a percent of goal or limit

Eat more of these:
- Whole grains: 15%
- Vegetables: 59%
- Fruits: 42%
- Dairy: 52%
- Seafood: 44%
- Oils: 61%

- Fiber: 40%
- Potassium: 56%
- Vitamin D: 28%
- Calcium: 75%

Eat less of these:
- Calories from SoFAS*: 280%
- Refined grains: 200%
- Sodium: 149%
- Saturated fat: 110%

Percent of goal or limit

*SoFAS-solid fats and added sugars.

culinary students. Even though restaurants sell a lot of big hamburgers and fries, the National Restaurant Association reported in 2011 that 71 percent of adults are trying to eat healthier at restaurants than two years earlier. Research from the NPD Group shows that consumers have indeed been ordering fewer foods high in sugar content (such as carbonated soft drinks) and high in fat (such as hot dogs). At the same time, more consumers are selecting healthier foods such as grilled chicken, fruit, and yogurt. Consumers who seek healthy options are looking for fresh and nutritious ingredients.

When professional chefs, all members of the American Culinary Federation, were asked to identify the top 20 trends in foods, beverages, cuisines, and culinary themes for the National Restaurant Association's "What's Hot in 2012" survey for full-service restaurants, nutrition/health was one of the top trends. Other related trends included the following:

- Gluten-free/food-allergy conscious
- Sustainability and locally grown and sourced foods
- Children's nutrition and healthful kids' meals (Figure 1-2) with whole grains, fruits, and vegetables
- Restaurant gardens

FIGURE 1-2 Feeding children nutritious meals is a trend in foodservice.

Courtesy of Monkey Business Images/Shutterstock.

Surveys for quickservice restaurants showed similar trends. This is significant, considering that the average consumer spends 49 percent of their food budget in restaurants.

Understanding good nutrition is important for you on both a personal and a professional level. This introductory chapter explores why we choose the foods we eat and then explains several important nutrition concepts that build a foundation for the remaining chapters. You will learn more about nutrients, kcalories, empty-kcalorie foods, characteristics of a nutritious diet, and how to recognize whole foods, processed foods, and organic foods in a grocery store.

SUMMARY

1. Nutrition is a science that studies nutrients and other substances in foods and in the body and the way those nutrients relate to health and disease. Nutrition also explores why you choose particular foods and the type of diet you eat. Nutrients are the nourishing substances in food that provide energy and promote the growth and maintenance of your body.
2. Eating healthy can help reduce your risk for heart disease, high blood pressure, diabetes, and several types of cancer, as well as help you maintain a healthy body weight.
3. Americans eat too few whole grains, fruits, vegetables, seafood, and dairy, and we eat too many foods high in solid fat and sugar (such as many desserts) as well as processed foods high in sodium.
4. Many adults are trying to eat healthier at restaurants by ordering fewer foods high in sugar and/or fat. Consumers are looking for fresh and nutritious ingredients such as grilled chicken, fruit, and yogurt.

WHY DO YOU EAT THE FOODS YOU DO?

Think about what you ate for your last meal yesterday. Did you eat at your job, at home, or out with friends? Were you making food choices based on cost or convenience, taste, or simply what foods are familiar to you? As you can see from this list, many factors influence what you eat:

- Flavor
- Other aspects of food (such as cost, convenience, nutrition)
- Demographics
- Culture and religion
- Health
- Social and emotional influences
- Marketing and the media
- Environmental concerns

Now we will look at these factors in depth.

FLAVOR

The most important consideration when choosing something to eat is the taste of the food (Figure 1-3). You may think that taste and flavor are the same thing, but taste is actually a component of flavor. **Flavor** is an attribute of a food that includes its taste, smell, feel in the mouth or texture, temperature, and even the sounds made when it is chewed. Flavor is a combination of all five senses: taste, smell, touch, sight, and sound. The taste buds in your mouth and the smell receptors in your nose work together to deliver signals to the brain that are translated into the flavor of food.

Flavor An attribute of a food that includes its taste, smell, feel in the mouth, texture, temperature, and even the sounds made when it is chewed.

FIGURE 1-3 The most important consideration when choosing something to eat is taste.

Courtesy of Anna Omelchenko/Shutterstock.

Taste Sensations perceived by the taste buds on the tongue.

Taste buds Clusters of cells found on the tongue, cheeks, throat, and roof of the mouth. Each taste bud houses 60 to 100 receptor cells that bind food molecules dissolved in saliva and alert the brain to interpret them.

Umami A taste often referred to as "savory" that is characteristic of monosodium glutamate and is associated with meats, mushrooms, tomatoes, Parmesan cheese, and other foods. It is a basic taste along with sweet, sour, salty, and bitter.

Taste

Taste comes from 10,000 **taste buds**—clusters of cells that resemble the sections of an orange. Taste buds, found on the tongue, cheeks, throat, and roof of the mouth, house 60 to 100 receptor cells each. The body regenerates taste buds about every three days.

These taste cells bind food molecules dissolved in saliva and alert the brain to interpret them. Although the tongue is often depicted as having regions that specialize in particular taste sensations—for example, the tip is said to detect sweetness—researchers know that taste buds for each sensation (sweet, salty, sour, bitter, and umami) are actually scattered around the tongue. In fact, a single taste bud can have receptors for all five sensations. We also know that the back of the tongue is more sensitive to bitter, and that food temperature can influence taste. For example, sugar seems sweeter at warmer temperatures whereas salt tastes stronger at colder temperatures.

Taste buds are most numerous in children under age six, and this might explain why youngsters are such picky eaters (Figure 1-4). We know that many children do not like bitter taste, thereby interfering with vegetable consumption. However their heightened sensitivity to bitter tastes will decrease with age and they will eventually eat more vegetables as long as they are presented with them. As for older adults, it is normal for smell and taste to gradually decline. By age 50, the number of taste buds begins to decrease, which may explain why some older people like saltier and spicier foods. Smoking and some medications also reduce the ability to taste food normally.

Umami, the fifth basic taste, differs from the traditional sweet, sour, salty, and bitter tastes by providing a savory, sometimes meaty, sensation. Umami is a Japanese word, and the taste is evident in many Japanese ingredients and flavorings, such as seaweed, dashi stock, soy sauce, and mushrooms, as well as other foods. The umami taste receptor is very sensitive to glutamate, an amino acid found in protein that occurs in foods such as meat, fish, and milk, and it is often added to processed foods in the form of the flavor enhancer monosodium glutamate (MSG). MSG is an inexpensive, intensely umami ingredient with no off-flavors. Despite the frequent description of umami as meaty, many foods, including mushrooms, tomatoes, and Parmesan cheese, have a higher level of glutamate than an equal amount of beef or pork. This explains why foods that are cooked with mushrooms or tomatoes seem to have a fuller, rounder taste than when cooked alone.

Umami flavor is strengthened when sodium is present, which explains why tomatoes have a strong taste after adding salt. Many popular sauces for cooking combine savory and salty tastes—think of ketchup, soy sauce, or fish sauce.

FIGURE 1-4 Taste buds are most numerous in children under age six, which might explain why they can be picky eaters.

Courtesy of Zurjeta/Shutterstock.

FIGURE 1-5 The heat of hot peppers is felt not by the taste buds but by pain receptors in the mouth that sense heat.

Reprinted with permission of John Wiley & Sons, Inc.

When incorporating umami ingredients such as Parmesan cheese and tomato products into recipes, chefs can reduce the fat and salt content of foods without sacrificing flavor. Chefs can also build umami flavor through cooking techniques. Any process that breaks down protein—such as drying, aging, curing, and slow cooking—increases umami because glutamate is released from protein.

If you like to eat hot chili peppers, you may wonder what kind of taste it has. The heat of chili peppers is not felt by the taste buds. The substance that makes a chili so hot is called capsaicin. Capsaicin actually binds with pain receptors in the mouth and throat that are responsible for sensing heat (Figure 1-5). When you eat, you perceive heat or cooling in the food.

Smell

If you could only taste sweet, salty, sour, bitter, and umami, how could you taste the flavor of cinnamon, chicken, or any other food? This is where smell comes in. Your ability to identify the flavors of specific foods requires smell.

The ability to detect the strong scent of a fish market, the antiseptic odor of a hospital, the aroma of a ripe melon or a glass of wine, and thousands of other smells is possible thanks to a yellowish patch of tissue the size of a quarter high up in your nose (Figure 1-6). This patch is actually a layer of 12 million specialized cells, each sporting 10 to 20 hairlike growths that bind with the smell and send a message to the brain. Our sense of smell may not be as refined as that of dogs, which have billions of olfactory cells, but we can distinguish among about 10,000 scents.

Of course, if you have a bad cold and mucus clogs up your nose, you lose some sense of smell and everything tastes bland. With a cold, you can still taste salty and sweet, but you will have a hard time distinguishing the difference between flavors.

You can smell foods in two ways. If you smell coffee brewing while you are getting dressed, you smell it directly through your nose. But if you are drinking coffee, the smell of the coffee goes to the back of your mouth and then up into your nose. To some extent, what you smell (or taste) is determined by your genetics and also your age.

FIGURE 1-6 The sense of smell and detecting the aromas in wine is the way wine is tasted.

Courtesy of Chiyacat/Shutterstock.

Touch

All foods have **texture**—think of a tender cookie or a smooth soup. The human body is very adept at evaluating a food's texture. We use not only the sense of feeling in our mouths—or **mouthfeel**, as food professionals refer to it—but also our other senses to evaluate the texture of foods. Textures can range from moist to dry, tender to tough, fluid to solid, thick to thin, gritty/rough to smooth, coarse to fine, hard to soft, crunchy to soggy. Even carbonated drinks have texture—they tingle in your mouth as you drink them.

Texture Those physical properties of food that can be felt with the tongue, mouth, teeth, or fingers—such as tender, juicy, or firm.

Mouthfeel How the texture of a food is perceived in the mouth.

Textures that most people like include crispy, crunchy, juicy, creamy, tender, and firm. Consumers generally don't like foods that are tough, crumbly, lumpy, soggy, or watery.

Whichever the texture, it influences whether you like the food and also can tell you whether the food is fresh. Think about eating a crispy cracker—when it is stale, it loses its crispiness and loses some appeal. Green beans that are overcooked lose their al dente texture and become floppy and undesirable.

The natural texture of a food may not be the most desirable texture for a finished dish, and so a cook may create a different texture. For example, a fresh apple may be too crunchy to serve at dinner, so it is baked or sautéed for a softer texture. Or a cream soup may be too thin, so a thickening agent is used to increase the viscosity of the soup or, simply stated, make it harder to pour.

FIGURE 1-7 Eye appeal is especially important for cold foods.

Reprinted with permission of John Wiley & Sons, Inc.

Sight and Sound

Appearance creates the first impression of food and strongly influences which foods you choose to eat. Color is very important—think of the eye appeal of a juicy red tomato or a nicely browned loaf of bread. Color gives us a clue about the quality of a food as well as its flavor. We make decisions about foods before we actually eat them. For example if the skin of an apple looks wrinkly, then the apple will probably have lost its crunchy bite. Qualities such as color, size, shape, consistency, and arrangement all contribute to eye appeal. Eye appeal is especially important for cold foods because they lack the come-on of an appetizing aroma (Figure 1-7). Just the sight of something delicious to eat can start your digestive juices flowing. It is certainly true that "you eat with your eyes."

The sound made when a food is eaten, such as the snap of a carrot, is also part of the enjoyment of eating. Think of the sizzle of fajitas or the crunch of a crispy cookie. These sounds also help stimulate appetite.

OTHER ASPECTS OF FOOD

Food cost is a major consideration in what foods you eat. Cost is a factor in many purchasing decisions, whether one is buying dry beans at $1.99/pound or fresh salmon at $16.99 per pound at the supermarket, or choosing where to eat out. We continually feel the pinch at the grocery store as food prices go up for a wide variety of reasons such as drought, tight supplies, or increased fuel costs.

Convenience is also a major consideration. You may not have the time or inclination to prepare meals from scratch. Instead, you can choose ready-to-eat foods, frozen dinners, precut fruits and vegetables, and baked goods. Of course, convenience foods are more expensive than their raw counterparts, and not every budget can afford them. Take-out meals are also more expensive, but very popular.

Your food choices are also affected by availability, familiarity, and habits. Whether it is a wide choice of foods at an upscale supermarket or a choice of only two eating places within walking distance of where you work or go to school, you can eat only what is available. Fresh fruits and vegetables are perfect examples of foods that are most available (and at their lowest prices) when in season. Of course, you are more likely to eat fruits and vegetables, or any food for that matter, with which you are familiar and which you have eaten before and enjoyed.

So what are some of your food habits? Do you eat cold cereal for breakfast, pizza out with friends, popcorn at the movies? Do you love ramen noodles? These are some typical eating habits of college students. Yours might be similar or very different.

The nutritional content of a food can be an important factor in deciding what to eat. You probably have watched people reading nutrition labels on a food package, or perhaps you have read nutrition labels yourself. Some consumers are looking out for kcalories, others for sodium or saturated fat content. Current estimates show that over 60 percent of Americans use nutrition information labels. Older people tend to read labels more often than younger people do.

DEMOGRAPHICS

Demographic factors that influence your food choices include age, gender, educational level, income, and cultural background (discussed next). Women and older adults tend to consider nutrition more often than do men or young adults when choosing what to eat. Older adults are probably more nutrition-minded because they have more health problems, such as heart disease and high blood pressure, and are more likely to have to change their diet for health reasons. Older adults also have more concerns with poor dental health, swallowing problems, and digestive problems. People with higher incomes and educational levels tend to think about nutrition more often when choosing what to eat.

CULTURE AND RELIGION

Culture can be defined as the behaviors and beliefs of a certain social, ethnic, or age group. A culture strongly influences the eating habits of its members. Each culture has norms about which foods are edible, which foods have high or low status, how often foods are consumed, what foods are eaten together, when foods are eaten, and what foods are served at special events and celebrations (such as weddings).

Culture The behaviors and beliefs of a certain social, ethnic, or age group.

In short, your culture influences your attitudes toward and beliefs about food. For example, some French people eat horsemeat, but Americans do not consider horsemeat acceptable to eat. Likewise, many common American practices seem strange or illogical to persons from other cultures. For example, what could be more unusual than boiling water to make tea and adding ice to make it cold again, sugar to sweeten it, and then lemon to make it tart? When immigrants come to live in the United States, their eating habits gradually change, but they are among the last habits to adapt to the new culture.

For many people, religion affects their day-to-day food choices. For example, many Jewish people abide by the Jewish dietary laws, called the Kashrut. They do not eat pork, nor do they eat meat and dairy products together. Muslims also have their own dietary laws. Like Jews, they will not eat pork. Their religion also prohibits drinking alcoholic beverages. For other people, religion influences what they eat mostly during religious holidays and celebrations. Religious holidays such as Passover are observed with appropriate foods. Table 1-2 explains the food practices of different religions.

HEALTH

Have you ever dieted to lose weight? Most Americans are trying to lose weight or keep from gaining it. You probably know that obesity and overweight can increase your risk of heart disease, diabetes, and other health problems. What you eat influences your health. Even if you are healthy, you may base food choices on a desire to prevent health problems and/or improve your appearance.

A knowledge of nutrition and a positive attitude toward nutrition might or might not translate into nutritious eating practices. Just knowing that eating lots of fruits and vegetables may prevent heart disease does not mean that someone will automatically start eating more of those foods. For most people, knowledge is not enough and change is difficult. Many circumstances and beliefs prevent change, such as a lack of time or money to eat right. But some people manage to change their eating habits, especially if they take small, realistic steps toward their goal and feel that the advantages (such as losing weight or preventing diabetes) outweigh the disadvantages.

SOCIAL AND EMOTIONAL INFLUENCES

Your food choices are influenced by the social situations you find yourself in, such as when you are eating with others. Peer pressure no doubt influences many of your food choices. Have

TABLE 1-2	Food Practices of World Religions

RELIGION	DIETARY PRACTICES
Judaism	Kashrut: Jewish dietary law of keeping kosher. 1. Meat and poultry. Permitted: Meat of animals with a split hoof that chew their cud (includes cattle, sheep, goats, deer); a specific list of birds (includes chicken, turkey, goose, pheasant, duck). Not permitted: Pig and pork products, mammals that don't have split hooves and chew their cud (such as rabbit), birds not specified (such as ostrich). All animals require ritual slaughtering. All meat and poultry foods must be free of blood, which is done by soaking and salting the food or by broiling it. Forequarter cuts of mammals are also not eaten. 2. Fish. Permitted: Fish with fins and scales. Not permitted: Shellfish (scallops, oysters, clams), crustaceans (crab, shrimp, lobster), fishlike mammals (dolphin, whale), frog, shark, and eel. Do not cook fish with meat or poultry. 3. Meat and dairy are not eaten or prepared together. Meals are dairy or meat, not both. It is also necessary to have two sets of cooking equipment, dishes, and silverware for dairy and meat. 4. All fruits, vegetables, grains, and eggs can be served with dairy or meat meals. 5. A processed food is considered kosher only if the package has a rabbinical authority's name or insignia.
Roman Catholicism	1. Abstain from eating meat on Fridays during Lent (the 40 days before Easter). 2. Fast (one meal is allowed) and abstain from meat on Ash Wednesday (beginning of Lent) and Good Friday (the Friday before Easter).
Eastern Orthodox Christianity	Numerous feast days and fast days. On fast days, no fish, meat, or other animal products (including dairy products) are allowed. They also abstain from wine and oil, except for certain feast days that may fall during a fasting period. Shellfish are allowed. Wednesdays and Fridays are also fast days throughout the year.
Protestantism	1. Food on religious holidays is largely determined by a family's cultural background and preferences. Some churches within Protestantism prohibit alcohol. 2. Fasting is uncommon.
Mormonism	1. Prohibit tea, coffee, and alcohol. Some Mormons abstain from anything containing caffeine. 2. Eat only small amounts of meat and base diet on grains. 3. Some Mormons fast once a month.
Seventh-Day Adventist Church	1. Many members are lacto-ovo vegetarians (eat dairy products and eggs but no meat or poultry). 2. Avoid pork and shellfish. 3. Prohibit coffee, tea, and alcohol. 4. Drink water before and after meals, not during. 5. Avoid highly seasoned foods and eating between meals.
Islam	1. All foods are permitted (halal) except for swine (pigs), four-legged animals that catch prey with the mouth, birds of prey that grab prey with their claws, animals (except fish and seafood) that have not been slaughtered according to ritual, and alcoholic beverages. Use of coffee and tea is discouraged. 2. Celebrate many feast and fast days. On fast days, they do not eat or drink from sunup to sundown.
Hinduism	1. Encourages eating in moderation. 2. Meat is allowed, but the cow is sacred and is not eaten. Also avoided are pork and certain fish. Many Hindus are vegetarian. 3. Many Hindus avoid garlic, onions, mushrooms, and red foods such as tomatoes. 4. Water is taken with meals. 5. Some Hindus abstain from alcohol. 6. Hindus have a number of feast and fast days.
Buddhism	1. Dietary laws vary, depending on the country and the sect. Many Buddhists do not believe in taking life, and so they are lacto-ovo vegetarians (eat dairy products and eggs but no meat or poultry). 2. Celebrate feast and fast days.

you ever noticed that there are a lot of college students who become vegetarians? Perhaps social influences during the college years have something to do with that. Even as adults, we tend to eat the same foods that our friends and neighbors eat. This is due to cultural influences as well.

Food is often used to convey social status. For example, getting your morning coffee beverage from Starbucks is certainly more upscale than picking up coffee at a convenience store. Social class, and how much money you have, definitely can influence what you choose to eat.

Emotions are closely tied to some of our food selections. As a child, you may have been given something sweet to eat, such as cake or candy, whenever you were unhappy or upset. As an adult, you may gravitate to those kinds of foods, called comfort foods, when in a stressful situation such as studying and taking final exams. Carbohydrates, such as in cake or candy, tend to have calming effects. Eating in response to emotions can lead to overeating and overweight.

MARKETING AND THE MEDIA

The food industry very much influences what you choose to eat. After all, the food companies decide what foods to produce and where to sell them. They also use advertising, product labeling and displays, and websites to sell their products.

On a daily basis, the media (television, newspapers, magazines, radio, Internet) portray food in many ways: paid advertisements, articles on food in magazines and newspapers, foods eaten on television shows, or simply a tweet on Twitter that gives you the location of your favorite food truck.

Much research has been done on the impact of television food commercials on children. Quite often, the commercials succeed in getting children to eat foods such as cookies, candies, and fast food. Television commercials are likely contributing to higher kcalorie and fat intakes.

ENVIRONMENTAL CONCERNS

There is growing concern about the environmental impacts of our food. Conventional farming/food production and transportation use considerable amounts of energy and create undesirable waste. Consumers are clamoring for sustainable food choices and environmentally friendly restaurants, and indeed the restaurant industry is responding to this with practices such as purchasing locally, using organic produce and sustainable seafood, growing gardens, and implementing green practices in the facility. See the Hot Topic at the end of this chapter for more information on some of the environmental concerns of commercial food production and how it influences food choices.

Now that you have a better understanding of why we eat the foods we do, we can look at some basic nutrition concepts and terms.

SUMMARY

1. Flavor is an attribute of a food that includes its taste, smell, mouthfeel, texture, temperature, and even the sounds made when it is cooked or chewed. Flavor is a combination of all five senses: taste, smell, touch, sight, and sound.

2. Taste buds in your mouth sense any of five basic tastes: sweet, salty, sour, bitter, and umami. Umami is a taste often referred to as "savory" that is characteristic of monosodium glutamate and is associated with meats, mushrooms, tomatoes, Parmesan cheese, and other foods.

3. Taste buds are more numerous in children under age six, which might explain why they can be picky eaters. By age 50 the number of taste buds begins to decrease, which might explain why older people like saltier and spicier foods.

4. Textures that most people like include crispy, crunchy, juicy, creamy, tender, and firm. Consumers generally don't like foods that are tough, crumbly, lumpy, soggy, or watery.

5. Qualities such as color, size, shape, consistency, and arrangement all contribute to the eye appeal of foods.

6. Table 1-3 lists factors that influence what you choose to eat.

TABLE 1-3	Factors Influencing What You Eat

- Flavor
  - Taste
  - Smell
  - Touch/texture
  - Sight and sound
- Other aspects of food
  - Food cost
  - Convenience
  - Availability
  - Familiarity/food habits
  - Nutrition
- Demographics
- Culture and religion
- Health
- Social and emotional influences
- Marketing and the media
- Environmental concerns

WHAT ARE KILOCALORIES?

Kilocalories A unit of measure used to express the amount of energy found in different foods.

Food energy, as well as the energy needs of the body, is measured in units of energy called **kilocalories**. The number of kilocalories in a particular food can be determined by burning a weighed portion of that food and measuring the amount of heat (or kilocalories) it produces. A kilocalorie raises the temperature of 1 kilogram of water 1 degree Celsius.

When you read in a magazine that a cheeseburger has 350 calories, understand that it is actually 350 kilocalories. The American public has been told for years that an apple has 80 calories, a glass of regular milk has 150 calories, and so on, when the correct term is not calories but kilocalories. This has been done in part to make the numbers easier to read and to ease calculations. Imagine adding up your calories for the day, and having most numbers be six digits long, such as 350,000 calories for a cheeseburger. This book uses the term *kilocalorie* and its abbreviations, kcalorie and kcal, throughout each chapter.

The number of kcalories you need is based on three factors: your energy needs when your body is at rest and awake (referred to as **basal metabolism**), your level of physical activity, and the energy you need to digest and absorb food. For every 100 kcalories you eat, about 10 are used for digestion, absorption, and metabolism of nutrients, our next topic.

Basal metabolism The minimum energy needed by the body for vital functions when at rest and awake.

Basal metabolic needs include energy needed for vital bodily functions when the body is at rest but awake. For example, your heart is pumping blood to all parts of your body, your cells are making proteins, and so on. Your basal metabolic rate (BMR) depends on the following factors:

1. **Gender.** Men have a higher BMR than women do because men have a higher proportion of muscle tissue (muscle requires more energy for metabolism than fat does).
2. **Age.** As you get older, you generally gain fat tissue and lose muscle tissue. BMR declines about 2 percent per decade after age 30.
3. **Growth.** Children, pregnant women, and lactating women have higher BMRs.
4. **Height.** Tall people have more body surface than shorter people do and lose body heat faster. Their BMR is therefore higher.
5. **Temperature.** BMR increases in both hot and cold environments, to keep the temperature inside the body constant.

TABLE 1-4   Kcalories per Hour Expended in Common Physical Activities

MODERATE PHYSICAL ACTIVITIES	IN 1 HOUR	IN 30 MINUTES
Hiking	370	185
Light gardening/yard work	330	165
Dancing	330	165
Golf (walking and carrying clubs)	330	165
Bicycling (less than 10 miles per hour)	290	145
Walking (3½ miles per hour)	280	140
Weight training (general light workout)	220	110
Stretching	180	90
VIGOROUS PHYSICAL ACTIVITIES	**IN 1 HOUR**	**IN 30 MINUTES**
Running/jogging (5 miles per hour)	590	295
Bicycling (more than 10 miles per hour)	590	295
Swimming (slow freestyle laps)	510	255
Aerobics	480	240
Walking (4½ miles per hour)	460	230
Heavy yard work (chopping wood)	440	220
Weight lifting (vigorous effort)	440	220
Basketball (vigorous)	440	220

Source: www.choosemyplate.gov 2012.

6. **Fever and stress.** Both of these increase BMR. The body reacts to stress by secreting hormones that speed up metabolism so that the body can respond quickly and efficiently.

7. **Exercise.** Exercise increases BMR for several hours afterward.

8. **Smoking and caffeine.** Both cause increased energy expenditure.

9. **Sleep.** Your BMR is at its lowest when you are sleeping.

The basal metabolic rate also decreases when you diet or eat fewer kcalories than normal. The BMR accounts for most of the energy you burn—about two-thirds of your energy if you are not very active.

Your level of physical activity strongly influences how many kcalories you need. Table 1-4 shows the kcalories burned per hour for a variety of activities. The number of kcalories burned depends on the type of activity, how long and how hard it is performed, and your size. The larger your body is, the more energy you use in physical activity. Aerobic activities such as walking, jogging, cycling, and swimming are excellent ways to burn calories if they are brisk enough to raise heart and breathing rates. Physical activity accounts for 25 to 40 percent of your total energy needs (Figure 1-8).

FIGURE 1-8 Physical activity accounts for 25 to 40 percent of total energy needs.

Courtesy of Rafal Olechowski/Shutterstock.

SUMMARY

1. The number of kcalories (a measure of the energy in food) you need is based on three factors: your energy needs when your body is at rest and awake (basal metabolism), your level of physical activity, and the energy you need to digest and absorb food (about 10 percent of kcalories).

2. Men have a higher BMR than women due to more muscle tissue. Tall people have higher BMRs due to more body surface. BMR declines after age 30. BMR is higher during growth and in hot and cold environments. Exercise, fever, stress, smoking, and caffeine increase BMR.

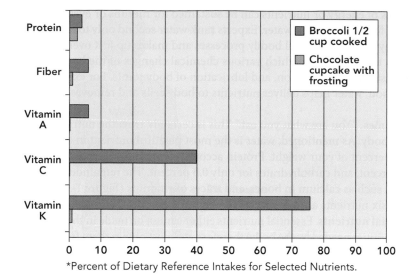

FIGURE 1-14 Broccoli is more nutrient dense than a chocolate cupcake. Broccoli contains more protein, fiber, and vitamins.

*Percent of Dietary Reference Intakes for Selected Nutrients.

Nutrient density A measure of the nutrients provided in a food per kcalorie of that food.

Empty-kcalorie foods Foods with added solid fats or sugar that provide few nutrients for the number of kcalories they contain.

Adequate diet A diet that provides enough kcalories, essential nutrients, and fiber to maintain health.

Moderate diet A diet that avoids excessive amounts of kcalories or any particular food/food group or nutrient.

Balanced diet A diet in which foods are chosen to provide kcalories, essential nutrients, and fiber in the right proportions.

Varied diet A diet in which you eat a wide selection of foods to get necessary nutrients.

The **nutrient density** of a food depends on the amount of nutrients it contains and the comparison of that to its kcaloric content. In other words, nutrient density is a measure of the nutrients provided per kcalorie of a food. As Figure 1-14 shows, broccoli offers many nutrients for its few kcalories. Broccoli is considered to have a high nutrient density because it is high in nutrients relative to its kcaloric value.

Nutrient-dense foods and beverages are lean or low in solid fats, and have little or no sugar or sodium added to them. All vegetables, fruits, whole grains, seafood, eggs, beans and peas, unsalted nuts and seeds, fat-free and low-fat milk and milk products, and lean meats and poultry—when prepared without adding solid fats or sugars—are nutrient-dense foods. Regular soft drinks and cakes are examples of **empty-kcalorie foods** because the kcalories they provide are "empty" (that is, they deliver few nutrients).

Understanding the concept of nutrient density is helpful when learning about nutritious diets. A nutritious diet has four characteristics. It is:

1. Adequate
2. Moderate
3. Balanced
4. Varied

Your diet must provide enough nutrients, but not too many. This is where adequate and moderate diets fit in. An **adequate diet** provides enough kcalories, essential nutrients, and fiber to keep you healthy, whereas a **moderate diet** avoids taking in excessive amounts of kcalories or eating more of one food or food group than is recommended. In the case of kcalories, for example, consuming too many leads to obesity. The concept of moderation allows you to choose appropriate portion sizes of any food as well as to indulge occasionally in empty-kcalorie foods such as french fries and premium ice cream.

Next, you need a **balanced diet**. Eating a balanced diet means eating more servings of nutrient-dense foods such as whole grains, fruits, and vegetables and fewer servings of foods such as cakes, cookies, and chips, which supply few nutrients. For example, if you drink a lot of soft drinks, you will be getting too much sugar and possibly not enough calcium, a mineral found in milk. This is a particular concern for children, whose bones are growing and who are more likely than ever before to be obese. The typical American diet is unbalanced. We eat more fried foods and fatty meats than we need, and we drink too much soda. At the same time we eat too few fruits, vegetables, and whole grains. A balanced diet is also likely to be adequate and moderate.

Last, you need a **varied diet**—in other words, you need to eat a wide selection of foods to get the necessary nutrients. If you imagine everything you eat for one week piled in a grocery cart, how much variety is in that cart from week to week? Do you eat the same bread, the same cereal, the same types of fresh fruit, and so on, every week? Do you constantly eat favorite foods? Do you try new foods? A varied diet is important because it makes it more likely that you will get the essential nutrients in the right amounts.

1. Foods with high nutrient density, such as fruits and vegetables, are high in nutrients relative to their kcaloric value.
2. Empty-kcalorie foods provide few nutrients for the kcalories you take in. Solid fats and sugars are usually found in empty-kcalorie foods such as cookies and regular soft drinks.
3. A nutritious diet is adequate, moderate, balanced, and varied.

HOW TO RECOGNIZE WHOLE, PROCESSED, FORTIFIED, AND ORGANIC FOODS

When people talk about food, you may hear some terms that you are not familiar with or are unsure of. **Whole foods** (besides being the name of a chain of natural and organic grocery stores) are foods pretty much as we get them from nature. Examples include eggs, fresh fruits and vegetables, beans and peas, whole grains, and fish (Figure 1-15). Whole foods are generally not processed or refined and do not have any added ingredients. Some whole foods, such as milk, are minimally processed to make it safe to drink. Fresh meat is also minimally processed so that consumers can buy just what they want.

Processed foods (Figure 1-16) have been prepared using a certain procedure: milling (wheat is milled to make white flour), cooking and freezing (such as frozen pancakes or dinners), canning (canned vegetables), dehydrating (dried fruits), or culturing with bacteria (yogurt). In some cases, processing removes nutrients, such as when whole wheat is milled to make white flour. In other cases, processing helps retain nutrients, such as when freshly picked vegetables are frozen.

Whereas the food supply once contained mostly whole farm-grown foods, today's supermarket shelves are stocked primarily with processed foods. Many processed foods contain parts of whole foods and often have added ingredients such as sugars and fats. For instance, cookies are made with eggs and flour. Then sugar and fat are added. Highly processed foods, such as many breakfast cereals, cookies, crackers, sauces, canned or frozen soups, baking mixes, frozen entrees, snack foods, and condiments, are staples nowadays.

When processing adds nutrients, the resulting food is either an enriched or a fortified food. For example, white flour must be enriched with several vitamins and iron to make up for some of the nutrients lost during milling. A food is considered **enriched** when nutrients are added to it to replace the same nutrients that are lost in processing.

Milk is often fortified with vitamin D because there are few good food sources of this vitamin. A food is considered **fortified** when nutrients are added that were not present

Whole foods Foods as we get them from nature; some may be minimally processed.

Processed foods Foods that have been prepared using a certain procedure such as canning, cooking, freezing, dehydration, or milling.

Enriched A food to which nutrients are added to replace the same nutrients that were lost in processing.

Fortified A food to which nutrients are added that were not present originally or nutrients are added that increase the amount already present.

FIGURE 1-15 Whole foods are generally not processed or refined.

Snow peas and papaya courtesy of CookingDistrict.com; Bluefish and wild rice reprinted with permission of John Wiley & Sons, Inc.

FIGURE 1-16 Processed foods are prepared using various techniques such as milling, cooking, freezing, or canning.

Courtesy of Joe Belanger/Shutterstock.

Natural Meat or poultry products that contain no artificial ingredient or added color and are only minimally processed; for other foods, natural means that there are no added colors, artificial flavors, or synthetic ingredients.

originally or nutrients are added that increase the amount already present. For example, orange juice does not contain calcium, so when calcium is added to orange juice, the product is called calcium-fortified orange juice (Figure 1-17). Probably the most notable fortified food is iodized salt. Iodized salt was introduced in 1924 to decrease iodine deficiency in Americans.

Another term you see on food labels is **natural**. When you are buying meat or poultry, natural products contain no artificial ingredient or added color and are only minimally

FIGURE 1-17 Fortified foods provide additional sources of nutrients.

Courtesy of B. Calkins/Shutterstock.

processed. For other foods, natural means that there are no added colors, artificial flavors, or synthetic ingredients.

Organic food is produced by farmers who emphasize the use of renewable resources and the conservation of soil and water to improve the environment for future generations. Organic meat, poultry, eggs, and dairy products come from animals that are given no antibiotics or growth hormones. Organic food is produced without using most conventional pesticides, fertilizers made with synthetic ingredients, bioengineering, or ionizing radiation. Before a product can be labeled *organic,* a government-approved certifier inspects the farm where the food is grown to make sure the farmer is following all the rules necessary to meet USDA organic standards. Companies that handle or process organic food before it gets to your local supermarket or restaurant must be certified, too.

Products labeled "100 percent organic" must contain only organically produced ingredients. Products labeled "organic" must consist of at least 95 percent organically produced ingredients. Products meeting the requirements for "100 percent organic" and "organic" may display the USDA Organic seal (Figure 1-18).

Processed products that contain at least 70 percent organic ingredients can use the phrase "made with organic ingredients" and list up to three of the organic ingredients or food groups on the label. Processed products that contain less than 70 percent organic ingredients cannot use the term *organic* other than to identify the specific ingredients that are organically produced in the ingredients statement (Figure 1-19).

Some studies show that organic foods may be higher in certain vitamins and/or minerals compared with conventionally grown foods. However, there is no solid body of research yet. The nutrient composition of any food grown in soil will vary due to many factors, such as variations in the soil quality, the amount of sunshine, and the amount of rain. Vitamins in plants are created by the plants themselves as long as they get adequate sunshine, water, carbon dioxide, and fertilizer. Minerals must come from the soil.

Many chefs feel that organic foods taste better than their conventional counterparts. Whether organic foods taste better is to some extent a matter of personal taste. Also, taste will vary among any fresh produce, depending on their freshness, the seeds used, where they were grown, and so on.

Sometimes when consumers see a box of organic crackers or cookies, they mistakenly think that the product must be healthier and lower in kcalories than a nonorganic product. The truth is that an organic cookie does contain organic ingredients, but it could still use white flour, butter, and sugar just like the nonorganic cookie and contain an equal number of kcalories.

FIGURE 1-18 The USDA Organic seal signifies that the product is at least 95 percent organic.

Courtesy of the U.S. Department of Agriculture.

Organic foods Food produced without antibiotic or growth hormones, most conventional pesticides, fertilizers made with synthetic ingredients or sewage sludge, bioengineering, or ionizing radiation.

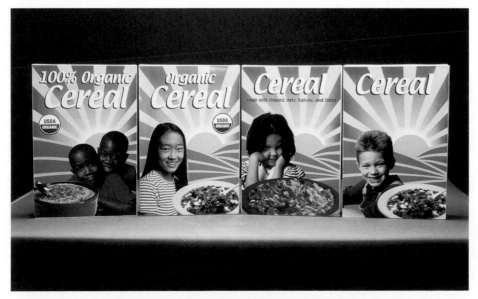

FIGURE 1-19 The sample cereal boxes show the four labeling categories.

Courtesy of the U.S. Department of Agriculture.

1. Whole foods, such as eggs, fresh fruits, and vegetables, are generally not processed or refined and do not have any added ingredients. Some whole foods (e.g., milk) are minimally processed to make them safe for consumption.

2. Processed foods, such as frozen meals, canned soups, and many snack foods, have been prepared using a certain procedure such as canning, cooking, freezing, dehydration, or milling.

3. Enriched foods contain nutrients that were added to replace those lost in processing. Fortified foods contain added nutrients that either were not present originally or that increase the amount already present.

4. Natural meat or poultry cannot contain any artificial ingredients or added color and can only be minimally processed. For other foods, *natural* means there are no added colors, artificial flavors, or synthetic ingredients.

5. Organic meat, poultry, eggs, and dairy products come from animals that are given no antibiotics or growth hormones. Organic food is produced without using most conventional pesticides, fertilizers made with synthetic ingredients or sewage sludge, bioengineering, or ionizing radiation.

6. Organic foods don't necessarily contain more nutrients and often contain the same number of kcalories as their nonorganic counterpart.

DIETARY REFERENCE INTAKES

Dietary Reference Intake (DRIs) Nutrient standards set for healthy Americans and Canadians that estimate how much you need daily of various nutrients, as well as when you might be taking in too much of certain nutrients.

Recommended Dietary Allowance (RDA) The dietary intake value that represents what you need to take in on a daily basis.

Adequate Intake (AI) The dietary intake that is used when there is not enough scientific research to support an RDA.

Tolerable Upper Intake Level (UL) The maximum intake level of a nutrient above which you may feel adverse health effects.

The **Dietary Reference Intakes (DRIs)** estimate how much you need daily of various nutrients as well as when you might be taking in too much of a nutrient. The DRI is only useful for healthy individuals, and was developed for use in the United States and Canada. They include the following.

- **Recommended Dietary Allowance (RDA)**. The RDAs represent how much of a nutrient you should be getting on a daily basis. If there is not enough scientific evidence to set an RDA, an **Adequate Intake** is given.

- **Tolerable Upper Intake Level (UL)**. The UL is the maximum you should take in of a nutrient—if you take more than the UL, you risk side effects that can be bad for your health. For most nutrients, the UL includes how much you take in from foods and supplements. A UL cannot be established for some nutrients, due to inadequate research.

The DRIs (see Appendix B) vary, depending on age and gender, and there are separate DRIs for pregnant and lactating women. The DRIs are meant to help healthy people maintain health and prevent disease. They are not designed for seriously ill people, whose nutrient needs may be much different.

TABLE 1-5	Acceptable Macronutrient Distribution Ranges		
AGE	AMDR FOR CARBOHYDRATES	AMDR FOR FATS	AMDR FOR PROTEINS
1 to 3 years old	45–65%	30–40%	5–20%
4 to 18 years old	45–65%	25–35%	10–30%
Over 18 years old	45–65%	20–35%	10–35%

Another important concept related to the DRIs is **Acceptable Macronutrient Distribution Ranges (AMDR).** As you recall, macronutrients refer to carbohydrates, fats, and proteins—the only nutrients that provide kcalories. AMDR is simply the percent of kilocalories you should eat from carbohydrates, fats, or proteins to reduce your risk of chronic disease while providing adequate intake and nutrients. For example, adults (and children over 1 year old) should obtain 45 to 65 percent of their total kcalories from carbohydrates (Table 1-5). The AMDR for adults is 20 to 35 percent of total kcalories from fat and 10 to 35 percent of total kcalories from protein. The wide range allows for more flexibility in dietary planning for healthy people.

Acceptable Macronutrient Distribution Range (AMDR) The percent of total kilocalories coming from carbohydrates, fats, or proteins that is associated with a reduced risk of chronic disease while providing adequate intake.

SUMMARY

1. The Dietary Reference Intakes provide standards for healthy Americans and Canadians that estimate how much you need daily of various nutrients as well as when you might be taking in too much of certain nutrients.
2. The Recommended Dietary Allowance (RDA) represents how much of a nutrient you should be getting on a daily basis. If there is not enough scientific evidence to justify setting an RDA, an Adequate Intake is given.
3. Tolerable Upper Intake Level (UL) is the maximum intake level of a nutrient above which you may feel adverse health effects.
4. The DRIs vary depending on age and gender, and there are separate DRIs for pregnant and lactating women.
5. Acceptable Macronutrient Distribution Range is defined as the percent of total kilocalories in your diet that should come from carbohydrates, fats, or proteins—the percentages are based on reducing your risk of disease.

WHAT HAPPENS WHEN YOU EAT?

To become part of the body, food must be digested and absorbed. **Digestion** is the process by which food is broken down into its components in the mouth, stomach, and small intestine, with the help of digestive **enzymes** and fluids. For example, when you eat a hamburger, the bun and meat are broken apart by your teeth and stomach acid, which helps to release nutrients and other components. Nutrients such as carbohydrates, fats, and proteins must be broken down into their components before they can be absorbed from the stomach or small intestines into the body. Water, vitamins, and minerals do not need to be broken down any further.

Before the body can use any nutrients that are present in food, the nutrients must pass through the walls of the stomach or intestines into the body's tissues, a process called **absorption.** Much digestion occurs in the small intestine, so that nutrients such as carbohydrates, fats, and proteins can be absorbed. Nutrient are then transported in the blood through the body to the cells.

Within each cell, **metabolism** takes place. Metabolism refers to all the chemical processes by which nutrients are used to support life. Metabolism has two parts: the building up of substances and the breaking down of substances. Within each cell, nutrients such as glucose are split into smaller units to release energy. The energy is used to make heat to maintain body temperature or perform work within the cell. Substances such as protein are built from their building blocks in every cell.

Once we have smelled and tasted food, our meal goes on a journey through the **gastrointestinal tract** (also called the digestive tract), a hollow tube that runs down the middle of your body (Figure 1-20). The top of the tube is your mouth, which is connected, in turn, to your pharynx, esophagus, stomach, small intestine, large intestine, rectum, and anus, where solid wastes leave the body. The gastrointestinal tract is such a busy place that the cells lining it are replaced every few days.

Digestion The process by which food is broken down into its components in the mouth, stomach, and small intestine with the help of digestive enzymes and fluids.

Enzymes Compounds that speed up the breaking down of food so that nutrients can be absorbed. Enzymes also perform other functions in the body.

Absorption The passage of digested nutrients through the walls of the intestines or stomach into the body's tissues. Nutrients are then transported through the body to the cells in the blood.

Metabolism All the chemical processes by which nutrients are used to support life.

Gastrointestinal tract A hollow tube running down the middle of the body in which digestion of food and absorption of nutrients take place.

FIGURE 1-20 The human digestive tract digests and absorbs nutrients and removes solid wastes.

Mouth: Tastes food.
Chews food.
Makes saliva.

Pharynx: Directs food from mouth to esophagus.

Esophagus: Passes food to stomach.

Stomach: Makes enzyme that breaks down protein.
Makes hydrochloric acid.
Churns and mixes food.
Acts like holding tank.

Small intestine: Makes enzymes.
Digests most of food.
Absorbs nutrients across villi into blood and lymph.

Large intestine: Passes waste to be excreted.
Reabsorbs water and some minerals.
Absorbs vitamins made by bacteria.

Rectum: Stores feces.

Anus: Keeps rectum closed.
Opens for elimination.

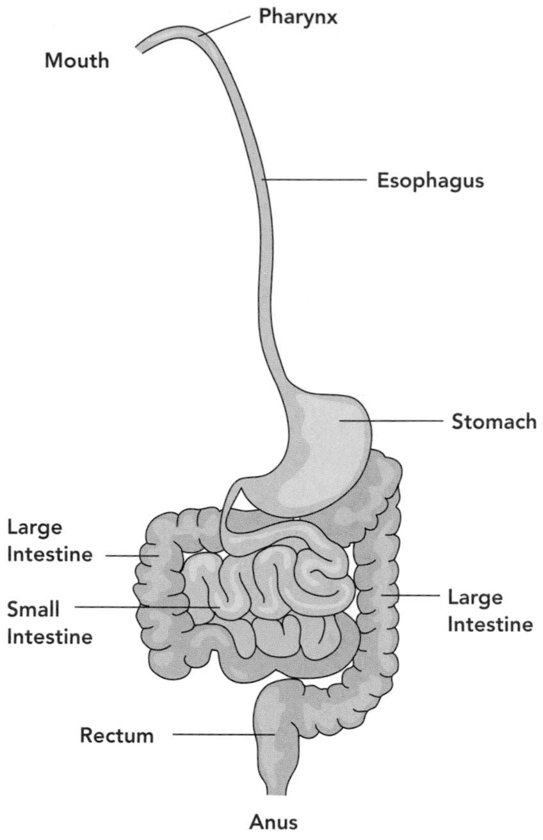

Mouth
Pharynx
Esophagus
Stomach
Large Intestine
Small Intestine
Large Intestine
Rectum
Anus

Saliva A fluid secreted into the mouth from the salivary glands that contains important digestive enzymes and lubricates the food so that it may readily pass down the esophagus.

Pharynx A passageway that connects your mouth to the esophagus.

Esophagus The muscular tube that connects the pharynx to the stomach.

Peristalsis Involuntary muscular contraction that forces food through the entire digestive system.

Stomach A muscular sac that holds about 4 cups of food when full and helps in digestion. Some alcohol is absorbed through the stomach.

Small intestine The digestive tract organ that extends from the stomach to the opening of the large intestine. The site of most digestion and absorption.

Villi Tiny fingerlike projections in the wall of the small intestines that are involved in absorption.

The digestive system starts with the mouth. Your tongue and teeth help with chewing. The tongue, which extends across the floor of the mouth, moves food around the mouth during chewing. Your 32 permanent teeth grind and break down food. Chewing is important because it breaks up the food into smaller pieces so that it can be swallowed. **Saliva**, a fluid secreted into the mouth from the salivary glands, contains important digestive enzymes and lubricates the food so that it may pass readily down the esophagus. Digestive enzymes help break down food into forms of nutrients that can be used by the body. The tongue rolls the chewed food into a ball to be swallowed.

When you swallow, food enters the **pharynx** (just 5 inches long) and then the **esophagus**, a muscular tube about 10 inches long that leads to the stomach. Food is propelled down the esophagus by **peristalsis**, which are rhythmic contractions of muscles in the wall of the esophagus. You might think of this contraction that forces food through the entire digestive system as squeezing a marble through a rubber tube. Peristalsis also helps break up food into smaller and smaller particles.

Food passes from the esophagus into the stomach. The **stomach**, a muscular sac, holds about 4 cups (or 1 liter) of food when full. The stomach makes an enzyme that helps in protein digestion and an acid that destroys harmful bacteria.

The stomach has the strongest muscles and thickest walls of all the organs in the gastrointestinal tract. The food now has a semiliquid consistency and is passed into the first part of the small intestine in small amounts (the small intestine can't process too much food at one time). Liquids leave the stomach faster than solids do, and carbohydrate or protein foods leave faster than fatty foods do. The stomach absorbs few nutrients, but it does absorb alcohol. It takes 1.5 to 4 hours after you have eaten for the stomach to empty.

The **small intestine** receives the digested food from the stomach as well as enzymes from other organs in the body, such as the liver. The small intestine itself produces digestive enzymes.

Most digestion and absorption are completed in the first half of the small intestine. On the folds of the intestinal wall (and throughout the entire small intestine) are tiny, fingerlike projections called **villi**. Nutrients are absorbed across the villi into the body where they are transported to your cells through the blood.

The **large intestine** (also called the **colon**) is located between the small intestine and the rectum. One of the functions of the large intestine is to receive the waste products of digestion and pass them on to be eliminated. Waste products are the materials that were not absorbed into the body. The large intestine does absorb water, some minerals, and a few vitamins made by bacteria residing there. Bacteria are normally found in the large intestine and are necessary for a healthy intestine. Intestinal bacteria make some important substances, such as vitamin K. They also can digest some components of food, such as fiber.

The **rectum** stores the waste products until they are released as solid feces through the **anus**, which opens to allow elimination.

Two common digestive problems are heartburn and peptic ulcers. It is the acid in the stomach that contributes to heartburn. **Heartburn** is a painful burning sensation in the esophagus, just below or behind the breastbone. Heartburn occurs when your stomach contents back up (also called reflux) into the esophagus. This partially digested material is usually acidic and can irritate the esophagus. Frequent, ongoing heartburn may be a sign of gastroesophageal reflux disease (GERD). Ways to treat heartburn and GERD include eating small meals, avoiding foods and beverages such as fatty foods that aggravate heartburn, losing weight (if overweight), and possibly taking medications.

Ulcers are a common digestive problem that can affect part of the small intestine or the stomach. A **peptic ulcer** is a sore or defect on the lining of the stomach or small intestine. Peptic ulcers are common: One in ten Americans develops an ulcer at some time in his or her life. One cause of peptic ulcer is bacterial infection, but some ulcers are caused by long-term use of nonsteroidal anti-inflammatory agents (NSAIDs), such as aspirin and ibuprofen. Taking antibiotics, quitting smoking, limiting consumption of caffeine and alcohol, and reducing stress can speed healing and prevent ulcers from recurring.

Large intestine (colon) The part of the gastrointestinal tract between the small intestine and the rectum. It absorbs some water, vitamins, and minerals, as well as passes on waste products.

Rectum The last section of the large intestine, in which feces, the waste products of digestion, are stored until elimination.

Anus The opening of the digestive tract through which feces travel out of the body.

Heartburn A painful, burning feeling in your chest or throat. It happens when stomach acid backs up into your esophagus.

Peptic ulcer A sore on the lining of the stomach or duodenum.

S U M M A R Y

1. Before the body can use the nutrients in food, the food must be digested and the nutrients absorbed through the walls of the stomach and/or intestine to be distributed through the blood to the cells. Digestion is the process by which food is broken down into its components in the mouth, stomach, and small intestine with the help of digestive enzymes.
2. Within each cell, metabolism (all the chemical processes by which nutrients are used to support life) takes place. Metabolism has two parts: building substances such as bone or protein and breaking down substances such as sugar to produce energy.
3. Figure 1-20 summarizes food digestion and absorption.
4. Two common digestive problems are heartburn and peptic ulcers.

CHECK-OUT QUIZ

1. Nutrients are the nourishing substances in food that provide energy and promote the growth and maintenance of the body.

   a. true
   b. false

2. With regards to eating, flavor and taste are the same.

   a. true
   b. false

3. Juicy is a more desirable texture than dry.

   a. true
   b. false

4. Your energy needs are based on your basal metabolism and the energy you use to digest and absorb food.

   a. true
   b. false

5. As you get older, your basal metabolism speeds up.

   a. true
   b. false

6. Vitamins provide energy for the body.

   a. true
   b. false

7. Carbohydrates provide the same amount of energy per gram as fats.

   a. true
   b. false

8. Fats are an example of macronutrients.

   a. true
   b. false

9. The main structural component of all the body's cells is carbohydrate.

   a. true
   b. false

10. The most important nutrient is water.

    a. true
    b. false

11. White flour is an example of a whole food.

    a. true
    b. false

12. Milk with vitamin D is considered an enriched food.

    a. true
    b. false

13. Organic food is produced without using most conventional pesticides, fertilizers made with synthetic ingredients, bioengineering, or ionizing radiation.

    a. true
    b. false

14. An Adequate Intake is given when there is not enough research to support a Recommended Dietary Allowance.

    a. true
    b. false

15. In digestion, the nutrients pass through the walls of the stomach or intestines.

    a. true
    b. false

16. Name and briefly describe five factors that influence what you eat.

17. Is flavor the same as taste? Explain why or why not.

18. Name four factors that increase basal metabolism and two factors that slow it down.

19. Which nutrients yield energy? How much energy?

20. Briefly describe each of the six categories of nutrients.

21. Give an example of a whole food and a processed food made from it.

22. Describe two functions of the mouth, esophagus, stomach, and small and large intestines in digestion and absorption.

NUTRITION WEB EXPLORER

For a complete list of websites for the following activities, please visit the companion book page at www.wiley.com/college/drummond.

NUTRITION.GOV

From this government site, you can access many nutrition topics. Navigate around the website to get an idea of the resources available. Pick any nutrition article and summarize it in one paragraph.

CENTER FOR YOUNG WOMEN'S HEALTH: COLLEGE EATING AND FITNESS 101

Read "College Eating and Fitness 101." List five suggestions made to help you get/stay fit at college.

ALCOHOL CALORIE CALCULATOR

Fill in the "Average Drinks per Week" column and then press "Compute." You will see how many kcalories you take in each month and in one year from alcoholic beverages.

DIGESTION TUTORIAL AT SHEPPARD SOFTWARE

At this website, do the Digestion Game (Level 1 and Level 2) and also complete the Digestion Quiz. List the parts of the gastrointestinal tract from the mouth to the anus.

GREEN RESTAURANT ASSOCIATION

Click on "Green Your Restaurant for Restaurateurs." Next, click on "Certification Standards." Points are awarded for each of the following categories. List one thing restaurants can do in each category to run a more sustainable operation.

- Water Efficiency
- Waste Reduction and Recycling
- Sustainable Furnishings and Building Materials
- Sustainable Food
- Energy
- Disposables
- Chemical and Pollution Reduction

NATIONAL RESTAURANT ASSOCIATION

Use the National Restaurant Association's website to identify and list six tips for effective recycling in restaurants.

CENTER FOR SCIENCE IN THE PUBLIC INTEREST EATING GREEN

Click on "Eating Green Calculator" and fill in how much in the way of animal products you eat each week. Then click on "Calculate Impact" and write up the environmental impact of your eating habits including: acres of grain and grass needed for animal feed, pounds of fertilizer used to grow animal feed, and pounds of manure created by the animals you eat. Click on "Click Here to Improve Your Diet and Protect the Environment." See how changing your eating habits can impact the environment.

Also do "Score Your Diet." Write up your health, environmental, and animal welfare scores. How did you rate overall?

IN THE KITCHEN

Recipes

Broccoli Cheese Soup
Grilled Chicken on Garden Greens with Herb Sherry Vinaigrette
Penne and Cheese with Vegetables
Warm Baked Apples with Dried Cherries and Raisins
Fruit and Oatmeal Bars

You will be given one of these recipes to prepare. Along with the recipe, you will be given the kcalories and nutrients that one serving of your recipe contains. The nutrient list includes the amount of carbohydrate, fat, protein, and some vitamins and minerals. The information for the vitamins and minerals will include the percent of the RDA or AI each serving provides. During class, you will use a worksheet to help you determine which recipe is highest (or lowest) in selected nutrients.

HOT TOPIC

SUSTAINABLE FOOD SYSTEMS

As you drive by many farms across the United States, you might be inclined to think that we eat a lot of corn, soybeans, and grains. However, it is not Americans who are eating most of these foods: it is our livestock such as beef cattle, dairy cattle, hogs, chickens, and turkeys. Eventually, these livestock (except dairy cattle, which give us milk) will be slaughtered to produce meat and poultry. The typical American eats about 8 ounces of meat a day (including beef, poultry, and fish). This amount is about double the global average.

Producing large quantities of meat in America uses many resources and has serious environmental consequences, such as the following:

1. When forests have been cut down to create pastures for livestock, there are negative effects. The trees help balance the oxygen–carbon dioxide balance of the earth by absorbing carbon dioxide from the environment and releasing oxygen. Fewer trees lead to the accumulation of greenhouse gases, such as carbon dioxide and methane, in the atmosphere. The accumulation of these greenhouse gases leads to global warming. Trees also absorb rainfall by soaking up moisture through their roots, thus preventing runoff and the accompanying soil erosion and flooding.

2. Livestock farms are major air and water polluters. Cattle naturally produce methane, a strong greenhouse gas that contributes to global climate change. Livestock production systems also produce other greenhouse gases such as nitrous oxide and carbon dioxide. People who live near or work at these farms breathe in hundreds of gases, which are formed as manure decomposes. The stench can be unbearable. And, of course, there is the problem of what to do with millions of tons of manure each year.

3. Enormous quantities of water, fuel, fertilizers, and pesticides are required to grow the feed for livestock, utilizing many acres of farmland. In drier climates, huge amounts of irrigation water are used to produce feed grains such as corn. Fertilizers require a lot of energy to produce, and along with pesticides, they often wind up polluting waterways and drinking water.

To produce 100 kcalories of plant foods requires only about 50 kcalories from fossil fuels, but to get the equivalent amount of kcalories from beef requires almost 1,600 kcalories. Basically, the money you use to purchase a hamburger in the United States doesn't even start to cover the environmental costs of producing it.

Commercial farming, although producing an abundance of relatively inexpensive food, depletes natural resources such as topsoil and water. In addition, pesticides enter groundwater systems and fertilizers threaten ecosystems. Emissions from nitrogen fertilizers add to greenhouse gas emissions and chemical runoff from farms cause algae blooms in fresh-water lakes and rivers.

Our current method of producing food is not sustainable. In other words, at some point the damage being done to the environment will make it impossible to continue to farm. These harmful environmental consequences will become yet more worrisome as the world's population grows and demand for food increases. One answer to this dilemma can be found in sustainable agriculture.

Sustainable agriculture produces food without depleting the Earth's resources (water, soil, fuel) or polluting its environment. It is agriculture that follows the principles of nature to develop systems for raising crops and livestock that are, like nature, self-sustaining (Figure 1-21). In recent decades, sustainable farmers and researchers around the world have used a variety of techniques, such as those found in organic farming, to farm with nature. Sustainable practices lend themselves to smaller farms. These farms, in turn, tend to find their best niches in local markets, within local food systems, often selling directly to consumers in farmers' markets or to local restaurants. Some farms offer to consumers the possibility of buying shares of the harvest ahead of time. Then the consumers are entitled to pick up a bag or box of produce at certain time intervals, such as weekly,

FIGURE 1-21 Sustainable agriculture produces abundant food without depleting the Earth's resources or polluting its environment.

Courtesy of the U.S. Department of Agriculture.

from the farm during the growing season. This is referred to as community-supported agriculture. You can find a farm in your area that offers shares at the Local Harvest website.

Sustainable agriculture is part of a *sustainable food system.* Sustainable food systems involve not only growing crops and producing livestock in a sustainable manner but also processing, packaging, and distributing foods without depleting the Earth's resources or causing excessive pollution. In a sustainable food system, food should be affordable and workers, such as farm workers, should make a living wage.

So what are restaurants doing to embrace sustainable food systems? Lots! Let's take a look first at what chefs are doing in the kitchen and then see what chefs and managers are doing to run a sustainable facility. Chefs are using at least seven ways to purchase sustainably produced foods:

1. **Sourcing local foods.** By working with local farmers and sourcing local foods, less fuel is used to transport food to the restaurant. The average distance food travels from farm to plate—referred to as "food miles"—is 1,500 miles but

it is only 56 miles for locally produced items. Although food miles account for 11 percent of the food system's greenhouse gas emissions (food processing accounts for 83 percent of greenhouse gas emissions), buying local and choosing efficient modes of transportation does help save fuel and reduce emissions.

2. **Buying organic foods.** *Organic food* is produced by farmers who emphasize the use of renewable resources and the conservation of soil and water to enhance environmental quality for future generations. Organic meat, poultry, eggs, and dairy products come from animals that are given no antibiotics or growth hormones. Organic food is produced without using most conventional pesticides, fertilizers made with synthetic ingredients or sewage sludge, genetic engineering, or irradiation. Before a product can be labeled *organic,* a government-approved certifier inspects the farm where the food is grown to make sure the farmer is following all the rules necessary to meet US Department of Agriculture organic standards. Companies that handle or process organic food before it gets to your local supermarket or restaurant must be certified, too.

3. **Buying sustainable fish.** Fishing practices worldwide are depleting fish populations, destroying habitats, and polluting the water. Sustainable seafood comes from species of fish that are managed in a way that provides for today's needs without damaging the ability of the species to be available for future generations. Most fish and shellfish caught in US federal waters are harvested under fishery management plans that must meet standards to ensure fish stocks are maintained. Unfortunately, over 80 percent of the seafood eaten in the United States is imported, and some US seafood is overfished or caught or farmed in ways that harm the environment.

- For information on specific seafood to purchase, visit the websites of the Environmental Defense Fund (United States), Monterey Bay Aquarium Seafood Watch (United States) or SeaChoice (Canada). Also see Essential Fatty Acids in Chapter 4.

- Blue Ocean Institute and Chefs Collaborative have partnered to create Green Chefs, Blue Ocean: a comprehensive, interactive online sustainable seafood training program and resource center.

- The Fish Choice website helps seafood buyers source sustainable seafood.

4. **Starting a garden to grow herbs, vegetables, and other foods.** More chefs are growing some of their own produce and herbs (Figure 1-22).

5. **Serving meals that are lower on the food chain.** An animal-based diet requires more fertilizer, water, energy, and pesticides than a vegetarian diet. Recent studies show that changing from beef to chicken, fish, eggs, or vegetable-based entrees can reduce greenhouse gases. Beef is more costly to produce than any other meat or poultry. Chefs have been adding more vegetarian options to the

FIGURE 1-22 More chefs are growing some of their produce and herbs.

Courtesy of Yuri Arcurs/Shutterstock.

menu and using smaller portions of meat.

6. **Buying coffee and tea from sustainable operations.** To keep up with demand, some growers have been using mass production methods involving excessive chemicals and pesticides. Several certification programs exist, such as from Rainforest Alliance, and you can also buy organic coffee.

7. **Reducing bottled beverages.** Encourage the use of reusable cups and "bottleless" beverage options, and source bottles from companies that use less plastic or glass in their bottles.

Here are nine things chefs and managers are doing to maintain a sustainable dining room and kitchen:

1. **Saving energy** by turning off appliances when not in use. You can also save energy by choosing compact fluorescent or LED bulbs, using a programmable thermostat, and using variable speed exhaust hoods. Most kitchen hoods run constantly at the same speed. Variable speed fans can speed up when a lot of heat is being generated or slow down when there is less heat to remove. Using energy-efficient appliances and replacing old equipment with Energy Star certified equipment reduces energy consumptions by 25 to 60 percent. Keeping refrigerators, freezers, and appliances clean and well maintained helps reduce energy usage.

2. **Saving water** by running dishwashers to capacity and using devices in the dishwasher, such as low-flow spray valves, to maintain performance while saving water and energy. You can also save water by serving water on request, fixing leaks, and using tankless water heaters.

3. **Using washable, reusable plates, cups, and silverware.** When you must buy disposables, buy those made from bio-based materials or materials that have been previously recycled and made into these new products. You can also select containers that can be recycled and provide clearly marked recycling bins for everyone to use.

4. **Reducing the amount of trash produced.** About half of restaurant trash is food. Take a serious look at what foods are always being discarded to determine how to reduce the amount of waste. Use a pulper or disposer to break down food waste into a liquid that can be drained away. If this is not an option, hire a commercial composter to take away the waste and convert it to soil. Also consider donating unused product—find out how to do so at the Food Donation Connection website. Cutting back on bottled beverages also reduces waste.

5. **Setting up recycling bins and communicating to staff and customers.** A majority of customers prefer to patronize restaurants with recycling programs and are willing to sort recyclables into bins.

6. **Using green cleaning products** approved by third parties such as Green Seal.

7. **Taking nonchemical preventative measures** to eliminate the need for pesticides.

8. **Training employees** about energy-saving procedures and why they are important.

9. **Remodeling or building green.** New and remodeled restaurants use low-VOC (volatile organic compounds) or no-VOC paints, recycled flooring and other materials, skylights, sustainable furnishings, and energy-efficient windows to be more environmentally friendly.

Many factors are driving the green movement in restaurants, from meeting consumer demand for eco-friendly products to conserving resources and dollars to joining a global effort to protect and preserve our natural environment. The changes won't be accomplished overnight, but will take time. Sustainable food programs, waste reduction, recycling, and energy-conservation are major endeavors that take time, planning, and money.

USING DIETARY RECOMMENDATIONS, FOOD GUIDES, AND FOOD LABELS TO PLAN MENUS

CHAPTER 2

- Distinguish between dietary recommendations and food guides and give an example of a food guide.

- Discuss four nutrition messages that accompany MyPlate and identify how much food from each food group is allowed on a 2,000 kcalorie level using MyPlate.

- Identify what counts as 1 cup of vegetables or 1 cup of fruit. Give two benefits of eating lots of vegetables and fruit, and three tips to help you eat more vegetables and fruit.

- List serving sizes for grains, name 3 whole-grain foods, and explain the benefits of whole grains and how many you should eat daily.

- Identify foods/beverages and serving sizes in the dairy group and give the number of cups of dairy adults need each day and the nutrients provided.

- Identify foods and serving sizes for 1 ounce of protein foods including lean choices and choices high in saturated fat and cholesterol, and guidelines for eating seafood.

- Explain the concept of empty kcalorie foods as related to MyPlate, give five examples of foods containing solid fats and/or added sugar as well as healthier options, and explain how MyPlate treats oils.

- Discuss the two overarching concepts of the *Dietary Guidelines for Americans, 2010*.

- Use BMI to determine if someone is overweight or obese, explain how kcalorie imbalance can cause overweight and obesity, and list five tips to help overweight/obese individuals manage their weight.

- Identify foods and food components that are consumed in excessive amounts and foods/nutrients to increase.

- Identify foods high in sodium, and explain how to reduce your consumption of sodium and why it is important.

- Give examples of how you can replace foods high in saturated fat and/or trans fats with foods rich in monounsaturated and polyunsaturated fat and why it is important to do.

- Define moderate alcohol consumption and give two examples of nutrients of concern in the American diet.

- Plan and evaluate menus using MyPlate and the *Dietary Guidelines for Americans, 2010*.

- Read and interpret information on a food label including the Nutrition Facts label, discuss the relationship between portion size on food labels and portions in MyPlate, and identify everyday objects that can help you visualize portion sizes.

INTRODUCTION TO DIETARY RECOMMENDATIONS AND FOOD GUIDES

Dietary recommendations
Guidelines that discuss food groups, foods, and nutrients to eat for optimal health.

Food guides Guidelines that tell us the kinds and amounts of foods to make a nutritionally adequate diet. They are typically based on current dietary recommendations, the nutrient content of foods, and the eating habits of the targeted population.

Dietary recommendations have been published for the healthy American public for many years. Early recommendations centered on encouraging Americans to eat certain foods and nutrients to prevent nutrient deficiencies and fight disease. Although deficiency diseases have been virtually eliminated, they have been replaced by diseases such as heart disease, cancer, and diabetes, which touch the lives of most Americans. More recent dietary recommendations, therefore, have centered on changing what we eat, in most cases cutting back on certain foods, to reduce risk factors for diseases such as heart disease and diabetes.

Dietary recommendations are quite different from Dietary Reference Intakes (DRIs). Whereas DRIs deal with specific nutrients, dietary recommendations discuss specific foods and food groups that will help individuals meet the DRIs. DRIs also tend to be written in a technical style. Dietary recommendations are generally written in easy-to-understand terms.

Whereas dietary recommendations discuss specific foods to eat for optimum health, **food guides** tell us the amounts of foods we need to eat to have a nutritionally adequate diet. Their primary role, whether in the United States or around the world, is to communicate an optimum diet for the overall health of the population. Food guides are typically based on current dietary recommendations, the nutrient content of foods, and the eating habits of the targeted population.

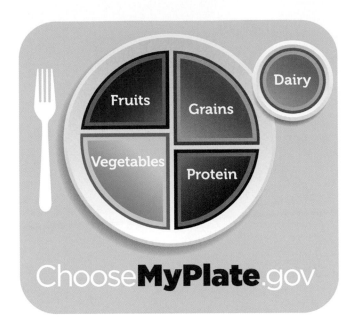

FIGURE 2-1 MyPlate was developed by the US Department of Agriculture.

Courtesy of the US Department of Agriculture.

MyPlate, a food guide developed by the US Department of Agriculture (USDA), is based on the Dietary Guidelines for Americans and DRIs. MyPlate (Figure 2-1) includes what and how much to eat from five food groups for a variety of kcalorie levels and also for a variety of eating styles, including nonvegetarian, lacto-ovo vegetarian, and vegan (see Appendix C.)

Other healthful food guides include the traditional Asian and Mediterranean diets (Figures 2-2 and 2-3). The Asian diet is high in plant foods, including rice, vegetables, fruits, beans, legumes, nuts, and vegetable oil. The diet varies of course from country to country and region to region. For example, people who live near the sea tend to eat more seafood. People who eat an Asian style diet are likely to have less heart disease, obesity, and certain cancers.

Like the Asian diet, no single set of criteria exists for what constitutes a traditional Mediterranean eating pattern. In general terms, it can be described as an eating pattern that emphasizes vegetables, fruits, nuts, olive oil, and grains—often whole grains. Only small amounts of meats and full-fat milk and milk products are usually consumed. Wine is often included with meals. Individuals following Mediterranean diets tend to have less chronic disease such as cardiovascular disease.

MyPlate A food guide developed by the US Department of Agriculture as a guide to the amounts of different types of foods needed to provide an adequate diet and comply with current nutrition recommendations.

SUMMARY

1. Whereas dietary recommendations discuss specific foods to eat for optimum health, food guides tell us the amounts of foods we need to eat to have a nutritionally adequate diet. More recent dietary recommendations in the US center on modifying the diet, in most cases cutting back on certain foods, to reduce risk factors for chronic disease such as heart disease and cancer.

2. MyPlate, a food guide developed by the US Department of Agriculture, is based on the Dietary Guidelines for Americans and nutrient recommendations. MyPlate (Figure 2-1) is accompanied by recommendations on what and how much to eat from five food groups.

3. Asian and Mediterranean diets all emphasize plant foods with small amounts of animal foods. Individuals following Asian and Mediterranean diets tend to have less chronic disease such as cardiovascular disease.

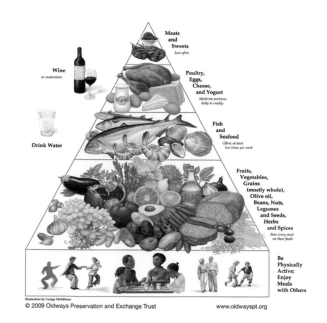

MYPLATE: INTRODUCTION AND VEGETABLE AND FRUIT GROUPS

MyPlate (Figure 2-1), developed by the US Department of Agriculture in 2011, has replaced the original Food Guide Pyramid. MyPlate translates the principles of the 2010 *Dietary Guidelines* and other nutritional standards to help you make healthier food choices.

As you can see from MyPlate, it includes five food groups: Fruits, Vegetables, Grains, Protein, and Dairy. Half of your plate should consist of fruits and vegetables, with slightly more vegetables than fruits. The other half of the plate includes grains and proteins, with slightly more grains than protein. Dairy is depicted as a circle—signifying a glass of milk—off to the side of the plate. Desserts are not shown—they are part of your daily "Empty Kcalorie" allowance, which is explained later in this discussion.

One very important point to remember about choosing foods from the five food groups is that your food choices are expected to be in their most nutrient-dense forms—lean or low-fat, with no added sugars. If choices that are not nutrient dense are routinely eaten, you will eat too many total kcalories due to too much fat and sugar.

Using Table 2-1, you can estimate what your kcalorie needs are daily by using your age, gender, and physical activity level. Once you know your kcalorie level, you can see how much you need to eat from each food group using Table 2-2. For example, if you need 2,000 kcalories a day to maintain a healthy weight, you can eat:

- 6 ounces or the equivalent of grains
- 2.5 cups of vegetables

TABLE 2-1	Estimated Kcalorie Needs per Day by Age, Gender, and Physical Activity Level[1]			

PHYSICAL ACTIVITY LEVEL[2]

Gender	Age (Years)	Sedentary	Moderately Active	Active
Child (female and male)	2–3	1,000–1,200	1,000–1,400	1,000–1,400
Female (3)	4–8	1,200–1,400	1,400–1,600	1,400–1,800
	9-13	1,400–1,600	1,600–2,000	1,800–2,200
	14–18	1,800	2,000	2,400
	19–30	1,800–2,000	2,000–2,200	2,400
	31–50	1,800	2,000	2,200
	51+	1,600	1,800	2,000–2,200
Male	4–8	1,200–1,400	1,400–1,600	1,600–2,000
	9–13	1,600–2,000	1,800–2,200	2,000–2,600
	14–18	2,000–2,400	2,400–2,800	2,800–3,200
	19–30	2,400–2,600	2,600–2,800	3,000
	31–50	2,200–2,400	2,400–2,600	2,800–3,000
	51+	2,000–2,200	2,200–2,400	2,400–2,800

[1]The estimates are rounded to the nearest 200 kcalories. An individual's kcalories needs may be higher or lower than these average estimates. The kcalorie ranges shown are to accommodate needs of different ages with the group. For children and adolescents, more kcalories are needed at older ages. For adults, fewer kcalories are needed at older ages.

[2]Sedentary means a lifestyle that includes only the light physical activity associated with typical day-to-day life. Moderately active means a lifestyle that includes physical activity equivalent to walking about 1.5 to 3 miles per day at 3 to 4 miles per hour, in addition to the light physical activity associated with typical day-to-day life. Active means a lifestyle that includes physical activity equivalent to walking more than 3 miles per day at 3 to 4 miles per hour, addition to the light physical activity associated with typical day-to-day life.

[3]Estimates for females do not include women who are pregnant or breastfeeding.

Source: US Department of Agriculture and US Department of Health and Human Services, *Dietary Guidelines for Americans, 2010*, 7th ed. (Washington, DC: US Government Printing Office, December 2010).

TABLE 2-2 | USDA Food Patterns

Calorie Level	1,200	1,400	1,600	1,800	2,000	2,200	2,400	2,600	2,800	3,000
Fruits	1 cup	1½ cup	1½ cup	1½ cup	2 cups	2 cups	2 cups	2 cups	2½ cups	2½ cups
Vegetables	1½ cup	1½ cup	2 cups	2½ cups	2½ cups	3 cups	3 cups	3½ cups	3½ cups	4 cups
Grains	4 oz.	5 oz.	5 oz.	6 oz.	6 oz.	7 oz.	8 oz.	9 oz.	10 oz.	10 oz.
Protein Foods	3 oz.	4 oz.	5 oz.	5 oz.	5.5 oz.	6 oz.	6.5 oz.	6.5 oz.	7 oz.	7 oz.
Dairy	2½ cups	2½ cups	3 cups	3 cups	3 cups	3 cups	3 cups	3 cups	3 cups	3 cups
Oils	17 g	17 g	22 g	24 g	27 g	29 g	31 g	34 g	36 g	44 g
Empty Calories	121	121	121	161	258	266	330	362	395	459

Source: US Department of Agriculture and US Department of Health and Human Services, *Dietary Guidelines for Americans, 2010*, 7th ed. (Washington, DC: US Government Printing Office, December 2010).

- 2 cups of fruit
- 3 cups of milk or the equivalent
- 5.5 ounces or the equivalent of lean meat and beans
- 6 teaspoons of oil
- 258 empty kcalories

The exact amounts of foods in these plans do not need to be achieved every day, but on average, over time.

MyPlate has several nutrition messages:

Balance Kcalories

- Enjoy your food, but eat less.
- Avoid oversized portions.

Foods to Increase

- Make half your plate fruits and vegetables.
- Make at least half of your grains whole grains.
- Switch to fat-free or low-fat (1 percent) milk.

Foods to Reduce

- Compare sodium in high-sodium foods like soup, bread, and frozen meals—and choose the foods with lower numbers.
- Drink water instead of sugary drinks.

Now, let's take a look at each food group (vegetables, fruits, grain, dairy, and protein) and then see how to use your allowance of oils and empty kcalories so you can use MyPlate. The nutrient contribution of each food group is summarized in Table 2-3. Complete serving size information for each group is in Appendix D.

TABLE 2-3	Nutrient Contribution of Food Groups
Vegetables	Dietary fiber, vitamin A, vitamin C, folate, potassium, magnesium
Fruits	Dietary fiber, vitamin C, folate, potassium, magnesium
Grains	Carbohydrate, dietary fiber, thiamin, riboflavin niacin, folate, iron, magnesium, selenium
Dairy	Protein, vitamin A, vitamin D, calcium, potassium
Protein	Protein, thiamin, riboflavin, niacin, B6, vitamin E, iron, zinc, magnesium

VEGETABLE GROUP

If you eat 2,000 kcalories a day, you should be eating 2.5 cups of vegetables daily. The recommendations from the vegetable group are given in cups. In general, 1 cup of raw or cooked vegetables or vegetable juice can be considered as 1 cup from the vegetable group. Two cups of raw leafy greens are considered 1 cup from the vegetable group. Appendix D gives additional information on serving sizes. Any vegetable or 100 percent vegetable juice counts toward your vegetable intake.

Vegetables are organized into five groups, based on their nutrient content. Some commonly eaten vegetables in each subgroup are as follows:

1. Dark green vegetables: dark green leafy lettuce, romaine lettuce, spinach, broccoli, bok choy, collard greens, turnip greens

2. Red and orange vegetables: carrots, tomatoes, sweet potatoes, acorn squash, butternut squash

3. Dry beans and peas: kidney beans, pinto beans, navy beans, split peas, lentils, soybeans, black-eyed peas

4. Starchy vegetables: potatoes, corn, peas, lima beans

5. Other vegetables: onions, celery, cucumbers, green peppers, iceberg lettuce, mushrooms, wax beans

A weekly intake of specific amounts from each of the five vegetable groups is recommended for adequate nutrient intake. Each subgroup provides somewhat different nutrients. The following weekly amounts are suggested from each subgroup.

Dark green vegetables	1½ cups/week
Red and orange vegetables	5½ cups/week
Beans and peas	1½ cups/week
Starchy vegetables	5 cups/week
Other vegetables	4 cups/week

Most vegetables are naturally low in kcalories and fat and none contain cholesterol. Vegetables are also important sources of many nutrients, including potassium, dietary fiber, folate, vitamins A and C.

FRUIT GROUP

If you eat 2,000 kcalories a day, 2 cups of fruits are recommended daily. In general, 1 cup of fruit or 100 percent fruit juice or ½ cup of dried fruit can be considered as 1 cup from the fruit group. You can also count the following as 1 cup: 1 small apple, 1 large banana, 32 seedless grapes, 1 medium pear, 2 large plums, or 1 large orange. Appendix D gives information on additional serving sizes. Consumption of whole fruits (fresh, canned, or dried) rather than fruit juice for the majority of the total daily amount is suggested to ensure adequate fiber intake.

There are three reasons to support the recommendation to eat lots of vegetables and fruits.

1. Eating at least 2½ cups of vegetables and fruits per day is associated with a reduced risk of cardiovascular disease, including heart attack and stroke. Some vegetables and fruits may be protective against certain types of cancer.

2. Most vegetables and fruits, when prepared without added fats or sugars, are relatively low in kcalories. Eating them instead of higher kcalorie foods can help adults and children achieve and maintain a healthy weight.

3. Most vegetables and fruits contain nutrients that Americans often don't get enough of— including folate, magnesium, potassium, dietary fiber, and vitamins A, C, and K.

Very few Americans consume the amounts of vegetables recommended as part of healthy eating patterns. Figure 2-4 gives tips to add more vegetables and fruits to your meals.

liven up your meals with vegetables and fruits

ChooseMyPlate.gov

10 tips to improve your meals with vegetables and fruits

Discover the many benefits of adding vegetables and fruits to your meals. They are low in fat and calories, while providing fiber and other key nutrients. Most Americans should eat more than 3 cups—and for some, up to 6 cups—of vegetables and fruits each day. Vegetables and fruits don't just add nutrition to meals. They can also add color, flavor, and texture. Explore these creative ways to bring healthy foods to your table.

1 fire up the grill
Use the grill to cook vegetables and fruits. Try grilling mushrooms, carrots, peppers, or potatoes on a kabob skewer. Brush with oil to keep them from drying out. Grilled fruits like peaches, pineapple, or mangos add great flavor to a cookout.

2 expand the flavor of your casseroles
Mix vegetables such as sauteed onions, peas, pinto beans, or tomatoes into your favorite dish for that extra flavor.

3 planning something Italian?
Add extra vegetables to your pasta dish. Slip some peppers, spinach, red beans, onions, or cherry tomatoes into your traditional tomato sauce. Vegetables provide texture and low-calorie bulk that satisfies.

4 get creative with your salad
Toss in shredded carrots, strawberries, spinach, watercress, orange segments, or sweet peas for a flavorful, fun salad.

5 salad bars aren't just for salads
Try eating sliced fruit from the salad bar as your dessert when dining out. This will help you avoid any baked desserts that are high in calories.

6 get in on the stir-frying fun
Try something new! Stir-fry your veggies—like broccoli, carrots, sugar snap peas, mushrooms, or green beans—for a quick-and-easy addition to any meal.

7 add them to your sandwiches
Whether it is a sandwich or wrap, vegetables make great additions to both. Try sliced tomatoes, romaine lettuce, or avocado on your everday sandwich or wrap for extra flavor.

8 be creative with your baked goods
Add apples, bananas, blueberries, or pears to your favorite muffin recipe for a treat.

9 make a tasty fruit smoothie
For dessert, blend strawberries, blueberries, or raspberries with frozen bananas and 100% fruit juice for a delicious frozen fruit smoothie.

10 liven up an omelet
Boost the color and flavor of your morning omelet with vegetables. Simply chop, saute, and add them to the egg as it cooks. Try combining different vegetables, such as mushrooms, spinach, onions, or bell peppers.

USDA Center for Nutrition Policy and Promotion

Go to www.ChooseMyPlate.gov for more information.

DG TipSheet No. 10
June 2011
USDA is an equal opportunity provider and employer.

FIGURE 2-4 Ten Tips to Improve Your Meals with Vegetables and Fruits.
Courtesy of the US Department of Agriculture.

1. MyPlate includes five food groups, as seen in Figure 2-1. You should select a variety of foods within each group and foods should be in their most nutrient-dense forms to avoid taking in too many kcalories.
2. A 2,000 kcalorie eating pattern includes 6 ounce equivalents of grains; 2.5 cups of vegetables; 2 cups of fruit; 3 cups of milk or the equivalent; and 5.5 ounces or the equivalent of lean meat, poultry, fish, beans or peas, eggs, nuts and seeds, or soy.
3. MyPlate's nutrition messages are: enjoy your food but eat less, avoid oversized portions, make half your plate fruits and vegetables, make at least half of your grains whole grains, switch to fat-free or low-fat (1 percent) milk, compare sodium in foods and choose foods lower in sodium, and drink water instead of sugary drinks.
4. Vegetables are organized into five groups based on their nutrient content. One cup of raw or cooked vegetables or vegetable juice can be considered as 1 cup from the vegetable group. Two cups of raw leafy greens are considered 1 cup.
5. One cup of fruit or 100 percent fruit juice or ½ cup of dried fruit can be considered as 1 cup from the fruit group. Eating whole fruits rather than fruit juice for the majority of the total daily amount is suggested to get enough fiber.
6. Almost all fruits and vegetables are naturally low in kcalories, fat, and sodium. Eating at least 2½ cups of vegetables and fruits/day is associated with a reduced risk of cardiovascular disease. Eating vegetables and fruits in place of higher kcalorie foods can help you achieve and maintain a healthy weight. Most vegetables and fruits contain nutrients that many of us don't get enough of: folate, magnesium, potassium, dietary fiber, and vitamins A, C, and K.
7. Figure 2-4 gives tips to add more vegetables and fruits to your meals.

MYPLATE: GRAIN, DAIRY, AND PROTEIN GROUPS

We have discussed two of MyPlate's five food groups: fruits and vegetables. This section covers the remaining three food groups: grain, dairy, and protein groups.

GRAIN GROUP

Any food made from wheat, rice, oats, cornmeal, barley, or another cereal grain is a grain product. Bread, pasta, oatmeal, breakfast cereals, tortillas, and grits are examples of foods made from grains.

Grains are divided into two subgroups: whole grains and refined grains. **Whole grains** contain the entire grain kernel: the fiber-rich bran, the vitamin-rich germ, and the starchy endosperm (see Table 3-4). Examples of whole grains include whole-wheat flour, bulgur (partially cooked cracked wheat), oatmeal, brown rice, and whole cornmeal.

Refined grains have been milled, a process that removes the bran and the germ. This is done to give grains a finer texture and improve their shelf life, but it also removes dietary fiber and many vitamins and minerals. Some examples of refined grain products are white flour, white bread, white rice, and degermed cornmeal. Most refined grains are enriched. This means that certain B vitamins and iron are added back after processing. Fiber, along with a number of vitamins and minerals, are not added back to enriched grains.

At the 2,000-kcalorie level, you need to eat 6 ounces (or their equivalent) daily, and at least half of these should be whole grain. At all kcalorie levels, you should eat at least half your

Whole grains Grains and grain products made from the entire grain seed, usually called the kernel, which consists of the bran, germ, and endosperm.

Refined grains Grains and grain products missing the bran, germ, and/or endosperm; any grain product that is not a whole grain.

1. 3 ounces of 100% whole grains and 3 ounces of refined-grain products

a. Each one-ounce slice of bread represents a 1 ounce-equivalent of grains: 1 one-ounce slice bread; 1 ounce uncooked pasta or rice; ½ cup cooked rice, pasta, or cereal; 1 tortilla (6" diameter); 1 pancake (5" diameter); 1 ounce ready-to-eat cereal (about 1 cup cereal flakes). The figure uses an example for a person whose recommendation is 6 ounces of total grains with at least 3 ounces from whole grains per day.
b. Partly whole-grain products depicted are those that contribute substantially to whole-grain intake. For example, products that contain at least 51% of total weight as whole grains or those that provide at least 8 grams of whole grains per ounce-equivalent.

2. 2 ounces of 100% whole grains, 2 ounces of partly whole-grain products,[b] and 2 ounces of refined-grain products

3. 6 ounces of partly whole-grain products

FIGURE 2-5 Three ways to make at least half of your total grains whole grains.

Courtesy of the US Department of Agriculture.

FIGURE 2-6 1 cup milk = 1 cup soy milk, 2 cups cottage cheese, 1 cup yogurt, 1½ ounces hard cheese, or 1/3 cup shredded cheese.

Photo by Peter Pioppo.

grains as whole grains. In general, 1 slice of bread, 1 cup of ready-to-eat cereal, 1 small muffin, half an English muffin, or ½ cup of cooked rice, cooked pasta, or cooked cereal can be considered as 1 ounce-equivalent from the grains group. Appendix D gives additional information on serving sizes.

Grains are important sources of many nutrients, including dietary fiber, several B vitamins, and minerals such as iron. Whole grains contain more dietary fiber than refined grains, as well as more vitamins and minerals. Moderate evidence indicates that whole-grain intake may reduce the risk of heart disease and is associated with a lower body weight.

Figure 2-5 shows three ways to make at least half of your grain choices whole grain. If a food contains at least 8 grams of whole grains per ounce (it is often listed on the label), then it is considered to be at least 50 percent whole grain.

DAIRY GROUP

The amount of milk/dairy products you need is based on age. Anyone 9 years and older needs 3 cups daily. The following can count as 1 cup (Figure 2-6). Remember that most choices should be fat-free or low-fat.

1 cup of milk or calcium fortified soymilk

1 cup yogurt

1½ ounces hard cheese

1/3 cup shredded cheese

2 ounces processed (American) cheese

½ cup ricotta cheese

2 cups cottage cheese

Appendix D gives additional information on serving sizes. Cream cheese, cream, and butter are not included in the dairy group because they contain little to no calcium.

If you choose milk or yogurt that is not fat-free or cheese that is not low-fat, the fat in the product counts as part of your empty kcalorie allowance (to be discussed in a moment). If sweetened milk products are chosen, such as chocolate milk and some yogurts, the added sugars also count as part of the empty kcalorie allowance.

Foods in the dairy group provide nutrients vital to health, such as calcium, vitamin D, potassium, and protein.

- Calcium is used for building bones and teeth and maintaining bone mass. Dairy products are the primary source of calcium in American diets. Diets that provide 3 cups or the equivalent of milk products daily help your bones stay strong. Dairy products are especially important during childhood and adolescence when bones are getting bigger and stronger.

- Vitamin D also helps build bones. Most milk and soymilk are fortified with vitamin D because there are not many foods that provide vitamin D.

PROTEIN FOODS GROUP

All foods made from meat, poultry, fish, dry beans or peas, eggs, nuts and seeds, and soy products are considered part of this group. Your protein choices should be lean or low-fat, such as the following. Note that all meats should be trimmed of fat and cooked with dry heat.

- Beef: bottom round steak (outside), bottom round roast, flank steak, eye round roast, top sirloin steak, tenderloin filet, top round roast or steak, 90/10 or 95/5 ground beef (See Figure 5-8)

- Lamb or Veal: loin or rib chop, top round

- Pork: pork tenderloin, top loin chops, whole loin

- Poultry: breast or thigh (skinless or skin removed after cooking)

- Fish: all fish and shellfish

Choose seafood at least twice a week. An intake of 8 or more ounces per week (less for young children) of a variety of seafood is recommended, so choose seafood in place of meat or poultry twice a week.

Dry beans and peas are the mature forms of **legumes** such as kidney beans, pinto beans, lima beans, black-eyed peas, and lentils. They can be counted either as vegetables or in the meat, poultry, fish, dry beans, eggs, and nuts (meat and beans) group. Beans and peas are excellent sources of plant protein, and also provide other nutrients such as iron and zinc, making them similar to meats, poultry, and fish. However, they are also considered vegetables because they are excellent sources of dietary fiber and nutrients such as potassium.

At the 2,000-kcalorie level, you should eat 5.5 ounces daily. In general, 1 ounce is equal to 1 ounce of meat, poultry, or fish; ¼ cup cooked dry beans; 1 egg; 1 tablespoon of peanut butter; ½ ounce of nuts or seeds; ¼ cup of tofu; 2 tablespoons of hummus; or ½ soy or bean burger patty. Appendix D gives additional information on serving sizes.

The protein foods group supplies many nutrients, including protein, vitamins, and iron. Proteins function as building blocks for bones, muscles, skin, and blood. Iron is used to carry oxygen in the blood.

Some food choices in this group are high in saturated fat, which can contribute to developing heart disease. Food choices that are high in saturated fat include fatty cuts of beef, pork, and lamb; regular ground beef (75 to 85 percent lean); regular sausages, hot dogs, and bacon; some luncheon meats, such as regular bologna and salami; duck; and eggs. You should eat these foods in moderation.

Figure 2-7 gives you tips for making wise choices in the protein foods group.

Legumes Plant species that have seed pods that split along both sides when ripe. Some of the more common legumes are beans, peas, and lentils.

10 tips
Nutrition Education Series

with protein foods, variety is key

10 tips for choosing protein

ChooseMyPlate.gov

Protein foods include both animal (meat, poultry, seafood, and eggs) and plant (beans, peas, soy products, nuts, and seeds) sources. We all need protein—but most Americans eat enough, and some eat more than they need. How much is enough? Most people, ages 9 and older, should eat 5 to 7 ounces* of protein foods each day.

1 vary your protein food choices
Eat a variety of foods from the Protein Foods Group each week. Experiment with main dishes made with beans or peas, nuts, soy, and seafood.

2 choose seafood twice a week
Eat seafood in place of meat or poultry twice a week. Select a variety of seafood—include some that are higher in oils and low in mercury, such as salmon, trout, and herring.

3 make meat and poultry lean or low fat
Choose lean or low-fat cuts of meat like round or sirloin and ground beef that is at least 90% lean. Trim or drain fat from meat and remove poultry skin.

4 have an egg
One egg a day, on average, doesn't increase risk for heart disease, so make eggs part of your weekly choices. Only the egg yolk contains cholesterol and saturated fat, so have as many egg whites as you want.

5 eat plant protein foods more often
Try beans and peas (kidney, pinto, black, or white beans; split peas; chickpeas; hummus), soy products (tofu, tempeh, veggie burgers), nuts, and seeds. They are naturally low in saturated fat and high in fiber.

6 nuts and seeds
Choose unsalted nuts or seeds as a snack, on salads, or in main dishes to replace meat or poultry. Nuts and seeds are a concentrated source of calories, so eat small portions to keep calories in check.

7 keep it tasty and healthy
Try grilling, broiling, roasting, or baking—they don't add extra fat. Some lean meats need slow, moist cooking to be tender—try a slow cooker for them. Avoid breading meat or poultry, which adds calories.

8 make a healthy sandwich
Choose turkey, roast beef, canned tuna or salmon, or peanut butter for sandwiches. Many deli meats, such as regular bologna or salami, are high in fat and sodium—make them occasional treats only.

9 think small when it comes to meat portions
Get the flavor you crave but in a smaller portion. Make or order a smaller burger or a "petite" size steak.

10 check the sodium
Check the Nutrition Facts label to limit sodium. Salt is added to many canned foods—including beans and meats. Many processed meats—such as ham, sausage, and hot dogs—are high in sodium. Some fresh chicken, turkey, and pork are brined in a salt solution for flavor and tenderness.

* What counts as an ounce of protein foods? 1 ounce lean meat, poultry, or seafood; 1 egg; ¼ cup cooked beans or peas; ½ ounce nuts or seeds; or 1 tablespoon peanut butter.

Center for Nutrition Policy and Promotion

Go to www.ChooseMyPlate.gov for more information.

DG TipSheet No. 6
June 2011
USDA is an equal opportunity provider and employer.

FIGURE 2-7 Ten Tips for Choosing Protein.
Courtesy of the US Department of Agriculture.

1. At all kcalorie levels, you should be eating at least half your grains as whole grains. Whenever the fiber-rich bran and vitamin-rich germ are left on the endosperm of a grain, the grain is a whole grain such as whole wheat or brown rice. When the bran and germ are removed from the endosperm, the grain is a refined or milled grain. White flour is a refined grain and is enriched with iron and some B vitamins. White flour is still not nearly as nutritious as whole-wheat flour.

2. A one ounce-equivalent of grains would be 1 slice of bread, 1 cup of ready-to-eat cereal, 1 small muffin, or ½ cup of cooked rice, cooked pasta, or cooked cereal.

3. Key nutrients in grains are B vitamins, iron, and fiber.

4. Anyone over 9 years old needs 3 cups or equivalent of dairy daily.

5. One cup of milk or calcium-fortified soymilk is equivalent to 1 cup yogurt, 1½ ounces hard cheese, ½ cup shredded cheese, 2 ounces processed cheese, or ½ cup ricotta cheese. Most choices should be fat-free or low-fat.

6. Dairy foods provide calcium and vitamin D (for healthy bones), potassium, and protein.

7. One ounce equivalent of protein foods is equal to 1 ounce of meat, poultry, or fish; ¼ cup cooked dry beans; 1 egg; 1 tablespoon of peanut butter; ½ ounce of nuts or seeds; ¼ cup of tofu; 2 tablespoons of hummus; or ½ soy or bean burger patty. Most meat and poultry choices should be lean, such as lean beef and pork cuts and skinless poultry.

8. Vegetarians should count beans and peas in the protein group. People who eat meat, poultry, and fish should count beans and peas in the vegetable group.

9. The protein foods group supplies protein, vitamins, and iron.

10. Food choices that are high in saturated fat include fatty cuts of beef, pork, and lamb; regular ground beef (75 to 85 percent lean); regular sausages, hot dogs, and bacon; some luncheon meats, such as regular bologna and salami; and duck. These foods should be used in moderation because their saturated fat content contributes to heart disease.

MYPLATE: EMPTY KCALORIES AND OILS

Many of the foods and beverages we eat and drink contain empty kcalories—kcalories from solid fats and/or added sugars that add few or no nutrients. Solid fats and added sugars contribute about 35 percent of daily kcalories—without contributing to overall nutrient adequacy of the diet.

Solid fats are found naturally in foods, such as beef or whole milk, or are added when foods are prepared at home or processed by food companies. Examples of solid fats include butter, stick margarine, shortening, and beef and chicken fat (found mostly in the skin). Most solid fats contain saturated fat, which contributes to heart disease if you eat too much.

Added sugars are sugars like white sugar that are added to a food for sweetening and flavor (such as soda and cookies) or added to foods or beverages at the table (such as sugar added to coffee). Increased consumption of added sugars is associated with a higher risk of obesity and is likely one of the many factors involved in rising obesity rates among adults and children. Sugars that are found naturally in fruits and dairy products are not "added sugars." You don't need to completely eliminate solid fats or added sugars from your diet, but you do need to moderate your intake of them.

Solid fats and added sugars make foods and beverages more appealing. The following foods and beverages are the major sources of empty kcalories (Figure 2-8):

- Cakes, cookies, pastries, and donuts (contain both solid fat and added sugars)
- Sodas, energy drinks, sports drinks, and fruit drinks (contain added sugars)
- Cheese and pizza (contains solid fat)
- Ice cream (contains both solid fat and added sugars)
- Sausages, hot dogs, bacon, and ribs (contain solid fat)

Solid fats Fats found in most animal foods but also can be made from vegetable oils through hydrogenation. Some common solid fats include butter, beef fat, stick margarine, and shortening. Solid fats contain more saturated fat than vegetable oils.

Added sugars Sugars, syrups, and other kcaloric sweeteners that are added to foods during processing, preparation, or consumed separately; do not include naturally occurring sugars such as those in fruit or milk.

FIGURE 2-8 Empty kcalorie foods.

Photo by Peter Pioppo.

Many of these foods can be found in forms with less or no solid fat or added sugars (such as low-fat cheese or sugar-free soda).

Empty kcalories from solid fats and added sugars can also be found in foods that contain important nutrients. Here are some examples of foods that provide nutrients, shown in forms with and without empty kcalories:

Foods with Some Empty Kcalories	Foods with Few or No Empty Kcalories
Sweetened juice or applesauce (contains added sugars)	Unsweetened juice or applesauce
75 percent lean ground beef (contains solid fats)	90 percent lean ground beef
Fried chicken (contains solid fats from frying and skin)	Baked chicken breast without skin
Sugar-sweetened cereals (contains added sugars)	Unsweetened cereals
Whole milk (contains solid fats)	Fat-free milk

Making better choices, such as unsweetened applesauce or extra lean ground beef, can help you moderate your intake of added sugars and solid fats, and also save kcalories.

Each kcalorie level in MyPlate contains a maximum limit for empty kcalories. For example, if you can eat 2,000 kcalories/day—your empty kcalorie limit is 258 empty kcalories. Appendix D includes a guide to how to count empty kcalories.

Oils are not a food group, but are emphasized because they contribute essential fatty acids and vitamin E to the diet. Oils are naturally present in foods such as olives, nuts, avocados, and seafood. Many common oils are extracted from plants, such as canola, corn, olive, peanut, soybean, and sunflower oils. Foods that are mainly oil include mayonnaise, oil-based salad dressings, and soft (tube or squeeze) margarine with no trans fats.

You can use vegetable oils instead of solid fats in cooking and baking and use soft margarine instead of stick margarine. Because oils are a concentrated source of kcalories, you should use oils in small amounts. Coconut oil, palm kernel oil, and palm oil are high in saturated fats so they should be considered as solid fats.

A 2,000 kcalorie diet allows for 6 teaspoons of oil each day. Here is how to count the oils you eat.

1 tablespoon vegetable oil = 3 teaspoons oil

1 tablespoon margarine (trans fat free) = 2½ teaspoons oil

1 tablespoon mayonnaise = 2½ teaspoons oil

2 tablespoons Italian dressing = 2 teaspoons oil

1. In MyPlate, each kcalorie level has an allowance for empty kcalories—kcalories from solid fats and/or added sugars that add few or no nutrients.
2. Solid fats are found naturally in foods, such as beef, or are added when foods are prepared at home or processed by food companies. Examples of solid fats include butter, stick margarine, shortening, and beef and chicken fat. Added sugars are sugars like white sugar that are added to a food for sweetening and flavor (such as soda and cookies) or added to foods or beverages at the table.
3. Drinking fat-free milk, lean beef, skinless chicken, unsweetened or fresh fruit, and sugar-free soda are ways to consume less solid fats and added sugars.
4. Oils are not a food group, but are emphasized because they contribute essential fatty acids and vitamin E to the diet. Oils include vegetable oils, soft margarine, mayonnaise, and oil-based salad dressings. A 2,000 kcalorie diet allows for 6 teaspoons daily.

DIETARY GUIDELINES FOR AMERICANS: INTRODUCTION AND WEIGHT CONTROL

The most recent set of US dietary recommendations are the **Dietary Guidelines for Americans, 2010.** The US Department of Agriculture and the US Department of Health and Human Services jointly create each edition.

The intent of the *Dietary Guidelines* is to summarize current nutritional knowledge and then make recommendations for healthy eating to promote health, reduce the risk of chronic diseases, and reduce overweight and obesity. Taken together, the *Dietary Guidelines* recommendations encompass two overarching concepts:

1. *Maintain kcalorie balance over time to achieve and sustain a healthy weight.* People who are most successful at achieving and maintaining a healthy weight do so through continued attention to consuming only enough kcalories from foods and beverages to meet their needs and by being physically active.

2. *Focus on consuming nutrient-dense foods and beverages.* These include fruits and vegetables, beans and peas, whole grains, lean meats and poultry, and fat-free or low-fat milk and dairy products. We currently consume too much sodium and too many kcalories from solid fats, added sugars, and refined grains. These replace nutrient-dense foods and beverages and make it difficult for people to achieve recommended nutrient intake while controlling kcalorie and sodium intake.

The *Dietary Guidelines for Americans* include 23 key recommendations for the general population and six additional key recommendations for specific population groups, such as women who are pregnant (Table 2-4). Key recommendations are the most important messages within the Guidelines in terms of their implications for improving public health.

A basic premise of the *Dietary Guidelines* is that nutrient needs should be met primarily through eating foods. In some cases, fortified foods and dietary supplements may be useful in providing one or more nutrients that otherwise might be consumed in less than recommended amounts.

The *Dietary Guidelines* provide recommendations in these areas:

1. Balancing kcalories to manage weight
2. Foods and food components to reduce
3. Foods and nutrients to increase
4. Building healthy eating patterns

First, we will discuss balancing kcalories to manage weight. Then we will look at foods to reduce and foods to increase. Healthy eating patterns, such as MyPlate or the Mediterranean diet, have already been discussed.

Dietary Guidelines for Americans, 2010 A set of dietary recommendations for Americans that is updated every five years and designed to promote dietary changes to improve health, reduce the risk of chronic diseases, and reduce the number of people who are overweight or obese. The recommendations are intended for Americans ages 2 years and older, including those at increased risk of chronic disease.

BALANCING KCALORIES TO MANAGE WEIGHT

- Prevent and/or reduce overweight and obesity through improved eating and physical activity behaviors.
- Control total calorie intake to manage body weight. For people who are overweight or obese, this will mean consuming fewer calories from foods and beverages.
- Increase physical activity and reduce time spent in sedentary behaviors.
- Maintain appropriate calorie balance during each stage of life—childhood, adolescence, adulthood, pregnancy and breastfeeding, and older age.

FOODS AND FOOD COMPONENTS TO REDUCE

- Reduce daily sodium intake to less than 2,300 milligrams (mg) and further reduce intake to 1,500 mg among persons who are 51 and older and those of any age who are African American or have hypertension, diabetes, or chronic kidney disease. The 1,500 mg recommendation applies to about half of the US population, including children, and the majority of adults.
- Consume less than 10 percent of calories from saturated fatty acids by replacing them with monounsaturated and polyunsaturated fatty acids.
- Consume less than 300 mg per day of dietary cholesterol.
- Keep *trans* fatty acid consumption as low as possible by limiting foods that contain synthetic sources of *trans* fats, such as partially hydrogenated oils, and by limiting other solid fats.
- Reduce the intake of calories from solid fats and added sugars.
- Limit the consumption of foods that contain refined grains, especially refined grain foods that contain solid fats, added sugars, and sodium.
- If alcohol is consumed, it should be consumed in moderation—up to one drink per day for women and two drinks per day for men—and only by adults of legal drinking age.

FOODS AND NUTRIENTS TO INCREASE

Individuals should meet the following recommendations as part of a healthy eating pattern while staying within their calorie needs:

- Increase vegetable and fruit intake.
- Eat a variety of vegetables, especially dark-green and red and orange vegetables and beans and peas.
- Consume at least half of all grains as whole grains. Increase whole-grain intake by replacing refined grains with whole grains.
- Increase intake of fat-free or low-fat milk and milk products, such as milk, yogurt, cheese, or fortified soy beverages.
- Choose a variety of protein foods, which include seafood, lean meat and poultry, eggs, beans and peas, soy products, and unsalted nuts and seeds.
- Increase the amount and variety of seafood consumed by choosing seafood in place of some meat and poultry.
- Replace protein foods that are higher in solid fats with choices that are lower in solid fats and calories and/or are sources of oils.
- Use oils to replace solid fats where possible.
- Choose foods that provide more potassium, dietary fiber, calcium, and vitamin D, which are nutrients of concern in American diets. These foods include vegetables, fruits, whole grains, and milk and milk products.

BUILDING HEALTHY EATING PATTERNS

- Select an eating pattern that meets nutrient needs over time at an appropriate kcalorie level.
- Account for all foods and beverages consumed and assess how they fit within a total healthy eating pattern.
- Follow food safety recommendations when preparing and eating foods to reduce the risk of foodborne illnesses.

RECOMMENDATIONS FOR SPECIFIC POPULATION GROUPS

Women capable of becoming pregnant

- Choose foods that supply heme iron, which is more readily absorbed by the body, additional iron sources, and enhancers of iron absorption such as vitamin C-rich foods.
- Consume 400 micrograms (mcg) per day of synthetic folic acid (from fortified foods and/or supplements) in addition to food forms of folate from a varied diet.

Women who are pregnant or breastfeeding

- Consume 8 to 12 ounces of seafood per week from a variety of seafood types.

- Due to their high methyl mercury content, limit white (albacore) tuna to 6 ounces per week and do not eat the following four types of fish: tilefish, shark, swordfish, and king mackerel.
- If pregnant, take an iron supplement, as recommended by an obstetrician or other health care provider.

INDIVIDUALS AGES 50 YEARS AND OLDER

- Consume foods fortified with vitamin B, such as fortified cereals, or dietary supplements.

Source: US Department of Agriculture and US Department of Health and Human Services, *Dietary Guidelines for Americans, 2010*, 7th ed. (Washington, DC: US Government Printing Office, December 2010).

Before talking about obesity, it is important for you to understand and use the concept of **Body Mass Index (BMI)**. BMI is a number calculated from a person's weight and height and is a fairly reliable indicator of body fatness for most people. An individual with a BMI:

- less than 18.5 is considered underweight;
- between 18.5 and 24.9 is considered normal weight;
- between 25.0 to 29.9 is considered overweight;
- and over 30.0 is considered obese.

Now use Appendix E to find your BMI using your weight and height.

The prevalence of overweight and obesity in the United States is dramatically higher now than it was a few decades ago. This is true for all age groups and virtually all subgroups.

	In the 1970s	**In 2007–2008**
Children ages 6–11	4% were obese	20% were obese
Children 12–19	6% were obese	18% were obese
Adults	15% were obese	34% were obese

The current high rates of overweight and obesity shows that many Americans consume more kcalories than they use. Kcalories consumed must equal kcalories used to maintain the same body weight. Consuming more kcalories than used results in weight gain. To curb the obesity epidemic, Americans need to do the following:

- Decrease the total number of kcalories they consume from foods and beverages.
- Increase kcalorie expenditure, or burn more kcalories, through physical activity (Figure 2-9).

This is important because being overweight or obese puts you at an increased risk of many health problems, including cardiovascular disease (which includes heart attacks and strokes), type 2 diabetes, and certain types of cancer such as colorectal cancer.

The following steps can help you get to or stay at a healthy weight:

1. Increase intake of whole grains, vegetables, and fruits.
2. Reduce intake of sugar-sweetened beverages such as soda. Check for added sugars using the ingredients list. When a sugar is close to first on the ingredients list, the food is high in added sugars.
3. Focus on the total number of kcalories consumed. Eating 500 fewer kcalories per day is a common initial goal for weight loss, as it typically results in the loss of one pound per week. Check for kcalories and serving size on the Nutrition Facts label.
4. Monitor food intake. Monitoring intake has been shown to help you become more aware of what and how much you eat and drink. Also, monitoring body weight and physical activity can help prevent weight gain and improve outcomes when actively losing weight or maintaining body weight following weight loss.
5. Monitor kcalorie intake from alcoholic beverages. Heavier than moderate consumption of alcohol over time is associated with weight gain.

Body Mass Index A measure of weight relative to height, which is considered a reasonably reliable indicator of total body fat.

FIGURE 2-9 Burning kcalories through physical activity can help you lose weight.

Courtesy of A. Turner/Shutterstock.

6. Prepare, serve, and consume smaller portions of foods and beverages, especially those high in kcalories. When given larger portions, you tend to eat more. Stop eating when you are satisfied, not full.

7. Eat a nutrient-dense breakfast. Eating breakfast is associated with weight loss and keeping weight off, as well as improved nutrient intake.

8. When eating out, check posted kcalorie amounts and choose smaller portions or lower-kcalorie options. When possible, order a small-sized option, share a meal, or take home part of the meal.

9. Limit screen time. Also avoid eating while watching television or using the computer, which can result in overeating.

10. Be physically active. To achieve and maintain a healthy body weight, adults should do the equivalent of 150 minutes of moderate-intensity aerobic activity each week. For children and adolescents ages 6 years and older, 60 minutes or more of physical activity per day is recommended. Table 2-5 details current physical activity guidelines.

Appropriate eating behaviors and increased physical activity have tremendous potential to decrease the number of Americans who are overweight or obese.

TABLE 2-5	2008 Physical Activity Guidelines

AGE GROUP	GUIDELINES
6 to 17 years	Children and adolescents should do 60 minutes (1 hour) or more of physical activity daily. • Aerobic: Most of the 60 or more minutes a day should be either moderate or vigorous intensity aerobic physical activity, and should include vigorous-intensity physical activity at least three days a week. • Muscle-strengthening: As part of their 60 or more minutes of daily physical activity, children and adolescents should include muscle-strengthening physical activity on at least three days of the week. • Bone-strengthening: As part of their 60 or more minutes of daily physical activity, children and adolescents should include bone-strengthening physical activity on at least three days of the week. • It is important to encourage young people to participate in physical activities that are appropriate for their age, that are enjoyable, and that offer variety.
18 to 64 years	All adults should avoid inactivity. Some physical activity is better than none, and adults who participate in any amount of physical activity gain some health benefits. • For substantial health benefits, adults should do at least 2 hours and 30 minutes a week of moderate-intensity, or 1 hour and 15 minutes a week of vigorous-intensity aerobic physical activity, or an equivalent combination of moderate- and vigorous-intensity aerobic activity. Aerobic activity should be performed in episodes of at least 10 minutes, and preferably, it should be spread throughout the week. • For additional and more extensive health benefits, adults should increase their aerobic physical activity to 5 hours a week of moderate-intensity, or 2½ hours a week of vigorous-intensity aerobic physical activity, or an equivalent combination of moderate- and vigorous-intensity activity. Additional health benefits are gained by engaging in physical activity beyond this amount. • Adults should also include muscle-strengthening activities that involve all major muscle groups on two or more days a week.
65 years and older	Older adults should follow the adult guidelines. When older adults cannot meet the adult guidelines, they should be as physically active as their abilities and conditions will allow. • Older adults should do exercises that maintain or improve balance if they are at risk of falling. • Older adults should determine their level of effort for physical activity relative to their level of fitness. • Older adults with chronic conditions should understand whether and how their conditions affect their ability to do regular physical activity safely.

Source: Adapted from US Department of Health and Human Services, *2008 Physical Activity Guidelines for Americans* (Washington, DC: US Department of Health and Human Services, 2008). ODPHP Publication No. U0036. www.health.gov/paguidelines.

1. Two important concepts from the *Dietary Guidelines for Americans* are that it is important to maintain kcalorie balance to achieve and maintain a healthy weight and to focus on consuming nutrient-dense foods and beverages. The prevalence of overweight and obesity in the United States is dramatically higher now than it was a few decades ago.

2. A basic premise of the *Dietary Guidelines* is that nutrient needs should be met through eating foods—although fortified foods and dietary supplements may be useful in providing one or more nutrients.

3. Body Mass Index is a fairly reliable indicator of body fatness and is used to determine if you are normal weight, overweight, or obese.

4. Kcalorie imbalance is when you consume more kcalories than you use. Consuming more kcalories than used results in weight gain. Kcalories consumed must equal kcalories used to maintain the same body weight.

5. Ten tips are given on how to control kcalories to lose weight.

DIETARY GUIDELINES FOR AMERICANS: FOODS TO REDUCE AND FOODS TO INCREASE

The *Dietary Guidelines for Americans, 2010,* recommend that we eat more of certain foods and less of others:

1. Reduce sodium.

2. Reduce saturated and trans fats.

3. Reduce alcohol.

4. Reduce added sugar.

5. Reduce refined grains and increase whole grains.

6. Increase vegetables and fruit.

7. Increase milk and milk products.

8. Increase lean protein foods.

9. Address nutrients of concern.

The first three topics and the last topic are discussed in the following sections. The others have already been discussed.

REDUCE SODIUM

Sodium is an essential nutrient and is needed by the body in very small quantities, provided you are not sweating excessively. Sodium is primarily consumed as salt, and virtually all Americans consume considerably more sodium than needed. Table salt is 40 percent sodium by weight, and 1 teaspoon of salt contains 2,300 milligrams of sodium (Figure 2-10).

On average, the higher your sodium intake, the higher your blood pressure. Eating less salt is an important way to reduce the risk of high blood pressure, which then also lowers your risk of heart disease, stroke, and kidney disease.

FIGURE 2-10 Sodium is primarily consumed as salt.

Photo by Frank Pioppo.

The sodium Adequate Intake (AI) for individuals ages 9 to 50 years old is 1,500 mg per day. Older adults need less because they need fewer kcalories.

9 to 50 years old	1,500 mg sodium/day
51–70 years old	1,300 mg sodium/day
71 years +	1,200 mg sodium/day

For individuals 14 years and older, the Tolerable Upper Intake Level is 2,300 mg sodium/day.

The average intake of sodium for all Americans ages 2 years and older is about 3,400 mg per day, significantly above the Tolerable Upper Intake Level. Over 75 percent of the sodium that we eat is added to processed foods by manufacturers. The natural sodium content of unprocessed foods accounts for only 10 percent of total intake, while salt added at the table and in cooking provides another 5 to 10 percent of the total sodium intake. Examples of processed foods high in sodium include the following.

- Pizza
- Processed chicken, such as chicken nuggets
- Sausage, franks, bacon, and ribs
- Cold cuts
- Many canned and frozen foods, such as soups
- Cheese

Some of these foods can be purchased or prepared to be lower in sodium. For example, chicken naturally contains little sodium. By purchasing fresh or frozen chicken with no sodium added to it, you can prepare a meal lower in sodium.

You can reduce your consumption of sodium in a variety of ways:

- Use the Nutrition Facts label to purchase foods that are low in sodium.
- Eat more fresh foods and fewer processed foods that are high in sodium.
- Eat more home-prepared foods, where you have more control over sodium, and use little or no salt or salt-containing seasonings when cooking or eating foods.
- When eating at restaurants, ask that salt not be added to your food or order lower-sodium options.

It is challenging to keep your sodium intake below 2,300 mg/day.

If you like salty foods, you can retrain your taste buds to eat less salt. Sodium preferences in adults and children are influenced by exposure to sodium in the diet. Studies have demonstrated that reducing sodium intake over a time period of as little as three to four weeks can decrease your preference for salty foods.

African Americans; individuals with hypertension, diabetes, or chronic kidney disease; and individuals ages 51 and older make up about half of the US population over 2 years old. Although nearly everyone benefits from reducing sodium intake, the blood pressure of these individuals tends to be even more responsive to the blood pressure–raising effects of sodium than others.

REDUCE SATURATED AND TRANS FATS

Saturated fatty acids A type of fat found mostly in animal foods such as meat, chicken skin, eggs, milk, butter, and cheese, as well as some oils that have been hydrogenated to make stick margarine and shortening. Known to raise levels of bad cholesterol in the blood, which is a risk factor for cardiovascular disease.

Fats are found in both plant and animal foods. Fats are made of any of three types of fatty acids: saturated, monounsaturated, or polyunsaturated. You have probably heard about another type of fat called trans fats. Trans fatty acids are unsaturated fatty acids.

The *types* of fatty acids you eat are more important in influencing your risk of heart disease than is the *total amount* of fat in the diet. Animal fats tend to have a higher proportion of **saturated fatty acids** (seafood being the major exception), and plant foods tend to have a higher proportion of monounsaturated and polyunsaturated fatty acids (coconut oil, palm oil, and palm kernel oil being the exceptions).

Most fats with a high percentage of saturated or trans fatty acids are solid at room temperature, while those with more unsaturated fatty acids are usually liquid at room temperature (such as vegetable oils). Saturated fats are found in most animal foods but some are made when vegetable oils are used to make margarine or shortening. Foods high in saturated fats include the following. (Figure 2-11):

- Butter, stick margarine, and shortening
- Full-fat (regular) cheese
- Whole milk
- Cream
- Ice cream
- Well-marbled cuts of meats and regular ground beef
- Sausages, franks, bacon, and ribs
- Poultry skin
- Many baked goods (such as cookies, donuts, pastries)

This is important because diets high in saturated fats raise levels of LDL cholesterol (called bad cholesterol) in your blood that then increases your risk for cardiovascular disease. To lower your risk of cardiovascular disease, it is best to eat less than 10 percent of kcalories from saturated fatty acids by limiting foods high in saturated fat and replacing them with foods that are rich in mono-unsaturated and polyunsaturated fatty acids. For example, you can:

FIGURE 2-11 Foods high in solid fats.

Photo by Peter Pioppo.

- Buy fat-free or low-fat versions of milk and other dairy products.
- Prepare foods with vegetable oils rich in monounsaturated and polyunsaturated fatty acids such as canola or olive oil instead of using butter.
- Choose baked, steamed, or broiled rather than fried foods most often.
- Eat lean meats and moderate portions of cheese and baked goods.

You should also try to keep your **trans fat** intake as low as possible because, like saturated fat, it increases your levels of bad cholesterol in the blood. Trans fats can be found in some fried foods like French fries and doughnuts; baked goods including pastries, pie crusts, frozen pizzas, biscuits, ready-made frosting, and cookies; microwave popcorn, and some stick margarines and shortenings. You can determine the amount of trans fats by using the Nutrition Facts label, where it must be listed.

Trans fats Unsaturated fatty acids found mostly in foods that have been hydrogenated such as shortening or margarine. Act like saturated fats in the body to raise blood cholesterol levels.

REDUCE ALCOHOL

In the United States, about 50 percent of adults are regular drinkers. Alcohol consumption may have beneficial effects when consumed in moderation. Moderate alcohol consumption is associated with a lower risk of cardiovascular disease. Moderate alcohol consumption is defined as up to one drink per day for women and up to two drinks per day for men.

NUTRIENTS OF CONCERN

Because consumption of vegetables, fruits, whole grains, milk and milk products, and seafood is lower than recommended, intake by Americans of some nutrients is low enough to be of concern. These are potassium, dietary fiber, calcium, and vitamin D. In addition, women who are capable of becoming pregnant need to be concerned about iron and folate and adults 50 years and older need to consume adequate vitamin B_{12}. Table 2-6 shows good sources of these nutrients.

TABLE 2-7	Selected Daily Values
NUTRIENT	DAILY VALUE
Carbohydrate	300 grams
Fiber	25 grams
Fat	65 grams
Saturated fat	20 grams
Cholesterol	300 milligrams
Sodium	2,400 milligrams
Potassium	3,500 milligrams
Vitamin A	1,500 micrograms RAE
Vitamin C	60 milligrams
Folate	400 micrograms
Calcium	1,000 milligrams
Iron	18 milligrams

Source: US Food and Drug Administration.

sodium may increase your risk for certain chronic diseases, such as heart disease, some cancers, and high blood pressure. Americans often don't get enough of some of the other nutrients listed: dietary fiber, vitamins A and C, calcium, and iron. Eating enough of these nutrients can improve your health and help reduce the risk of some diseases and conditions. For example, a diet high in dietary fiber promotes healthy bowel function. You can use the food label to help limit those nutrients you want to cut back on and also increase those nutrients you want to consume in greater amounts.

Nutrient amounts are listed in two ways: in metric amounts (in grams) and as a percentage of the **Daily Value**. Developed by the Food and Drug Administration, Daily Values are recommended levels of intake specially developed for food labels (Table 2-7). DRIs can't be used on nutrition labels because they are set for specific age and gender categories, and you can't have several nutrition labels on one food for males and females in various age groups.

The table at the bottom of the food label shows the Daily Values for certain nutrients at both the 2,000- and 2,500-kcalorie levels. For example, if you eat a 2,000-kcalorie diet, you should get less than 65 grams of fat from all the foods you eat in a day. If you consume 2,500 kcalories per day, the amounts of cholesterol and sodium you eat in a day are not different from those eaten by others eating 2,000 kcalories per day. This table is found only on larger packages and does not change from product to product.

The percentage of the Daily Value (%DV) is based on a 2,000-kcalorie diet. Therefore, the Daily Value may be a little high, a little low, or right on target for you. The percentage of the Daily Value shows you how much of the Daily Value is in one serving of food. For example, in Figure 2-11, the %DV for total fat is 18 percent and for dietary fiber it is 0 percent. When one serving of macaroni and cheese contains 18 percent of the DV for Total Fat, that means you have 82 percent of your fat allowance left for all the other foods you eat that day.

When you are looking at the %DV on a food label, use this guide:

- Foods that contain 5 percent or less of the Daily Value for a nutrient are generally considered low in that nutrient.

- Foods that contain 10 to 19 percent of the Daily Value for a nutrient are generally considered good sources of that nutrient.

- Foods that contain 20 percent or more of the Daily Value for a nutrient are generally considered high in that nutrient. For example, one cup of macaroni and cheese contains 18 percent of the Daily Value for fat, which is just below 20 percent. Therefore, the macaroni and cheese is pretty high in fat, particularly if you eat 1½ to 2 cups.

You can use the %DV to help you make dietary tradeoffs with other foods throughout the day. You don't have to give up a favorite food to eat a healthy diet. When a food you like is high in fat, balance it by eating foods that are low in fat at other times of the day.

Daily Value (DV) A set of values developed by the Food and Drug Administration that are used as a reference for expressing nutrition content on nutrition labels.

The values listed for total carbohydrate include all carbohydrates, including dietary fiber and sugars listed below it. The sugar values include naturally present sugars, such as lactose in milk and fructose in fruits, as well as those added to the food, such as table sugar and corn syrup. The label can claim no sugar added but still include naturally occurring sugars. An example is fruit juice.

The values listed for total fat refer to all the fat in the food. Only total fat, saturated fat, and trans fat information is required on the label, because high intakes are linked to high blood cholesterol, which is linked to an increased risk of coronary heart disease. Trans fat is a specific type of fat that forms when liquid oils are made into solid fats such as margarine and shortening. Trans fat behaves like saturated fat in raising low-density lipoprotein ("bad") cholesterol. Listing the amount of polyunsaturated and monounsaturated fats in the food is voluntary.

The Daily Value for calcium is 1,000 milligrams (mg). Experts advise adolescents, especially girls, to consume 1,300 mg and postmenopausal women to consume 1,200 mg of calcium daily. The daily target for teenagers should therefore be 130%DV, and the daily target for postmenopausal women should be 120%DV.

NUTRIENT CLAIMS

Nutrient content claims, such as "good source of calcium" and "fat-free," can appear on food packages only if they follow legal definitions (Table 2-8). For example, a food that is a good source of calcium must provide 10 to 19 percent of the Daily Value for calcium in one serving. Phrases such as "sugar-free" describe the amount of a nutrient in a food but don't indicate exactly how much. These nutrient content claims differ from Nutrition Facts, which do list specific nutrient amounts.

If a food label contains a descriptor for a certain nutrient but the food contains other nutrients at levels known to be less healthy, the label has to bring that to consumers' attention. For example, if a food making a low-sodium claim is also high in fat, the label must state "see back panel for information about fat and other nutrients."

HEALTH CLAIMS

Health claims state that certain foods or components of foods (such as calcium) may affect or reduce the risk of a disease or health-related condition. Examples include calcium and osteoporosis and dietary saturated fat and the risk of coronary heart disease. Although food manufacturers may use health claims approved by the FDA to market their products, the intended purpose of health claims is to benefit you as the consumer by providing information on healthful eating patterns that may help reduce the risk of heart disease, cancer, osteoporosis, high blood pressure, dental cavities, and certain birth defects.

The FDA reviews and approves all health claims before they can be used on food labels. Table 2-9 lists food label health claims on which there is significant scientific agreement. Using the FDA's ranking system for health claims, these types of claims are given the highest grade, which is grade A.

A ranking system implemented in 2003 categorizes the quality and strength of the scientific evidence for every proposed health claim from A to D (Figure 2-14). Under this system, the grade of B is assigned to claims for which there is good supporting scientific evidence but the evidence is not entirely conclusive.

Health claims graded B, C, or D are referred to as **qualified health claims** because they require a disclaimer or other qualifying language to ensure that they do not mislead consumers. For example, supplements containing selenium (a mineral thought to possibly reduce the risk of some cancers) must include this disclaimer: "Selenium may reduce the risk of certain cancers. Some scientific evidence suggests that consumption of selenium may reduce the risk of certain forms of cancer. However, the FDA has determined this evidence is limited and not conclusive."

Nutrient content claims Claims on food labels about the nutrient composition of a food, such as "good source of fiber." Regulated by the US Food and Drug Administration.

Health claims Claims on food labels that state that certain foods or food substances—as part of an overall healthy diet—may reduce the risk of certain diseases.

Qualified health claims Health claims for which there is not yet well-established evidence and which must be accompanied by an explanatory statement to ensure they do not mislead consumers.

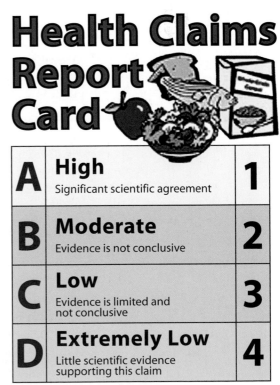

FIGURE 2-14 Ranking system for health claims.

Courtesy of the US Food and Drug Administration.

TABLE 2-8	Nutrient Content Claims: A Dictionary

NUTRIENT CONTENT CLAIMS	DEFINITION (PER SERVING)
Calories	
Calorie-free	Less than 5 kcalories/serving
Low-calorie	40 kcalories or less
Reduced or fewer calories	At least 25 percent fewer kcalories than the comparison food
Light	One-third fewer kcalories than the comparison food
Sugar	
Sugar-free	Less than 0.5 gram sugars
Reduced sugar or less sugar	At least 25 percent less sugars than the comparison food
No added sugar	No sugars added during processing or packing, including ingredients that contain natural sugars, such as fruit juice or purées
Fiber	
High-fiber	5 grams or more
Good source of fiber	2.5 to 4.9 grams
More or added fiber	At least 2.5 grams more than the comparison food
Fat and Cholesterol	
Fat-free (nonfat or no-fat)	Less than 0.5 gram fat
Percent fat-free	The amount of fat in 100 grams, may be used only if the products meets the definition of low-fat or fat-free
Low-fat	3 grams or less of fat
Light	50 percent or less of the fat than in the comparison food
Reduced or less fat	At least 25 percent less fat than the comparison food
Saturated fat-free	Less than 0.5 gram saturated fat and less than 0.5 gram trans fat
Low saturated fat	1 gram or less saturated fat and less than 0.5 gram trans fat
Reduced or less saturated fat	At least 25 percent less saturated fat than the comparison food
Trans fat-free	Less than 0.5 gram of trans fat and less than 0.5 gram of saturated fat
Cholesterol-free	Less than 2 milligrams cholesterol and 2 grams or less of saturated fat and trans fat combined
Low-cholesterol	20 milligrams cholesterol or less and 2 grams or less saturated fat and trans fat combined
Reduced or less cholesterol	25 percent or less cholesterol than the comparison food and 2 grams or less saturated fat and trans fat combined
Lean	Less than 10 grams of fat, 4.5 grams of saturated fat and trans fat combined, and 95 milligrams of cholesterol per serving and per 100 grams of meat, poultry, and seafood (about 3½ ounces)
Extra-lean	Less than 5 grams of fat, 2 grams of saturated fat and trans fat combined, and 95 milligrams of cholesterol per serving and per 100 grams of meat, poultry, and seafood (about 3½ ounces)
Sodium	
Sodium-free, salt-free	Less than 5 milligrams sodium
Very low sodium	35 milligrams or less sodium
Low sodium	140 milligrams or less sodium
Reduced or less sodium	At least 25 percent less sodium than the comparison food
Light in sodium	50 percent less sodium than the comparison food
General Claims	
High, rich in, excellent source of	Provides 20 percent or more of the Daily Value for a given nutrient
Good source, contains, provides	Provides 10 to 19 percent of the Daily Value for a given nutrient

TABLE 2-8	Nutrient Content Claims: A Dictionary (Continued)
More	Provides at least 10 percent or more of the Daily Value for a given nutrient than the comparison food
Fresh	Raw, unprocessed, or minimally processed, with no added preservatives
Healthy	Low in fat, saturated fat, cholesterol, and sodium and contains at least 10 percent of the Daily Value for vitamins A and C, calcium, iron, protein, and fiber (except for single ingredient fruits or vegetables)

Source: US Food and Drug Administration

PORTION SIZE COMPARISONS

Portion size is an important concept for anyone involved in preparing, serving, and consuming foods. Serving sizes vary from kitchen to kitchen, but American serving sizes have been increasing steadily. In comparison with MyPlate portion sizes as well as those served in many European countries, our serving sizes are huge. It wasn't that long ago that a "large" soft drink was typically 16 fluid ounces. Now that's the "small" size.

What you may consider one serving of the bread group may actually be three or four servings. For example, MyPlate considers 1 ounce of bread, about one slice, to be one serving. A typical New York–style bagel is about four ounces, or about four servings.

You may have noticed that the portion sizes in MyPlate do not always match the serving sizes found on food labels. This is the case because the purpose of MyPlate is not the same as that of nutrition labeling. MyPlate was designed to be very simple to use. Therefore, the USDA specified only a few serving sizes for each food group so that those sizes could be remembered easily. Food labels have a different purpose: to allow the consumer to compare the nutrients in equal amounts of foods. To compare the nutrient amounts in equal amounts of pasta, the portion size on the label is 2 ounces of uncooked pasta (about 56 grams), which will cook up

TABLE 2-9	Food Label Health Claims: The A List

A healthy diet with adequate **calcium** throughout life helps maintain good bone health and may reduce the risk of **osteoporosis**.

A diet **low in total fat** may reduce the risk of some **cancers**.

Diets low in **saturated fat** and **cholesterol**, and **as low as possible in trans fat**, may reduce the risk of **heart disease**.

Sugar alcohols used to sweeten foods do not promote **tooth decay**.

Low-fat diets rich in **fiber**-containing **grain products, fruit,** and **vegetables** may reduce the risk of **cancer**.

Healthful diets with adequate **folate** may reduce a woman's risk of having a child with a **brain or spinal cord defect**.

Low-fat diets rich in **fruits** and **vegetables** may reduce the risk of some **cancers**.

Diets low in **saturated fat** and **cholesterol** and rich in **fruits**, **vegetables**, and **grain** products that contain dietary **fiber**, particularly soluble fiber, may reduce the risk of **heart disease**.

Diets **low in sodium** may reduce the risk of **high blood pressure**.

Diets **low in saturated fat** and **cholesterol** that include **soluble fiber** from foods may reduce the risk of **heart disease**.

25 grams of **soy protein** a day, as part of a diet **low in saturated fat** and **cholesterol,** may reduce the risk of **heart disease**.

Diets **low in saturated fat** and **cholesterol** that include two servings of foods that provide a daily total of at least 3.4 grams of **plant stanol esters** in two meals may reduce the risk of **heart disease**.

Diets containing foods that are a good source of **potassium** and that are **low in sodium** may reduce the risk of **high blood pressure and stroke**.

Diets rich in **whole-grain** foods and other **plant foods** and low in **total fat, saturated fat,** and **cholesterol** may reduce the risk of **heart disease** and some **cancers**.

Drinking **fluoridated water** may reduce the risk of **tooth decay**.

Source: US Food and Drug Administration

FIGURE 2-15 Portion sizes.

SERVING SIZE CARD:

1 Serving Looks Like . . .

GRAIN PRODUCTS

1 cup of cereal flakes = fist

1 pancake = compact disc

½ cup of cooked rice, pasta, or potato = ½ baseball

1 slice of bread = cassette tape

1 piece of cornbread = bar of soap

1 Serving Looks Like . . .

VEGETABLES AND FRUIT

1 cup of salad greens = baseball

 1 baked potato = fist

1 med. fruit = baseball

½ cup of fresh fruit = ½ baseball

¼ cup of raisins = large egg

1 Serving Looks Like . . .

DAIRY AND CHEESE

 1½ oz. cheese = 4 stacked dice or 2 cheese slices

½ cup of ice cream = ½ baseball

FATS

1 tsp. margarine or spreads = 1 die

1 Serving Looks Like . . .

MEAT AND ALTERNATIVES

3 oz. meat, fish, and poultry = deck of cards

3 oz. grilled/baked fish = checkbook

 2 Tbsp. peanut butter = ping-pong ball

to about 1 cup of spaghetti or as much as 2 cups of a large shaped pasta such as ziti. Using MyPlate, the portion size for cooked pasta is only half a cup.

In many cases, the portion sizes are similar on labels and in the food guide, especially when expressed as household measures. For foods falling into only one major group, such as fruit juices, the household measures provided on the label (1 cup or 8 fluid ounces) can help you relate the label serving size to MyPlate serving size. For mixed dishes, MyPlate serving sizes may be used to visually estimate the food item's contribution to each food group as the food is eaten—for example, the amounts of bread, vegetables, and cheese contributed by a portion of pizza.

Use Figure 2-15 to help you visualize portion sizes. It may help to compare serving sizes to everyday objects. For example, ¼ cup of raisins is about the size of a large egg. Three ounces of meat or poultry is about the size of a deck of cards. Keep in mind that these size comparisons are approximations.

SUMMARY

1. Figure 2-12 shows required information on food labels.
2. The list of ingredients on food labels has the ingredient that is present in the largest amount by weight first, then other ingredient following in descending order of weight.
3. Food labels must clearly indicate the presence of the most common allergens such as milk, eggs, and peanuts.
4. Figure 2-13 shows a Nutrition Facts panel that includes serving size, kcalories, and selected nutrient information.
5. Daily Values are nutrient standards used on food labels to allow nutrient comparisons among foods.
6. Any nutrient claim on food labels must comply with Food and Drug Administration regulations and definitions as outlined in Table 2-8.
7. Health claims (Table 2-9) state that certain foods or components of foods (such as calcium) may reduce the risk of a disease or health-related condition. Health claims may be ranked A, B, C, or D (Figure 2-14). They must be approved by the FDA.
8. Qualified health claims (always ranked B, C, or D) require a disclaimer or other qualifying language to ensure that they do not mislead consumers.
9. Portion sizes in MyPlate don't always match the serving sizes on food labels because MyPlate only specifies a few serving sizes for each food group to keep it simple.
10. Use Figure 2-15 to help visualize portion sizes.

CHECK-OUT QUIZ

1. Besides MyPlate, other healthful food guides include the Asian diet and the *Dietary Guidelines for Americans*.
   a. True
   b. False

2. When using MyPlate, your food choices are expected to be in their most nutrient-dense forms with minimal fat or sugar.
   a. True
   b. False

3. One cup of raw leafy greens is counted as 1 cup from the vegetable group.
   a. True
   b. False

4. Fruits and vegetables are good sources of potassium and fiber and most are relatively low in kcalories.
   a. True
   b. False

5. If you need to eat 6 ounces of grains daily, 2 of them should be 100 percent whole grain.
   a. True
   b. False

6. Whole grains are good sources of fiber and several B vitamins.
   a. True
   b. False

7. One ounce of hard cheese is equivalent to 1 cup of milk or calcium fortified soymilk.
   a. True
   b. False

8. Meat and poultry choices in MyPlate include 80% lean ground beef and fried chicken.
   a. True
   b. False

9. Individuals who regularly eat meat, poultry, and fish should count beans and peas in the Protein Group.
   a. True
   b. False

10. Ice cream, pizza, cakes, and cookies all contain solid fats.
    a. True
    b. False

11. One tablespoon of vegetable oil contains 4 teaspoons of oil.
    a. True
    b. False

12. The current high rates of overweight and obesity demonstrate that many Americans are in kcalorie imbalance—that is, they consume more kcalories than they use.
    a. True
    b. False

13. One teaspoon of salt contains 300 milligrams of sodium.
    a. True
    b. False

14. Trans fat can be found in some commercial baked goods and many fried foods.
    a. True
    b. False

15. The percentage of the Daily Value on food labels is based on a 2,000 kcalorie diet.
    a. True
    b. False

16. Name the five food groups in MyPlate and two other categories in which you have an allowance. How much from each food group/category can you eat if you are on a 2,000 kcalorie diet?

17. Name five foods high in saturated fat and explain why you should limit saturated fat.

18. Using the *Dietary Guidelines for Americans, 2010*, which foods should you eat more of and which foods should you eat less of?

19. What are five tips to control kcalorie intake to lose weight?

20. What nutrients are Americans most likely to be taking in too little? What foods are they found in?

21. What are the rules about listing ingredients on a food label when the ingredient is a food allergen?

22. What is the difference between a nutrient content claim and a health claim on a food label?

NUTRITION WEB EXPLORER

For a complete list of websites for the following activities, please visit the companion book page at www.wiley.com/college/drummond.

ACADEMY OF NUTRITION AND DIETETICS (AND)

Visit the AND website and click on the tab "Public." Read one of the topics under "Latest Academy Food and Nutrition Information" and summarize what you learned.

DIETITIANS OF CANADA

Click on "Dietitians' Views" and pick a topic of interest to you. Read the information given on this topic and summarize what you learned in two paragraphs.

QUACKWATCH

Visit this website and click on "25 Ways to Spot It" under "General Observations–Quackery." List six ways to spot quackery.

USDA—DIETARY GUIDELINES FROM AROUND THE WORLD

Read about the dietary guidelines of another country. Compare and contrast them to the US Dietary Guidelines and MyPlate.

PURDUE UNIVERSITY—UNDERSTANDING NUTRITION INFORMATION

Complete Lessons 1-7 at this website to help you better evaluate nutritional claims made by the media or on packaging.

IN THE KITCHEN

MyPlate Box Lunch

You will be given ingredients to create and prepare a cold box lunch. You are to use MyPlate for guidance. Your box lunch will be evaluated as to how well it meets current dietary recommendations, as well as taste, appearance, and creativity. You will be given a worksheet to describe the ingredients and portion sizes of each item in your lunch, as well as if you have at least 1 serving from each food group. You will also need to identify any ingredients that contain solid fats or added sugars.

HOT TOPIC

FINDING RELIABLE NUTRITION INFORMATION

Food quackery has been defined by the US Surgeon General as "the promotion for profit of special foods, products, processes, or appliances with false or misleading health or therapeutic claims." Have you ever seen advertisements for supplements that are guaranteed to help you lose weight or herbal remedies to prevent serious disease? If a product's claim seems just too good to be true, it probably is too good to be true. The problem with quackery is not just loss of money—you can be harmed as well. Unproven remedies sometimes give patients false hope while important medical care is delayed.

Nutrition is brimming with quackery, in part because nutrition is such a young science. Questions on many fundamental nutrition issues, such as the relationship between sugar and obesity, are far from being resolved, yet the media publicize the results of research studies long before those results can be said to really prove a scientific theory. Unfortunately, because much research is only in its early stages, the public has been bombarded with conflicting ideas about issues that relate directly to two very important parts of their lives: people's health and their eating habits. This conflict leaves the public confused about the truth and vulnerable to dubious health products (most often nutrition products) and practices—on which people spend billions of dollars annually.

Much misinformation proliferates also because in some states anyone can call himself or herself a dietitian or nutritionist. In addition, one may even buy mail-order BS, MS, or PhD degrees in nutrition from "schools" in the United States. In all states, nutrition books that are entirely bogus can be published and sold in bookstores, dressed up to look like legitimate health books.

A quack is someone who makes excessive promises and guarantees for a nutrition product or practice that is said to enhance your physical and mental health by, for example, preventing or curing a disease, extending your life, or improving some facet of performance. Health schemes and misinformation proliferate because they thrive on wishful thinking. Many people want easy answers to their medical concerns, such as a quick and easy way to lose weight. Often, claims appear to be grounded in science. Here's how to recognize quacks.

1. Their products make claims such as:
   - Quick, painless, and/or effortless
   - Contains special, secret, foreign, ancient, or natural ingredients
   - Effective cure-all for a wide variety of conditions
   - Exclusive product not available through any other source
2. They use dubious diagnostic tests, such as hair analysis, to detect supposed nutritional deficiencies and illnesses. Then they offer you a variety of nutritional supplements, such as bee pollen or coenzymes, as remedies against deficiencies and disease.
3. They rely on personal stories of success (testimonials) rather than on scientific data for proof of effectiveness.
4. They use food essentially as medicine.
5. They often lack any valid medical or health-care credentials.
6. They come across more as salespeople than as medical professionals.
7. They offer simple answers to complex problems.
8. They claim that the medical community or government agencies refuse to acknowledge the effectiveness of their products or treatments.
9. They make dramatic statements that are refuted by reputable scientific organizations.
10. Their theories and promises are not written in medical journals using a peer-review process but appear in books written only for the lay public.

Keep in mind that there are few, if any, sudden scientific breakthroughs. Science is evolutionary—even downright slow—not revolutionary.

So, where can you find accurate nutritional information? In the United

States, over 50,000 registered dietitians (R.D.s) represent the largest and most visible group of professionals in the nutrition field. The medical profession recognizes registered dietitians as legitimate providers of nutrition care. They have specialized education in human anatomy and physiology, chemistry, medical nutrition therapy, foods and food science, the behavioral sciences, and foodservice management. Registered dietitians must complete at least a bachelor's degree from an accredited college or university, a program of college-level dietetics courses, a supervised practice experience, and a qualifying examination. Continuing education is required to maintain R.D. status. Registered dietitians work in private practice, hospitals, nursing homes, wellness centers, business and industry, and many other settings. Most are members of the American Dietetic Association, and most are licensed or certified by the state in which they live. The majority of states have enacted laws that regulate the practice of dietetics.

In addition to using the expertise of an R.D., you can ask some simple questions that will help you judge the validity of nutrition information seen in the media or heard from friends:

1. What are the credentials of the source? Does the person have academic degrees in a scientific or nutrition-related field?
2. Does the source rely on emotions rather than scientific evidence or use sensationalism to get a message across?
3. Are the promises of results for a certain dietary program reasonable or exaggerated? Is the program based on hard scientific information?
4. Is the nutrition information presented in a reliable magazine or newspaper, or is it published in an advertisement or a publication of questionable reputation?
5. Is the information someone's opinion or the result of years of valid scientific studies with possible practical nutrition implications?
6. How does this work fit with the body of existing research on the

subject? Even the most well-written article does not have enough space to discuss all relevant research on an issue. Yet it is extremely important for the article to address whether a study is confirming previous research and therefore adding more weight to scientific beliefs or whether the study's results and conclusions take a wild departure from current thinking on the subject.

Much nutrition information that we see or read is based on scientific research. It is helpful to understand how research studies are designed, as well as the pitfalls in each design, so that you can evaluate a study's results. The following three types of studies are used commonly in research.

Laboratory studies use animals such as mice or guinea pigs or tissue samples in test tubes to find out more about a process that occurs in people to determine if a substance might be beneficial or hazardous in humans or to test the effect of a treatment. A major advantage of using laboratory animals is that researchers can control many factors that they can't control in human studies. For instance, researchers can make sure that comparison groups are genetically identical and that the conditions to which they are exposed are the same. However, mice and other animals are not the same as humans, and so the results from these studies can't automatically be generalized to humans. For example, laboratory studies have indicated that the artificial sweetener saccharin causes cancer in mice, but this has never been proved for humans.

Another type of research, called epidemiological research, looks at how disease rates vary among different populations and also examines factors associated with disease. Epidemiological studies rely on observational data from human populations, and so they can only suggest a relationship between two factors; they cannot establish that a particular factor causes a disease. For example, a study looked at how much red meat adults ate over a number of years along with which of adults developed type 2 diabetes. The study showed

that the more red meat eaten, the more likely the individuals would get type 2 diabetes. This does not mean that eating red meat causes diabetes. It simply means that red meat consumption is associated with an increased risk of type 2 diabetes.

A third type of research goes beyond using animals or observational data and uses humans as subjects. Clinical trials are studies that assign similar participants randomly to two groups. One group receives the experimental treatment; the other does not. Neither the researchers nor the participants know who is in which group. For example, a clinical trial to test the effects of estrogen after menopause would randomly assign each participant to one of two groups. Both groups would take a pill, but for one group this would be a dummy pill, called a placebo. Clinical studies are used to assess the effects of nutrition-education programs and medical nutrition therapy. Unlike epidemiological studies, clinical studies can observe cause-and-effect relationships.

As you search online for nutrition information, you are likely to find websites for many health agencies and organizations that are not well known. By answering the following questions you should be able to find more information about these websites. A lot of these details can be found under the heading, "About Us" or "Contact Us."

- *Who sponsors the website? Can you easily identify the sponsor?* Websites cost money—is the funding source readily apparent? Sometimes the website address itself may help—for example:
  .gov identifies a government agency
  .edu identifies an educational institution
  .org identifies professional organizations (e.g., scientific or research societies, advocacy groups)
  .com identifies commercial websites (e.g., businesses, pharmaceutical companies, sometimes hospitals)
- *Is it obvious how you can reach the sponsor?* Trustworthy websites will have contact information for you to use. They often have a

toll-free telephone number. The website home page should list an e-mail address, phone number, or a mailing address where the sponsor and/or the authors of the information can be reached.

- *Who wrote the information?* Authors and contributors should be identified. Their affiliation and any financial interest in the content should also be clear. Be careful about testimonials. Personal stories may be helpful, but medical advice offered in a case history should be considered with a healthy dose of skepticism. There is a big difference between a website developed by a person with a financial interest in a topic and a website developed using strong scientific evidence. Reliable health information comes from scientific research that has been conducted in government, university, or private laboratories.

- *Who reviews the information? Does the website have an editorial board?* Click on the "About Us" page to see if there is an editorial board that checks the information before putting it online. Find out if the editorial board members are experts in the subject you are researching. For example, an advisory board made up of attorneys and accountants is not medically authoritative. Some websites have a section called "About Our Writers" instead of an editorial policy. Dependable websites will tell you where the health information came from and how it has been reviewed.

- *When was the information written?* New research findings can make a difference in making medically smart choices. So, it's important to find out when the information you are reading was written. Look carefully on the home page to find out when the website was last updated. The date is often found at the bottom of the home page. Remember: Older information isn't useless. Many websites provide older articles so readers can get an historical view of the information.

- *Is your privacy protected? Does the website clearly state a privacy policy?* This is important because, sadly, there is fraud on the Internet. Take time to read the website's policy—if the website says something like, "We share information with companies that can provide you with products," that's a sign your information isn't private.

- *Does the website make claims that seem too good to be true? Are quick, miraculous cures promised?* Be careful of claims that any one remedy will cure a lot of different illnesses. Be skeptical of sensational writing or dramatic cures. Make sure you can find other websites with the same information. Don't be fooled by a long list of links—any website can link to another, so no endorsement can be implied from a shared link.

Two organizations with great websites that will help you separate fact from fiction are Quackwatch and the National Council Against Health Fraud.

Table 2-10 lists organizations with websites with reliable nutrition information.

TABLE 2-10	Websites with Reliable Nutrition Information

American College of Sports Medicine—Provides fact sheets and more on a variety of health and fitness topics.

Center for Nutrition Policy and Promotion of the USDA—Promotes dietary guidance that links scientific research to the nutrition needs of consumers, and provides links to everything you need to know about the *Dietary Guidelines for Americans* and MyPlate.

EatRight—The website of the Academy of Nutrition and Dietetics, the world's largest organization of food and nutrition professionals.

Food and Nutrition Information Center of the USDA—Provides a directory to credible, accurate, and practical resources.

Healthfinder—A source of reliable health and nutrition information from the federal government.

Healthy Flavors, Healthy Kids at the Culinary Institute of America—An innovative website with important information and recipe solutions to address the problem of childhood obesity.

Mayo Clinic (Health Information tab)—More than 3,000 physicians, scientists, and researchers from Mayo Clinic share health information and expertise.

National Center for Complementary and Alternative Medicine—A federal agency that explores complementary and alternative medicine, and helps consumers stay informed to make wise health decisions.

Nutrition.gov—US federal guide offering access to all government websites with reliable and accurate information on nutrition and dietary guidance.

US Food and Drug Administration (Food tab)—Covers many topics from nutrition facts label programs to food safety to hot topics in food and nutrition.

Source: With permission, this Hot Topic used sections of "If It Sounds Too Good to Be True . . . It Probably Needs a Second Look" from *Food Insight,* published by the International Food Information Council Foundation, March/April 1999.

CARBOHYDRATES

CHAPTER 3

- Identify food sources of carbohydrates and distinguish between simple and complex carbohydrates.

- Compare and contrast glucose, fructose, sucrose, and lactose.

- Identify sugars on an ingredient label, foods high in added sugars, and the number of teaspoons of sugar in a food using a food label.

- Identify the simple sugar found in starch and fiber, list four foods rich in starch, and explain gelatinization and how starch is used in cooking.

- Identify examples of high-fiber foods and explain the difference between soluble and insoluble fiber, and between dietary fiber and functional fiber.

- Distinguish between a whole grain and a refined grain and explain why a whole grain is more nutritious.

- Summarize the functions of carbohydrates and describe how glycogen functions in the body.

- Describe how carbohydrates are digested and absorbed in the body, and explain how the body regulates the level of glucose in the blood.

- Identify foods with low to medium glycemic loads and how a low glycemic diet might affect your health.

- Discuss current recommendations for carbohydrate, sugar, fiber, and intake of fruits, vegetables, legumes, and whole grains.

- Explain the health effects (if any) of added sugars on dental cavities, obesity, diabetes, heart disease, hypoglycemia, and hyperactivity in children.

- Demonstrate how to select whole grains, and list two ways eating whole grains can improve your health.

- List three ways that a high fiber diet can improve your health.

- Define lactose intolerance and describe three strategies to manage it.

- Describe how to cook whole grains and legumes.

- Create an appetizer, entrée, side dish, salad, and snack using high-fiber carbohydrate foods.

- Read food labels to identify foods using alternative sweeteners.

INTRODUCTION TO CARBOHYDRATES

Carbohydrate A large class of nutrients, including sugars, starch, and fibers, that function as the body's primary source of energy.

You may, like many Americans, have a love-hate relationship with carbohydrates. You love to eat them, especially snack foods like cookies or pretzels or soda (and they are so easy to eat or drink), but you hate them because they have a reputation for being fattening. **Carbohydrates** include sugars, starches, and fiber. As you will learn in this chapter, carbohydrates containing fiber are your healthiest carbohydrates, including whole grains (such as whole wheat and products made from them), fruits, and vegetables.

We find carbohydrates in many of the foods we eat. Grains, such as oats and rice, are rich in carbohydrates. When grains such as wheat are ground, it produces flour used to bake breads, tortillas, crackers, cakes, and cookies (Figure 3-1). Fruits, vegetables, and milk also contain carbohydrates.

Photosynthesis A process in which plants use energy from sunlight to convert carbon dioxide and water to carbohydrate.

As you may have noticed, most carbohydrate foods are plant foods. Plants make their own carbohydrates in a process known as **photosynthesis**. In photosynthesis, plants use carbon dioxide from the air, water from the soil, and energy from the sun to make a carbohydrate called glucose. The plant then uses the energy in carbohydrates to grow and be healthy, much like we do.

Simple carbohydrates or sugars A form of carbohydrate that includes sugars occurring naturally in foods, such as fructose in fruits, as well as sugars that are added to foods, such as brown sugar in a cookie.

Carbohydrates are separated into two categories: simple and complex. **Simple carbohydrates** include sugars that occur naturally in foods, such as fructose in fruits or glucose in honey, as well as sugars that are added to foods, such as white or brown sugar in a chocolate chip cookie.

Complex carbohydrates Carbohydrates made of chains of sugars, including starches and fibers.

Carbohydrates are much more than just sugars, though, and include the **complex carbohydrates** starch and fiber. Starch and fiber are indeed more "complex" because they consist of long chains of many sugars. Starch and fiber are only found in plant foods such as fruits, vegetables, and grains.

FIGURE 3-1 Products made from wheat.

Courtesy of the US Department of Agriculture.

SUMMARY

1. Carbohydrates include sugars, starches, and fibers. They are found mostly in plant foods.
2. Carbohydrates are either simple or complex. Simple carbohydrates include sugars that occur naturally in food (such as fruit) and processed sugars (such as white sugar) that are added to foods.
3. Complex carbohydrates include starches and fibers. Starch and fiber are only found in plant foods such as fruits, vegetables, and grains.

SIMPLE CARBOHYDRATES (SUGARS)

Simple carbohydrates include monosaccharides and disaccharides. **Monosaccharides** are made of a single sugar unit ("mono" means one). **Disaccharides** are made of two sugar units ("di" means two). The chemical names of the six sugars to be discussed in this section all end in "-ose," which means sugar.

MONOSACCHARIDES (SIMPLE SUGARS)

Monosaccharides include these simple sugars.

1. Glucose
2. Fructose
3. Galactose

Monosaccharides consist of a single ring of atoms (Figure 3-2), and they are the building blocks of other carbohydrates, such as disaccharides and starch.

    Glucose is the most abundant sugar found in nature. Plants make their own glucose, which is burned for energy, and glucose is our primary energy source as well. Most of the carbohydrates

Monosaccharides Simple sugars, including glucose, fructose, and galactose, which are the building blocks for other carbohydrates, such as disaccharides and starch.

Disaccharides Pairs of monosaccharides linked together, including sucrose, maltose, and lactose.

Glucose The most important monosaccharide; the body's primary source of energy.

FIGURE 3-2 Monosaccharides and disaccharides.

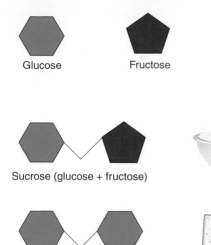

Glucose Fructose Galactose

Sucrose (glucose + fructose)

Maltose (glucose + glucose)

Lactose (glucose + galactose)

Blood glucose level (blood sugar level) The amount of glucose found in the blood; glucose is vital to the proper functioning of your body, as it provides most of your energy.

Fructose A monosaccharide found in fruits and honey; the sweetest natural sugar.

Galactose A monosaccharide found linked to glucose to form the sugar found in milk.

Sucrose (Sugar) A disaccharide formed by linking glucose and fructose; commonly called white sugar, table sugar, or simply sugar.

FIGURE 3-3 Granulated sugar is made from sugar cane and sugar beets.

Courtesy of the U.S. Department of Agriculture.

you eat, such as starch in bread, produce glucose for the body to use or store. The concentration of glucose in the blood, referred to as the **blood glucose level**, is vital to the proper functioning of your body. As glucose travels in the blood around the body, it enters cells where it is used as fuel. As a monosaccharide, glucose is found in fruits such as grapes, in honey, and in small amounts in many plant foods. Glucose is also part of all the disaccharides and most polysaccharides.

Fructose, the sweetest natural sugar, is found in fruits and also in honey. Fructose is also found in high fructose corn syrup, which is not found in nature but is manufactured from corn syrup. High fructose corn syrup is used as a sweetener in soft drinks, candy, and other foods.

The last single sugar, **galactose**, is almost always linked to glucose to make the sugar found in milk.

DISACCHARIDES (DOUBLE SUGARS)

Disaccharides, the double sugars, include sucrose, maltose, and lactose. They each contain glucose (Figure 3-2). **Sucrose** is the chemical name for what is commonly called white sugar, table sugar, granulated sugar, or simply sugar. White sugar is used to make other products such as brown sugar and powdered sugar. As Figure 3-3 indicates, sucrose is simply two common

FIGURE 3-4 Milk and most dairy products contain a sugar called lactose.

Photo by Peter Pioppo.

single sugars—glucose and fructose—linked together. Sucrose does occur naturally in small amounts in many fruits and vegetables.

Sucrose is found in sugar cane, a very tall grass that grows in tropical countries, and sugar beets, a root crop (Figure 3-3). The crude raw sugar made from these plants is not suitable for human consumption unless further refined or processed. White sugar is made from either sugar cane or sugar beets. White sugar is more than 99 percent pure sugar and provides virtually no nutrients for its 16 kcalories per teaspoon.

Maltose, present in germinating grains, is used to make beer.

The last disaccharide, **lactose**, is commonly called milk sugar because it occurs in milk and products made with milk (Figure 3-4). Although milk is not a food you think of as sweet, there is some sugar there. Lactose is one of the lowest-ranking sugars in terms of sweetness (Table 3-1). Lactose is the only sugar found naturally in animal products.

Lactose A disaccharide found in milk and milk products that is only slightly sweet.

TABLE 3-1	Relative Sweetness of Sugars and Alternative Sweetener
SWEETENER	**RATING**
Sugars	
Lactose	20
Glucose	70–80
Sucrose	100
High-fructose corn syrup	120–160
Fructose	140
Artificial Sweeteners	
Aspartame (Nutrasweet, Equal)	160–220
Acesulfame-K (Sunette)	200
Saccharin (Sweet 'N Low)	200–700
Sucralose (Splenda)	600
Neotame	7,000–13,000

Added sugars Sugars, syrups, and other kcaloric sweeteners that are added to foods during processing, preparation, or consumed separately. These do not include naturally occurring sugars such as those in fruit.

ADDED SUGARS

Added sugars are sugars like white sugar that are added to a food for sweetening and flavor (such as soda and cookies) or added to foods or beverages at the table (such as sugar added to coffee) (Table 3-2). Currently sugar-sweetened beverages are the primary source of added sugars in the American diet for both adults and children, followed by

TABLE 3-2	Sugars Used in Cooking and Baking
NAME	**DESCRIPTION**
Regular granulated sugar	Made from beet sugar or cane sugar. Refining removes the yellow-brown pigments of unrefined sugar. Also called table sugar, fine granulated, or sugar.
Very fine and ultrafine sugars	Finer than regular granulated. Can be used for making cakes and cookies because bakers can incorporate more fat into the recipe and the batter is more uniform.
Confectioner's sugar (powdered sugar)	Granulated sugar that has been crushed into a fine powder, and combined with a small amount of starch to prevent clumping. Available in several degrees of fineness, designated by the number of Xs following the name. 10X is the finest sugar and gives the smoothest texture in icings. 6X is the standard confectioners' sugar. Coarser types can be used for dusting.
Molasses	Molasses is concentrated sugar cane juice. If most of the sugar has been extracted from the juice, it is called sulfured molasses and has a bitter taste. Unsulfured molasses is less bitter because the sugar has not been taken out of it. Molasses contains large amounts of sucrose, other sugars, acids, and impurities.
Brown sugar	Sugar crystals contained in a molasses syrup with natural flavor and color—91 to 96% sucrose.
Maple sugar	A sugar made from the sap of sugar maple trees.
Turbinado sugar	Partially refined raw sugar that retains some of the natural molasses. Blond in color with large crystals and a mild brown sugar flavor. The large crystals do not readily dissolve. Has been steam-washed and spun in a turbine.
Demerara sugar	A light brown sugar with large crystals that is named after the region in South America from where it originally came. The minimal processing gives demerara sugar a unique flavor and texture. It is popular in countries such as the United Kingdom.
Fructose	Found naturally in fruits. Available in crystals or a light syrup. Sweeter than granulated sugar.
Corn syrup	A thick, sweet syrup made from cornstarch. Mostly glucose with some maltose. Only 75% as sweet as sucrose. Less expensive than sucrose. It helps to retain moisture and is used in some icings and candy.
High-fructose corn syrup	Corn syrup treated with an enzyme that converts some of the glucose to fructose. Sweeter than corn syrup and used in soft drinks, baked goods, jelly, fruit drinks, and other products.
Honey	Sweet syrupy fluid made by bees from the nectar collected from flowers. Honey is stored in nests or hives as food. Made mostly of glucose and fructose, plus other components that give it flavor. The flavor and color of honey vary, depending on the nectar source. Lighter shades are more mellow and darker shades richer and can be slightly bitter. One tablespoon of honey contains significantly more sugar and kcalories than granulated sugar (both provide kcalories with virtually no nutrients). Honey is associated with a form of botulism that has been linked with sudden infant death syndrome, so it is important never to feed it to an infant under one year of age.
Maple syrup	High-quality pure maple syrup can be made only by the evaporation of pure maple sap (from maple trees), and by weight may contain no less than 66 percent sugar. Maple syrup is classified according to its color, which is a rough guide to flavor intensity. The darker the syrup, the stronger the flavors.
Brown rice syrup	A sweetener made by culturing cooked brown rice with enzymes to break down the starches, then straining off the liquid and reducing it.
Agave nectar/syrup	Made from the sap of the agave plant. One tablespoon contains significantly more sugar and kcalories than granulated sugar. Very rich in fructose so it is sweet and has a lower glycemic index than granulated sugar.

FIGURE 3-5 Added Sugars. Examples of foods high in added sugars include pastries, cupcakes, donuts, cookies, candy, sweetened cereals, regular soft drinks, chocolate, and, of course, sugar.

Photo by Peter Pioppo.

grains-based desserts, sugar-sweetened fruit drinks, dairy-based desserts, and candy (Figure 3-5). In 2005–2008, half of Americans aged 2 and over drank sugar-sweetened drinks every day.

Added sugar is in soda, juice drinks, sweetened iced tea, and lemonade, as well as in many sports drinks, vitamin waters, and energy drinks. Fruit juices that are not 100 percent fruit juice usually have sugar added to them. Foods high in added sugars often contain few nutrients for the number of kcalories they provide. In other words, these foods are not nutrient-dense, as discussed in Chapter 2.

High-fructose corn syrup is corn syrup that has been treated with an enzyme to convert part of the glucose it contains to fructose, which is sweeter than glucose. High-fructose corn syrup is therefore sweeter, ounce for ounce, than corn syrup, and so smaller amounts can be used (making it cheaper). High-fructose corn syrup replaced sugar in many soft drinks and other sweetened foods starting in the 1980s. High-fructose corn syrup has been blamed recently in the media for the obesity epidemic, because obesity rates have gone up at the same time as the consumption of high-fructose corn syrup has increased. High-fructose corn syrup and sugar are equally harmful in excess—the two sweeteners are indistinguishable once absorbed into the body. In a moment we will discuss what we know about sugar affecting health issues such as obesity.

High-fructose corn syrup Corn syrup that has been treated with an enzyme that converts part of the glucose it contains to fructose; found in some sweetened foods.

In addition to sweetening foods, sugars prevent spoilage in jams and jellies. In cooking and baking, sugar performs many functions:

1. Sugar helps to balance the acidity of ingredients such as tomatoes and vinegar, resulting in balanced flavors.

2. In baking, sugars are responsible for browning the crust.

3. Sugar also helps retain moisture in baked goods so that they stay fresh longer.

4. Depending on how much sugar is used, sugars affect the texture of various foods. In general, increasing amounts of sugar makes baked products more tender.

5. Sugar also acts as a food for yeast in breads and other baked goods that use yeast for leavening.

When you look at "Sugars" on the Nutrition Facts panel, keep in mind that the number of grams given includes naturally occurring sugars and added sugars. Your body processes natural and added sugars in the same way. In fact, your body does not see any difference between a natural sugar, such as fructose in an apple, or high-fructose corn syrup in a soda.

To find out whether a food contains added sugar and how much, look first at the ingredient list. For example, here are the ingredients for an apple pie: "Apples, corn syrup, sugar, wheat flour, water, modified corn starch, dextrose, brown sugar, sodium alginate, spices, citric acid, salt, lecithin, dicalcium phosphate." The ingredients are listed in order by weight. The ingredient in the greatest amount by weight is listed first, and the one in the least amount is listed last. In this example, corn syrup is the second ingredient listed and sugar is the third, which means that combined, these two sugars are the main ingredients in the apple pie.

To determine exactly how much sugar is in the pie, keep in mind that one teaspoon of sugar weighs 4 grams (Figure 3-6). Therefore if the nutrition label says that one slice of pie has 40 grams of sugar, then a slice contains 10 teaspoons of sugar, as shown here.

$$\frac{40 \text{ grams of sugar}}{4 \text{ grams of carbohydrate in a teaspoon of sugar}} = 10 \text{ teaspoons sugar}$$

Table 3-3 shows the number of teaspoons of sugar in selected foods.

TABLE 3-3	Amount of Added Sugars in Selected Foods	
FOOD	**PORTION SIZE**	**AMOUNT OF SUGAR***
Foods high in sugar (4 or more teaspoons/serving)		
Cinnabon Classic Cinnamon Roll	1-7.8 oz.	14 teaspoons
Snapple Lemon Tea	16 fl oz.	10.5
Coca-Cola	12 fl. oz.	10
Ginger ale	12 fl. oz.	8
Starbucks Chocolate Chunks Cookie	3 oz.	8
Hershey Milk Chocolate Bar	16 oz.	5.5
Reese's Peanut Butter Cup	2	5.5
Jellybeans	10 large	5
Starbucks Tazo Shaked Ice Tea	16 fl oz.	5
Chocolate cupcake, crème-filled	1 cupcake	4.7
Sherbet, orange	½ cup	4.5
Honey	1 Tbsp.	4.3
Pancake syrup	¼ cup	4.3
Ice cream, vanilla	½ cup	4
Foods moderately high in sugar (1 to 3.9 teaspoons/serving)		
Frosted Flakes	1 cup	3.8
Vanilla ice cream	½ cup	3.5
Oreo Cookies	3	3.5
Syrups, chocolate, fudge-type	1 Tbsp.	2.8
Doughnuts, cake-type, plain	1 medium	2.6
Kellogg's Corn Pops	1 cup	2.6
Jams and preserves	1 Tbsp.	2.4
Barbecue sauce	2 Tbsp.	2.1
Catsup	2 Tbsp.	1.7
French dressing	2 Tbsp.	1.2

*4 grams sugar = 1 teaspoon

Source: US Department of Agriculture, Agricultural Research Service, 2010. *USDA Nutrient Database for Standard Reference, Release 23.* Nutrient Data Laboratory Home Page, www.nal.usda.gov/fnic/foodcomp. And Manufacturers' Food Labels.

FIGURE 3-6 One teaspoon of sugar or 2 sugar cubes contain 4 grams of sugar.

Photo by Peter Pioppo.

SUMMARY

1. Simple carbohydrates (also called simple sugars) include monosaccharides and disaccharides.
2. Monosaccharides include glucose, fructose, and galactose.
3. Most of the carbohydrates you eat are converted to glucose in the body. Your blood glucose level is vital to the proper functioning of the body because glucose provides energy. Glucose is found in fruits such as grapes, in honey, and in small amounts in many plant foods.
4. Fructose, the sweetest natural sugar, is found in fruits and honey. Galactose is part of the sugar in milk.
5. Disaccharides include sucrose (white sugar), maltose, and lactose (milk sugar). Sucrose is composed of 1 glucose and 1 fructose molecule.
6. Table 3-1 lists sources of added sugar, such as high-fructose corn syrup, found in the foods you buy. High-fructose corn syrup and sugar are equally harmful in excess.
7. Currently sugar-sweetened beverages are the primary source of added sugars in the American diet for both adults and children, followed by grains-based desserts, sugar-sweetened fruit drinks, dairy-based desserts, and candy.
8. Sugar performs many functions in cooking and baking: sweetness, balancing acidity of ingredients such as tomatoes, browning, retaining moisture in baked goods, giving a tender texture, and acting as a food for yeast.
9. "Sugars" on the Nutrition Facts label includes the number of grams of both naturally occurring sugars and added sugars. One teaspoon of sugar weighs 4 grams.

COMPLEX CARBOHYDRATES

The complex carbohydrates include starches and fibers found in plants.

STARCHES

Plants store glucose in the form of **starch**. Starch is made of many chains of hundreds to thousands of glucoses linked together. Starch is found only in plant foods (Figure 3-7). Grains, the seeds of cultivated grasses, are rich sources of starch and include wheat, corn, rice, rye, barley, and oats.

Starch A complex carbohydrate made up of a long chain of glucoses linked together; found in grains, legumes, vegetables, and some fruits.

TABLE 3-4   Whole Grains and Refined Grains

WHOLE GRAINS

Brown rice

Wild rice

Bulgur (cracked wheat)

Millet

Oatmeal

Rolled oats

Whole oats

Oatmeal

Whole wheat flour

Whole wheat bread, rolls, pasta, and cereals

Whole-grain sorghum

Whole-grain triticale

Whole-grain barley

Whole-grain corn

Whole rye

Buckwheat

Popcorn

Quinoa

REFINED GRAINS

White flour including all-purpose, cake, and pastry flour

White bread and rolls

White pasta

White rice

Degermed cornmeal

Corn flour

White rice flour

Endosperm In cereal grains, a large center area high in starch.

Germ In grains, the area of the kernels rich in vitamins and minerals that sprouts when allowed to germinate.

Bran In grains, the part that covers the grain and contains much fiber and other nutrients.

Whole Grains Grains and grain products made from the entire grain seed, usually called the kernel, which consists of the bran, germ, and endosperm.

crackers, cereals, baked goods, and many other foods. Rice, rye, barley, and oats are also grains (see Table 3-4).

All grains have a large center area high in starch known as the **endosperm**. The endosperm, about 85 percent of the kernel, also contains protein and B vitamins. At one end of the endosperm is the **germ**, the area of the kernel that germinates to grow into a new plant. The germ is rich in vitamins and minerals and contains some oil. The **bran**, containing much fiber and other nutrients, covers and protects the endosperm and germ. The seed contains everything needed to reproduce the plant: The germ is the embryo, the endosperm contains the nutrients for growth, and the bran protects the entire seed (Figure 3-9).

Most grains undergo some type of processing or milling after harvesting to allow them to cook more quickly and easily, make them less chewy, and lengthen their shelf life. Grains such as oats and rice have an outer husk or hull that is tough and inedible, and so it is removed. Other processing steps may include removing the bran and germ from the endosperm of the wheat kernel and further milling (as in white flour), cracking the grain (as in cracked wheat), or rolling and slightly steaming the grain (as in rolled oats) to shorten the cooking time.

A grain is called a **whole grain** whenever the fiber-rich bran and the vitamin-rich germ are left on the

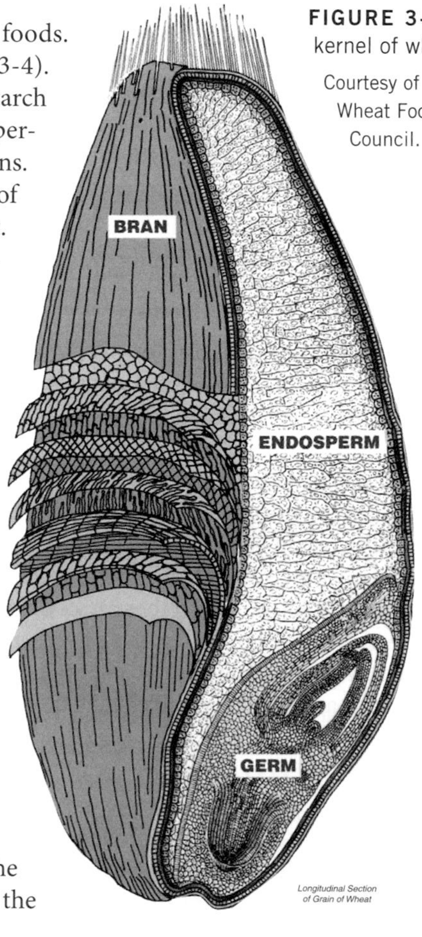

FIGURE 3-9 A kernel of wheat.

Courtesy of the Wheat Foods Council.

BRAN

ENDOSPERM

GERM

Longitudinal Section of Grain of Wheat

FIGURE 3-10 Whole grains. Top row: hard winter wheat berries, yellow popcorn, farro, rolled oats. Middle row: Quinoa, hulled buckwheat, rye berries, long-grain brown rice. Bottom row: Kasha, bulgur, whole wheat flour.

Photo by Peter Pioppo.

endosperm of a grain. Examples of whole grains include whole-wheat berries, bulgur (cracked wheat), whole-wheat flour, whole rye, whole oats, rolled oats (including oatmeal), whole corn-meal, whole hulled barley, popcorn, brown rice and wild rice (Figure 3-10). More exotic whole grains include amaranth, millet, and quinoa.

If the bran and germ are removed (or mostly removed) from the endosperm, the grain is called a **refined or milled grain**. Whereas whole-wheat flour is made from the whole grain, white flour (also called wheat flour) is made only from the endosperm of the wheat kernel. Whole-wheat flour does not stay fresh as long as white flours do. This is due to the presence of the germ, which contains oil. When the oil turns rancid or deteriorates, the flour will turn out a poor-quality product.

When you compare the nutrients in whole grains and refined grains, whole grains are always a far more nutritious choice (Figure 3-11). They surpass refined grains in their fiber, vitamin, and mineral content. When wheat is refined, over 20 nutrients and most of the fiber are removed. By federal law, refined grains are enriched with five nutrients that are lost

Refined grain Grains and grain products missing the bran, germ, and/or endosperm; any grain product that is not a whole grain.

FIGURE 3-11 On the right is whole wheat bread made from whole wheat flour. On the left is white bread, made from white flour that doesn't include the bran or germ. Whole wheat bread is more nutritious.

Photo by Peter Pioppo.

Phytochemicals A wide variety of compounds produced by plants that promote health.

in processing: thiamin, riboflavin, niacin, folate, and iron. With whole wheat you get more vitamin E, vitamin B$_6$, magnesium, zinc, potassium, copper, and, of course, fiber. Whole-grain foods also contain **phytochemicals**, substances in plants that have health-promoting properties.

SUMMARY

1. Starch is only found in plant foods such as whole grains, beans and peas, and certain vegetables such as potatoes.

2. Starches, such as cornstarch, are used extensively as thickeners in cooking. When heated in liquid, starches gelatinize meaning the granules absorb water and swell, making the liquid thicken.

3. Dietary fiber refers to the complex carbohydrates found in plants that are not digested by human digestive enzymes. Most dietary fibers are chains of glucose molecules and our digestive enzymes can't break the chemical bonds between the molecules, so most fiber is excreted. Some fiber is digested by bacteria in the colon.

4. Fiber is abundant in plants, so legumes, fruits, vegetables, whole grains, nuts, and seeds provide fiber. Fiber is not found in meat, poultry, fish, dairy products, or eggs.

5. Most foods contain a variety of soluble and insoluble fiber (based on how soluble they are in water).

6. Foods rich in soluble fiber include oats, barley, beans, many fruits (such as apples and pears), and many vegetables (such as carrots and sweet potatoes). Soluble fiber helps lower blood cholesterol levels and makes you feel full after eating.

7. Insoluble fiber does not swell in water and is found in wheat bran, whole grains, legumes, many vegetables and fruits, and seeds. Insoluble fiber helps prevent constipation.

8. Total fiber includes dietary fiber and functional fibers—fibers extracted from plants (such as oat fiber or vegetable gums) and added to a wide variety of foods. It is unclear if functional fiber provides the same health benefits as naturally occurring fiber.

9. Grains are the edible seeds of annual grasses such as wheat, oats, and rice. Grains contain an endosperm, germ, and bran.

10. Whenever the fiber-rich bran and vitamin-rich germ are left on the endosperm of a grain, the grain is a whole grain. When the bran and germ are removed from the endosperm, the grain is a refined or milled grain. Whereas whole-wheat flour is made from the whole grain, white flour is made only from the endosperm is not nearly as nutritious as whole wheat flour. Table 3-4 lists whole grains and refined grains.

11. By federal law, refined grains are enriched with thiamin, riboflavin, niacin, folate, and iron, to make up for some of the nutritional losses.

FUNCTIONS OF CARBOHYDRATES

Carbohydrates are the primary source of the body's energy (Figure 3-12), supplying about 4 kcalories per gram. Glucose, a simple carbohydrate, is the body's number-one source of energy. Most of the carbohydrates you eat are converted to glucose in the body.

Our cells can burn protein and fat for energy, but the body uses glucose first, in part because glucose is the most efficient energy source. The brain and other nerve cells are picky about their foods, and under most circumstances, they will only use glucose for energy.

Glycogen The storage form of glucose in the body; stored in the liver and muscles.

Some glucose from food is stored in your body in a form called **glycogen**. This way the body has a constant, available glucose source. Glycogen is stored in two places in the body: the liver and the muscles. When your blood sugar glucose starts to dip and more energy is needed, the liver converts glycogen into glucose, which then is delivered by the bloodstream to the cells. Much of the body's glycogen—about 75 percent—is stored in the muscles. Muscle glycogen does not supply glucose to the bloodstream but is used strictly as an energy source for the muscles during exercise.

Glycogen stores will give you energy for less than a day provided you are at rest. If you are training or exercising, your glycogen will last a few hours depending on variables such as the intensity of the activity and your fitness level.

If you run out of glycogen and do not eat any carbohydrates, the body will break down protein in muscles to some extent. Protein can be converted to glucose to maintain glucose levels in the blood and supply glucose to the central nervous system (brain and spinal cord). Carbohydrates in the diet spare protein from being burned for energy so that protein can be used to build and repair the body, its primary job.

Too little carbohydrate can also cause the body to convert some fat to glucose, but this is also not desirable. When fat is burned for energy without any carbohydrates present, the process is incomplete and can harm the body.

FIGURE 3-12 Carbohydrate is the body's number one source of energy.

Reprinted with permission of John Wiley & Sons, Inc.

You need at least 130 grams of carbohydrates daily to prevent protein and fat from being burned for fuel and to provide glucose to the central nervous system. This amount represents what you minimally need, not what is desirable (about two times more). You eat about 130 grams of carbohydrate in 1 cup of cereal, ¾ cup of milk, 1 cup of juice, and 1 bagel with 2 tablespoons jelly. We obtain about 50 to 60 percent of our energy intake from carbohydrates. Therefore, if you eat 2000 kcalories per day, you take in 1,000 to 1,200 kcalories of carbohydrates, which represents 250 to 300 grams—about twice the minimum of 130 grams carbohydrate.

Carbohydrates are part of various materials found in the body, such as connective tissues, some hormones and enzymes, and genetic material.

A generous intake of fiber lowers blood cholesterol, may reduce blood pressure modestly, decreases the risk of developing diabetes, helps to keep blood sugar at normal levels, promotes regularity, and is linked to lower body weights. The health effects of fiber are discussed in more detail in the Fiber and Its Health Effects section later in this chapter.

SUMMARY

1. Carbohydrates are the primary source of the body's energy and glucose is the body's number one source of energy. Under most circumstances, the brain and other nerve cells will only use glucose for energy.

2. Glycogen is a storage form of glucose in the liver and muscle that fills up after you eat carbohydrates. When your blood glucose level drops, the liver converts glycogen to glucose. Glycogen stores will give you energy for less than a day provided you are at rest. Muscle glycogen is used strictly to provide energy to muscles during exercise. Carbohydrates in the diet spare protein from being burned for energy.

3. Carbohydrates are also important to help the body use fat efficiently. When fat is burned for energy without any carbohydrates present, the process can harm the body.

4. You need at least 130 grams of carbohydrates daily to prevent protein and fat from being burned for fuel and to provide glucose to the nervous system. This amount is what is minimally needed—not what is desirable.

5. Carbohydrates are part of various materials in the body such as connective tissues, some hormones and enzymes, and genetic material.

6. A generous intake of fiber lowers blood cholesterol, may reduce blood pressure modestly, decreases the risk of developing diabetes, helps to keep blood sugar at normal levels, promotes regularity, and is linked to lower body weights.

DIGESTION, ABSORPTION, AND GLYCEMIC RESPONSE

Cooking carbohydrate-containing foods makes them easier to digest. As mentioned, starches gelatinize, making them easier to chew, swallow, and digest. Cooking also breaks down fiber in most raw fruits and vegetables.

Most of the digestion and absorption of carbohydrates take place in the small intestine. Before any carbohydrates can be absorbed, they must be broken down into one-sugar units.

During digestion, various enzymes break down starch and disaccharides into one-sugar units that are then absorbed and carried to the liver. In the liver, most fructose and galactose are converted to glucose.

The level of glucose in your blood is very closely regulated because glucose is the body's preferred fuel, especially for your brain and nervous system. When your blood glucose level is within the normal range, most of your body's cells will use glucose for their energy source.

Two hormones, **insulin** and **glucagon**, are responsible for hour-to-hour regulation of your blood glucose levels. The pancreas secretes both of these hormones. After a meal, blood glucose levels start to increase, and insulin is released. Insulin is necessary for glucose to get into most of your cells. Once the cells have the glucose they need, excess glucose is stored as glycogen or fat.

When blood glucose levels fall between meals and at night, insulin secretion slows and glucagon secretion increases. Glucagon's job is to prevent low blood sugar levels. Glucagon stimulates the liver to convert glycogen to glucose and release the glucose into the bloodstream. Figure 3-13 shows how these two hormones work.

Insulin A hormone that is necessary for glucose to leave the bloodstream and enter the body's cells.

Glucagon A hormone that stimulates the liver to convert glycogen to glucose and release the glucose into the bloodstream to bring the blood glucose level to normal levels.

FIGURE 3-13 How insulin and glucagon maintain normal blood glucose levels.

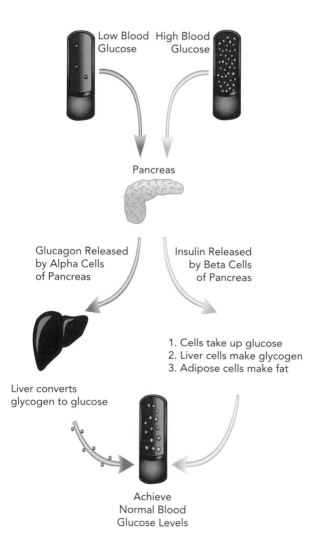

Low Blood Glucose High Blood Glucose

Pancreas

Glucagon Released by Alpha Cells of Pancreas

Insulin Released by Beta Cells of Pancreas

1. Cells take up glucose
2. Liver cells make glycogen
3. Adipose cells make fat

Liver converts glycogen to glucose

Achieve Normal Blood Glucose Levels

Fiber cannot be digested, or broken down into its components by human enzymes. Some bacteria in the large intestine digest soluble fiber and produce small fatty acids.

Soluble fiber slows the emptying of food from the stomach into the small intestine and also slows the absorption of glucose into the blood. This helps keep your blood glucose level stable. By delaying the emptying of food into the small intestine, soluble fiber also creates a feeling of fullness after eating.

Carbohydrate foods can be grouped according to whether they produce a high, moderate, or low **glycemic response.** Glycemic response refers to how quickly, how high, and how long your blood sugar rises after eating. Any number of factors influence how high your blood glucose will rise, such as the amount of carbohydrate eaten, the type of sugar or starch, and the presence of fat, protein, and fiber (which all slow down the emptying of the stomach). A low glycemic response (meaning that your blood sugar level rises slowly and modestly) is preferable to a high glycemic response (your blood sugar level rises quickly and high).

A more recent way to assess glycemic response is called **glycemic load**. It is more accurate because it takes into account actual portion size. Table 3-5 gives you an idea of which foods are considered low, medium, or high glycemic load. Again you see that most fruits and vegetables,

Glycemic response How quickly, how high, and how long your blood sugar rises after eating.

Glycemic load An index of the glycemic response that occurs after eating a specific food and takes into account portion size.

TABLE 3-5	Glycemic Load of Selected Foods

High Glycemic Load

- Chocolate cake with frosting (1 slice)
- Cornflakes or Rice Krispies (1 oz.)
- Boiled white long-grain rice (5 oz.)
- Spaghetti made from durum wheat (6 oz.)
- Baked potato or French fries (5 oz.)
- Coca-Cola (12 fl oz.)
- Jelly beans (1 oz.)
- Mars bar (2 oz.)

Medium Glycemic Load

- Puffed Wheat or Raisin Bran cereal (1 oz.)
- Steamed brown rice (5 oz.)
- Raw banana (1)
- Apple juice, unsweetened (1 cup)
- White bread (1 oz. slice)
- Pretzels (1 oz.)

Low Glycemic Load

- Whole wheat bread (1 oz. slice)
- All-Bran cereal (1 oz.)
- Raw apples, oranges, peaches, grapes, pears (1 piece)
- Cooked peas or carrots (5 oz.)
- Many legumes such as kidney beans, black beans, lentils, and chickpeas (about ¾ cup)
- Regular or fat-free milk (8 fl oz.)
- Yogurt with fruit (6 oz.)
- Peanuts (1 oz.)

legumes, and whole grains have low or medium glycemic loads. Choosing low-glycemic foods will help you to:

- Control your appetite and carbohydrate cravings, as they tend to help you feel full.
- Lose weight.
- Control your blood glucose levels and lower your risk of getting type 2 diabetes.

A low glycemic diet may also decrease your risk factors for cardiovascular disease.

SUMMARY

1. During digestion, various enzymes break down starch and disaccharides into monosaccharides that are then absorbed and carried to the liver. In the liver, most fructose and galactose are converted to glucose.
2. Two hormones, insulin and glucagon, are responsible for regulating your blood glucose levels. When you eat carbohydrates, your blood glucose rises and insulin helps glucose to enter cells and also stores some glucose as glycogen. Excess glucose is converted to fat.
3. When blood glucose is low, glucagon stimulates the liver to convert glycogen to glucose.
4. Fiber can't be digested by human enzymes, but soluble fiber can be digested by intestinal bacteria, producing small fatty acids that are either used for energy by the cells of the large intestine or they are absorbed.
5. Soluble fiber slows the emptying of food from the stomach and slows the absorption of glucose into the blood.
6. After eating, a low glycemic response (meaning that your blood sugar level rises slowly and modestly) is preferable to a high glycemic response (your blood sugar level rises quickly and high)—it may decrease your risk factors for cardiovascular disease and diabetes, as well as help you lose weight.
7. Glycemic load is more accurate because it takes into account portion size. Most fruits and vegetables, legumes, and whole grains have low or medium glycemic loads.

DIETARY RECOMMENDATIONS FOR CARBOHYDRATES

As you learned in Chapter 1, the Acceptable Macronutrient Distribution Range (AMDR) for carbohydrate is 45 to 65 percent of total kcalories for anyone over 1 year old. For example, if you eat 2,000 kcalories/day, you should get about 900 to 1,300 kcalories from carbohydrate, which is 225 to 325 grams. Americans typically do eat about 50 percent of total kcalories from carbohydrate, which is within the AMDR. Table 3-6 shows the amount of carbohydrate in common foods.

From 1971 to 2000, total kcalorie intake increased in the United States by an average of 168 to 335 kcalories—this was mostly due to increased carbohydrate intake. During this time period, per-person daily consumption of sugar-sweetened beverages (sodas, fruit drinks, sweetened iced tea, lemonade) increased 70 percent. Based on data from 2001 to 2004, the average intake of added sugars was 22 teaspoons per day (for adults and children), which equates to 350 kcalories.

Although Americans do eat enough carbohydrates, the choices are not nearly as healthy as they could be. Vegetables, fruits, whole grains, milk and milk products (preferably reduced in fat), and legumes all provide excellent sources of carbohydrate and fiber (except dairy), yet a large majority of Americans eat fewer of these foods than recommended

TABLE 3-6	Carbohydrate Content of Foods

Each of these foods contains approximately 5 grams of carbohydrate.

NONSTARCHY VEGETABLES

½ cup cooked vegetables

1 cup raw vegetables

Each of these foods contains approximately 12 grams of carbohydrate.

DAIRY

1 cup milk

²/₃ cup yogurt

Each of these foods contains approximately 15 grams of carbohydrate.

STARCHY FOODS

1 slice (1 oz.) white or whole grain bread

¼ large bagel

½ hamburger bun

½ of 6 in. pita bread

1 waffle or pancake (4 in. across)

¾ cup unsweetened cereal

¾ oz. pretzels

9–13 potato or tortilla chips

½ cup corn or 1 small baked potato

½ cup cooked beans, lentils, or split peas

FRUITS

1 extra small banana

1 small whole fruit

½ cup fruit juice, unsweetened

DESSERTS

2 small sandwich cookies

½ cup regular ice cream

1¼-in. square brownie

Each of these foods contains approximately 30 grams of carbohydrate.

Chocolate candy bar, 1¼ oz.

1 small frosted cupcake

1 glazed donut

1 cup lasagna

¼ of a 12-inch pizza—thin crust

1 small French fry

(Figure 3-14). Instead, we are eating more carbohydrates in the form of added sugars and refined grains (such as white flour found in many breads, cereals, and baked goods). Unfortunately, foods high in sugar and refined grains do not supply nearly the amount of fiber, vitamins, and minerals found in legumes, vegetables, fruits, and whole grains. Less than 5 percent of Americans consume the minimum recommended amount of whole grains, which for many is three servings per day.

Following are the dietary recommendations for carbohydrates.

1. The RDA for carbohydrates (Table 3-7) is 130 grams per day for children (from 1 year old) and adults. It is based on the minimum amount of carbohydrates needed to supply

SUMMARY

1. High-sugar foods, such as cookies, are teamed up with fat and high in kcalories. They are referred to as empty kcalorie foods.
2. Sugars and starches contribute to the development of dental caries or cavities. When you eat something sweet, the bacteria living on your teeth ferment, or digest, the carbohydrates for about 20 to 30 minutes—producing acid that dissolves the enamel surface of the tooth.
3. To decrease the chance of cavities, avoid sticky carbohydrate foods such as raisins and caramels, eat carbohydrates with meals rather than between meals, eat certain cheeses such as cheddar that help neutralize the acids produced by the bacteria, use carbonated beverages (both regular and diet) in moderation because of their acidity, and choose foods that do not seem to cause cavities such as meats, fish, cheese, peanuts, some vegetables, and sugar-free gum.
4. Increased consumption of added sugars is associated with a higher risk of obesity and is likely one of the many factors involved in rising obesity rates among adults and children. Strong evidence shows that children and adolescents who consume more sugar-sweetened beverages have higher body weight compared to those who drink less, and moderate evidence also supports this relationship in adults. Kcalories from sugar-sweetened beverages are not as filling as solid foods.
5. Being overweight or obese increases your risk for high blood pressure, high blood cholesterol, heart disease, stroke, type 2 diabetes, and other diseases. High consumption of added sugars also makes it difficult to get in enough vitamins and minerals that are necessary, which is a real concern for children.
6. Emerging research shows that higher consumption of sugar-sweetened beverages is associated with a higher risk of developing type 2 diabetes.
7. Diets high in sucrose, glucose, and fructose do increase fat levels in the blood, a risk factor for coronary heart disease.
8. Regular, well-balanced meals with moderate amounts of refined sugars and sweets, as well as protein, fiber, and fat, can moderate swings in blood glucose levels in people with hypoglycemia.
9. High sugar intake does not cause hyperactivity or attention deficit hyperactivity disorder (ADHD).

CHOOSING WHOLE GRAINS AND THEIR HEALTH EFFECTS

Choosing a loaf of bread in the grocery store can be hard. There are just so many different ones to pick from. If you are looking for 100 percent whole grain bread, you have to watch out for the many breads that are simply pretending to be 100 percent whole grain. Breads simply called "wheat breads" often contain as much as 75 percent enriched white flour with molasses or caramel added to make them look like whole wheat. Here are some tips for finding whole-grain breads, rolls, crackers, and cereals.

1. Read the ingredient list on the food label. For many whole-grain products, "whole" or "whole grain" will appear before the grain ingredient's name. The whole grain should be the first ingredient listed. Look for whole wheat, whole oats, whole rye, whole grain corn, brown rice, or whole grain barley. If the first ingredient is wheat flour, enriched flour, or degerminated cornmeal, the product is not a good source of whole grains.
2. Look on the package for the Whole Grains Council stamp (Figure 3-16). The stamp shows how many grams of whole grain are in one serving of the product. The *Dietary Guidelines for Americans* recommend half of your grains servings should be whole grains. Most people need to eat a minimum of three whole grains a day. A serving of whole grains contains 16 grams of whole grains, so strive to eat at least 48 (16 grams × 3 servings) grams of whole grains daily.

FIGURE 3-16 Whole Grains
Council stamp.
Courtesy Oldways and the Whole Grains
Council, wholegrainscouncil.org.

3. Look for the whole-grain health claim on food product labels: "Diets rich in whole-grain foods and other plant foods and low in saturated fat and cholesterol may help reduce the risk of heart disease." Foods that bear the whole-grain health claim must contain 51 percent or more whole grains by weight and be low in fat.

4. Look on the nutrition label for the amount of fiber in one serving. The product should provide at least 2 grams of fiber/serving if it contains a significant amount of whole grain ingredients. Be aware that you will find some products that contain no whole grain ingredients but have lots of fiber. This is because food manufacturers simply add functional fibers to foods ranging from ice cream and yogurt to baked products. Depending on functional fibers to meet your daily fiber needs is not nearly as healthful as eating a diet rich in whole grains, legumes, fruits, and vegetables.

At the store, you will also find whole grains that are labeled as stone ground. Grinding by stones is a technique for milling grains that allows them to be processed without the intense heat created by industrialized milling. The high heat causes more loss of flavor as well as nutrients. Stone-ground flours are often more flavorful and nutritious than other types of flours, although more expensive. Because there is no legal definition from the Food and Drug Administration for the term "stone ground," some stone-ground products may have only been partially stone ground. Stone-ground is also not a guarantee that the product is whole grain—you still need to read the label.

In baking, whole-grain flours produce breads that are denser and chewier. For example, breads made with only whole-wheat flour are heavier and more compact than breads made with only white flour. This is the case because the strands of gluten (a protein in wheat that gives bread structure) in the whole-wheat bread dough are cut short by the sharp edges of the bran flakes.

Traditional whole-wheat flour is a hard flour, meaning it is higher in protein than soft whole-wheat flour. Hard whole-wheat flour is good for any type of baking. However, whole-wheat pastry flour, which is made from soft wheat, is perfect for muffins, biscuits, scones, pancakes, piecrusts, and cakes—because the product will be more tender. Of course you can combine hard whole-wheat flour with whole-wheat pastry flour in a recipe to get a little more tenderness. If you add all-purpose white flour to whole-wheat flour, you will lighten the color and texture of the baked good—and increase the rise.

Research shows that a diet high in whole grains (at least three whole grains/day) is associated with less weight gain, and less risk of:

• Heart disease

• Type 2 diabetes

• Stroke

• High blood pressure

Components of whole grains, such as vitamins, minerals, antioxidants such as vitamin E that help keep body cells healthy, *phytochemicals* (substances in plants that promote health), and fiber all contribute to the health-promoting properties of whole grains. Refined grains are lower in vitamins such as vitamin E and minerals such as potassium.

SUMMARY

1. A whole-grain product will have a whole grain listed as the first ingredient, such as whole wheat, whole oats, whole grain corn, or brown rice. It should provide at least 2 grams of fiber/serving if it contains a significant amount of whole grain ingredients. A serving of whole grains contains 16 grams of whole grains.
2. Traditional whole-wheat flour is a hard flour, meaning it is higher in protein than soft whole-wheat flour. Hard whole-wheat flour is good for any type of baking. However, whole-wheat pastry flour, which is made from soft wheat, can be used to get more tenderness.
3. Moderate evidence shows that a diet high in whole grains is associated with less weight gain and less risk of heart disease, type 2 diabetes, stroke, and high blood pressure. Components of whole grains, such as vitamins, minerals, antioxidants such as vitamin E, phytochemicals, and fiber all contribute to the health-promoting properties of whole grains.

FIBER AND ITS HEALTH EFFECTS

What can fiber do for you? Lots—fiber promotes health in many ways, yet most American adults and children get less than half of what is recommended due to low intakes of fruits, vegetables, legumes, whole grains, and nuts. White flour (in breads, rolls, cereals, crackers, cookies, and pizza crust) and white potatoes provide the most fiber in the American diet—not because they are high in fiber, but because we eat a lot of them. Table 3-8 gives examples of the fiber content of common foods. Numerous studies have found that diets high in fiber reduce your risk of the following:

- Heart disease
- Type 2 diabetes
- Stroke
- High blood pressure
- Obesity
- Some gastrointestinal disorders

High-fiber foods also contain antioxidants, phytochemicals, and other substances that may offer protection against these diseases.

Findings on the health effects of fiber show that it plays a number of roles.

- *Improves blood cholesterol.* Studies show that a diet high in fruits, vegetables, beans, and whole grain products that contain *soluble fiber* can lower blood cholesterol levels and therefore lower your risk of heart disease. As it passes through the gastrointestinal tract, soluble fiber binds to cholesterol in the foods you just ate, helping the body eliminate the cholesterol. This reduces blood cholesterol levels, a major risk factor for heart disease. Soluble fibers in foods such as oatmeal, oat bran, barley, apples, and beans help lower blood cholesterol levels.

- *May reduce blood pressure modestly*, especially in people with high blood pressure and in older adults.

- *Decreases risk of developing diabetes.* Americans with the highest level of fiber intake, compared to those with the lowest intake, have a reduced risk of developing diabetes.

- *Helps to keep blood sugar at normal levels.* Soluble fibers cause a delay in the digestion and absorption of carbohydrates (and lipids). The slower absorption of glucose helps keep both blood glucose and insulin levels in the normal range.

- *Promotes regularity.* Insoluble fiber found in whole grains, beans, and many fruits and vegetables, is helpful in preventing **constipation** (infrequent passage of feces) because it increases fecal weight and improves transit time through the intestines. Fiber also helps

Constipation Infrequent passage of feces.

TABLE 3-8	Fiber Content of Selected Foods	
FOOD	**PORTION SIZE**	**AMOUNT OF FIBER (GRAMS)**
Foods High in Dietary Fiber (5 grams or more/serving)		
Kellogg's All-Bran	½ cup	9 g
Lentils, cooked	½ cup	8
Refried beans	½ cup	7
Beans, kidney, cooked	½ cup	7
Raspberries, frozen, red	½ cup	6
Cornmeal	½ cup	5
Foods Moderately High in Dietary Fiber (2.5 to 4.9 grams/serving)		
Bulgur, cooked	½ cup	4
Vegetables, mixed, cooked	½ cup	4
Prunes, stewed	½ cup	4
Kellogg's Raisin Bran	½ cup	4
Pear, raw	1 medium	4
Potato, baked	1 potato	4
Squash, winter, cooked	½ cup	3
Broccoli, cooked	½ cup	3
Raisins	½ cup	3
Blueberries, frozen	½ cup	3
Carrots, boiled	½ cup	3
Seeds, sunflower	¼ cup	3
Oatmeal, instant	1 pack	3
Peanuts	1 oz.	3
Apples, raw with skin	1 apple	3
Banana, raw	1 banana	3
Foods Moderate in Dietary Fiber (1 to 2.4 grams/serving)		
Oats, cooked	½ cup	2
Pears, canned	½ cup	2
Corn, sweet	½ cup	2
Rice, brown, cooked	½ cup	2
Mushrooms, cooked	½ cup	2
Bread, whole wheat	1 slice	2
Bread, rye	1 slice	2
Potatoes, mashed	½ cup	1.5
Carrots, raw	½ cup	1.5
Wild rice, cooked	½ cup	1.5
Applesauce, canned	½ cup	1.5
Squash, summer, cooked	½ cup	1.5

FOOD	PORTION SIZE	AMOUNT OF FIBER (GRAMS)
Foods Low in Dietary Fiber (Less than 1 gram/serving)		
Macaroni, cooked	½ cup	0.1
Doughnuts, cake-type	1 medium	0.7
Rice, white, cooked	½ cup	0.4
Cookies, oatmeal	1 cookie	0.7
Bread, white	1 slice	0.6
Orange juice	½ cup	0.25

Source: US Department of Agriculture, Agricultural Research Service, 2004. *USDA Nutrient Database for Standard Reference, Release 17*. Nutrient Data Laboratory Home Page, www.nal.usda.gov/fnic/foodcomp.

reduce the risk of diverticulosis, a condition in which small pouches form in the colon wall, usually from the pressure created within the colon by the small bulk and/or from straining during bowel movements. A diet high in insoluble fiber is also used to prevent or treat hemorrhoids, enlarged veins in the lower rectum. Larger, soft feces are easier to eliminate, and so there is less pressure and the rectal veins are less likely to swell.

- *Linked to lower body weights.* Because fiber-containing foods, such as fruits and vegetables, tend to be low in fat and sugar, they often contain fewer kcalories. And since fiber-containing foods require more chewing, slow digestion and provide bulk, they create a sense of fullness. Eating more high fiber foods, especially ones rich in soluble fiber that absorb water in the digestive tract, can increase **satiety** (feeling full) and decrease hunger.

Satiety The feeling of fullness and satisfaction after eating.

Research on fiber intake and cancer have had mixed results. More research is being done to determine if fiber can decrease the risk of cancer, such as colon cancer, but until then, healthcare professionals suggest eating plenty of whole grains, fruits, vegetables, legumes, and nuts.

So, as you just read, there are many good reasons to follow the recommendation to eat 14 grams of dietary fiber for every 1,000 kcal you eat.

If you want to increase the fiber in your diet, look at Table 3-9 for tips. It is important to add fiber gradually to your diet to give the gastrointestinal tract time to adapt. Also drink more fluids to soften the fiber. Think of fiber as a dry sponge absorbing water and digestive juices as it moves through your gastrointestinal system. You need to drink enough fluids for this to happen. If too much fiber is added too fast, you can experience intestinal discomfort, gas, and diarrhea.

SUMMARY

1. A diet high in fruits, vegetables, beans, and whole grain products that contain soluble fiber can lower blood cholesterol levels and therefore lower your risk of heart disease. Soluble fiber can bind with dietary cholesterol in the gastrointestinal tract so that some cholesterol is excreted.

2. A diet high in dietary fiber may reduce blood pressure modestly, decreases your risk of developing diabetes, helps to keep blood glucose at normal levels, and promotes regularity.

3. A diet high in dietary fiber is linked to lower body weights as it helps you feel full.

4. It is important to add fiber gradually to your diet to give the gastrointestinal tract time to adapt. Also drink more fluids to soften the fiber.

TABLE 3-9   How to Increase Fiber in Your Diet

BREAKFAST

Instead of: Choose:

Cereal made mostly of refined grains with less   Cereals made with whole grains and/or bran such as Wheaties,
than 1 gram of fiber/serving such as rice or cocoa   Shredded Wheat, Cheerios, and oatmeal
krispies or corn flakes

English muffin made with white flour   100% whole wheat English muffin

Waffles and pancakes made with white flour   Waffles and pancakes made with at least 50% whole grains

White flour bagel   Whole wheat bagel

LUNCH

Instead of: Choose:

White bread or white tortilla or white pita bread   Whole grain or whole wheat bread, tortilla, or pita bread

Meat/cheese sandwich fillings   Peanut butter, tahini, roasted vegetables, bean dip, or hummus on
                                                     whole grain bread or crackers

SNACKS

Instead of: Choose:

Crackers and pretzels made with white flour   Crackers and pretzels made with whole wheat or other whole grain
                                                     flours, popcorn

Cookies made with white flour   Granola bars with whole grains and nuts, whole grain cookies such as
                                                     whole grain fig bars

Fruit juice drinks   Fresh and canned fruits

DINNER

Instead of: Choose:

White spaghetti   Whole wheat pasta

White rice   Brown rice, bulgur, barley

Creamy soups   Soups with vegetables, beans, and lentils

Meat stews and chili   Meatless stews and chili using beans, vegetables, and whole grains

LACTOSE INTOLERANCE

Lactose (milk sugar) is a problem for certain people who have a deficiency of the enzyme **lactase**. Lactase is needed to split lactose into its components in the small intestine. If lactose is not split, it travels to the colon, where it attracts water and causes bloating and diarrhea. Intestinal bacteria also ferment the lactose and produce fatty acids and gas, making the stomach discomfort worse and contributing still more to the diarrhea. Symptoms usually occur within 30 minutes to 2 hours after eating or drinking milk products and clear up within 2 to 5 hours.

This condition, called **lactose intolerance**, seems to be an inherited problem that is especially prevalent among Asians, Native Americans, African Americans, and Latinos, as well as some other population groups. Most people experience a normal age-related decline in lactase between 2 to 20 years of age.

Treatment for lactose intolerance includes a diet that is limited in lactose, which is present in large amounts in milk, ice cream, ice milk, sherbet, cottage cheese, eggnog, and cream. Most individuals can drink small amounts of milk (½ cup milk) without any symptoms, especially if it is taken with food.

People who have difficulty digesting lactose report tremendous variation in which lactose-containing foods they can eat and even the time of day they can eat them. For example, one individual may not tolerate milk at all, whereas another can tolerate milk as part of a big meal. The ability to tolerate lactose is not an all-or-nothing phenomenon. As people with lactose

Lactase An enzyme needed to split lactose into its components in the intestines.

Lactose intolerance A condition caused by a deficiency of the enzyme lactase, resulting in symptoms such as flatulence and diarrhea after drinking milk or eating most dairy products.

deficiency usually decrease their intake of dairy products and thus their calcium intake, they should try different dairy products to see what they can tolerate.

Strategies to manage lactose intolerance include the following:

- Use lactose-free milk and milk products. A wide variety of lactose-free dairy products (e.g., reduced-fat, low-fat, fat-free, chocolate, and whole milk, ice cream, cottage cheese) are available in the supermarket (Figure 3-17). These lactose-free dairy products contain the same nutrients as their regular counterparts, just without the lactose, and they taste the same, too!

- Lactase is also available for home use in liquid and tablet or capsule form. The lactase can be consumed directly or added to regular milk.

- Consume small serving sizes of dairy with a meal containing fat and protein and introduce dairy slowly.

FIGURE 3-17 Lactose-free products.

Courtesy of the US Department of Agriculture.

- Choose dairy products naturally lower in lactose, such as yogurt. Yogurt is usually well tolerated because it is cultured with live bacteria that digest lactose. This is not always the case with frozen yogurt, because most brands do not contain nearly the number of bacteria found in fresh yogurt (there are no federal standards for frozen yogurt at this time)

- Try hard cheeses such as Cheddar, Colby, Monterey Jack, and Swiss. Hard cheeses are low in lactose because lactose is primarily in the whey, not the curds. When cheese is being made (with the exception of some soft cheeses that contain whey, like Ricotta) the whey is discarded and the lactose goes with it. Curds still have a little bit of lactose, but not much. As cheese ages and loses moisture and becomes hard, there is even less lactose left in the curds.

- Soy milk and rice milk are possible milk substitutes for people with lactose intolerance.

Milk is the leading food source of calcium, vitamin D, potassium, and magnesium in the American diet. Avoiding milk and dairy foods increases the risk of having vitamin and mineral deficiencies.

SUMMARY

1. In lactose intolerance, lactase is deficient so lactose (milk sugar) is not split into its components in the small intestines. Instead, it travels to the colon where it attracts water and causes bloating and diarrhea. In addition, intestinal bacteria ferment lactose and produce gas. Symptoms usually occur with 30 minutes to 2 hours and clear up within 2 to 5 hours.
2. Lactose intolerance is an inherited problem and is especially prevalent among Asians, Native Americans, African Americans, and Latinos.
3. Treatment for lactose intolerance includes a diet limited in lactose (present in dairy and added to some other foods), use of lactose-free milk and milk products and/or lactase, and consuming small servings of dairy with a meal and/or dairy products lower in lactose (such as hard cheeses and yogurt) as tolerated.

CULINARY FOCUS: GRAINS AND LEGUMES

Most Americans need to get more whole grains and legumes such as beans into their diet. Both whole grains and legumes are very nutritious, contain little fat, and are filling due to their fiber content. This section will help you highlight dishes with grains and legumes on the menu.

GRAINS: PRODUCT

Nutritionally, grains are a powerhouse of nutrients. They are:

- High in vitamins and minerals that perform thousands of job in your body every day to keep you strong and healthy
- High in fiber (if whole grain) that makes you feel full after eating and reduces your risk of heart disease, obesity, and type 2 diabetes
- Low in fat
- Moderate in protein
- Low or moderate in kcalories

In terms of protein content, amaranth and quinoa are a little higher in protein than other grains, while corn is lower in protein than most grains.

Whole grains are also a source of antioxidants and phytochemicals. Grains are moderately priced and can be quite profitable as protein accompaniments or main dish components. Both traditional grains, such as rice varieties, and grains such as quinoa, spelt, and wheatberries, are expected and accepted on menus by your clientele in salad bar selections, side dishes, appetizers, or main course options.

GRAINS: PREPARATION

Following are Chef's Tips for preparing and serving grains:

- See Table 3-10 for information on flavor and cooking times for a variety of grains.

TABLE 3-10	Grains: Cooking Time and Uses		
TO 1 CUP OF THIS GRAIN	ADD THIS MUCH WATER OR BROTH (CUPS)	BRING TO A BOIL, THEN SIMMER FOR	AMOUNT AFTER COOKING (CUPS)
Amaranth	2½ cups	25 minutes	3½ cups
Barley, pearl	3	35–40 minutes	3½
Barley, whole hulled	3	60–90 minutes	4
Buckwheat, whole white	2	20 minutes	2½
Buckwheat, roasted (kasha)	2	10–15 minutes	2½
Corn, whole hominy	2½	2½–3 hours	3
Corn, hominy grits	4	20–25 minutes	3
Millet	2	30–35 minutes	3
Oats, steel-cut	2	45–60 minutes	2
Quinoa	2	12–15 minutes	2½
Rice, regular-milled, long-grain	2	15–20 minutes	3
Rice, regular milled, medium or short grain	1½	20–25 minutes	3
Rice, parboiled	2–2½	20–25 minutes	3–4
Rice, brown	2½	40–50 minutes	4
Rice, wild	3	30–45 minutes	3½–4
Rice, basmati	1½	25 minutes	3
Jasmine rice	2	15–20 minutes	3
Texmati rice	2	15–20 minutes	3
Rye, whole berries	3	1½ hours	3
Wheat, bulgur	2½	20–25 minutes	2
Wheat, whole berries	3	1 hour	2

FIGURE 3-18 Grains: Clockwise from top: Multicolored popcorn, farro, rolled oats, black quinoa, pearled barley, kasha, red quinoa, quinoa, and hard winter wheat berries. In center circle from top: hulled buckwheat, rye berries, and bulgur.

Photo by Peter Pioppo.

- Figure 3-18 showcases many popular grains.
- Figure 3-19 highlights types of rice.
- Grains are a versatile addition to a main course, appetizer, salad, side dish, breakfast selection, or center of the plate applications. Methods and recipes such as pilaf, grain cakes, stuffings, pancakes, baked goods and salads provide the ability to incorporate grains into every meal period to create flavorful, interesting and beneficial selections. When preparing grains, the best way to achieve maximum flavor and texture is to cook each of the grains separately, adding seasonings such as bay leaf, large diced onion, and fresh thyme stem. Once cooked and strained, mix the grains with other ingredients, depending on the application or desired result.
- Toasted barley or quinoa, combined with other grains or small pastas (such as orzo or ditalini) then mixed with ingredients such as roasted vegetables, fresh herbs and dried fruits with a balanced dressing create attractive and flavorful dishes.
- Rice and beans, a staple in cultures around the world, has gained tremendous popularity with its versatility of combining many different grains and legumes. The traditional white rice and black beans can be substituted with purple rice or wild rice, adding beans such as cranberry, fava, or white kidney. Each of these mixtures can be seasoned according to cultural taste, local ingredients, or seasonal genre. The addition of a well-balanced dressing can be used for a cold salad as well as a hot application.

FIGURE 3-19 Rices. Basmati rice in middle. Clockwise from top center: short-grain brown rice, Bhutan red rice, wild rice, long-grain brown rice, purple Thai rice, and Arborio rice.

Photo by Peter Pioppo.

GRAINS: MENUING AND PRESENTATION

Over the past decade the addition of grains on menus has increased in popularity due to the health benefits as well as the creative flavorings and proper cooking techniques used in food service establishments globally. There are endless ways to use grains in every meal period.

Breakfast

Examples of ways to add these items to your menus with positive results include the following breakfast ideas:

- Serve whole-grain cereals hot and cold, from simple steel-cut oatmeal to cold favorites such as shredded wheat or kasha (whole grain buckwheat) dressed up with yogurt, fresh berries and other fresh fruits, dried fruits, roasted nuts, and sweet spices such as cinnamon, nutmeg, and mace. Steel-cut oats are whole-grain groats that have been cut into two or three pieces using steel discs. Gold in color, they look like little pieces of rice. Rolled oats are flakes that have been steamed, rolled, and toasted. Steel-cut oats have a firmer, chewier texture than rolled oats and also more oat flavor.
- Serve wheat berry or whole grain pancakes with fresh kiwi salsa, or the classic Bircher Muesli. This is a mixture of whole oats, fruits, skim milk, and nuts.

Then there are the more daring alternatives, such as Banh Cuon, a Vietnamese crêpe made with ground rice, rolled around a mixture of pork or shrimp filling served with orange-spiced fish sauce, or congee (also known as jook)—Chinese thin porridge rice soup eaten for breakfast. In Ethiopia, they prepare breakfast items such as Genfo—a barley-based hot cereal like porridge, Kinche—a cracked wheat and milk oatmeal dish, teff and whole grain cake served with apricots, apples and figs, and Fir Fir—a grain bread called injera, broken up and tossed with meat, spiced sauce, and vegetables or eggs.

Lunch

For lunch, examples include a warm whole-wheat burrito or wrap stuffed with a selection of grains, proteins, and vegetables; a meatless entree where grains take on the center of the plate such as a burger or millet and brown rice cake; or a complex salad mixed with a variety of greens and grains.

Dinner

Dinner choices include grains as the starch accompanying a main entrée; a flavorful stuffing used in a chicken breast or fish preparation; an ingredient stacked inside a Portobello, spinach, roasted pepper mozzarella napoleon; or as a bed for seared sea scallops, tomato confit, and broccolini. Each of these choices is well balanced, creative, and flavorful.

Other meal periods can include such favorites as homemade granola bars; oatmeal white raisin or other whole grain cookies; carrot cake with whole-wheat flour and egg whites; or a cheese platter with grapes, apples, and whole grain crackers. The ingredients are an exciting challenge for culinary professionals using a variety of methods and techniques.

LEGUMES: PRODUCT

Legumes include a variety of dried beans, peas, and lentils. Dried beans are among the oldest foods and are an important staple for millions of people around the world. Beans were once considered to be worth their weight in gold—the jeweler's carat owes its origin to a pealike bean on the east coast of Africa.

From a nutritional point of view, legumes are extremely valuable. They are:

- High in fiber
- Low in fat (only a trace, except in a couple of cases)
- Cholesterol-free
- A good source of many vitamins and minerals
- Low in sodium

Besides being so nutritious, like grains, they are extremely cost-effective, as well as complement a variety of foods, adding color, flavor, and texture.

Flours are made from a variety of beans, such as black beans, garbanzo beans, and soybeans. Compared to wheat flour, bean flours are gluten-free and higher in protein and dietary fiber. White beans or fava beans have a milder taste, making these flours suitable for use in most recipes calling for white flour. Substitute one-quarter of the white flour with bean flour. Bean flours made from kidney, pinto, garbanzo, or black beans have a stronger flavor and can be used, for example, as thickeners in soups and stews. The bean flours add body and flavor as well as protein and fiber to rice flour mixes that are often used in gluten-free baking.

LEGUMES: PREPARATION

Following are Chef's Tips for preparing and serving legumes:

- See Table 3-11 for information on cooking times for a variety of legumes.
- Figures 3-20 and 3-21 highlight many popular legumes.
- When choosing legumes for a dish, think color and flavor profile. Make sure the flavor profile and appearance of the legume you pick will complement the completed dish. Furthermore, think of other ingredients you will use to add flavor. In a salad, for example, black-eyed peas (black and white) go well with flageolets and red adzuki beans. To add a little more color and develop the flavor, you might add chopped tomatoes, fresh cilantro (Chinese parsley), thin-sliced haricots verts (green beans) or fresh roasted corn.
- Bigger beans, such as gigante white beans, hold their shape well and lend a hearty flavor to stews, ragouts, and salads.

TABLE 3-11	Cooking Times for 1 Cup of Legumes			
BEAN (1 CUP DRY)	SOAKING REQUIRED?	CUPS LIQUID FOR COOKING	COOKING TIME	YIELD IN CUPS
Adzuki beans	Yes	3	1–1½ hours	2
Anasazi beans	Yes	3	2 hours	2
Black beans (turtle beans)	Yes	4	1½ hours	2
Black-eyed peas (cowpeas, black-eyed beans)	No	3	50–60 minutes	2
Chickpeas (garbanzo beans, ceci)	Yes	4	1–1½ hours	4
Fava beans, whole	Yes	2½	3 hours	4
Great Northern beans	Yes	3½	1½ hours	2
Kidney beans	Yes	3	1–1½ hours	2
Lentils	No	2	30–45 minutes	2¼
Lima beans	Yes	2	1½ hours (large) 1 hour (baby)	1¼
Navy beans (pea beans)	Yes	3	1½ hours	2
Peas, split	No	3	30 minutes	2¼
Peas, whole	Yes	3	40 minutes	2¼
Pinto beans	Yes	3	1½ hours	2
Pink beans	Yes	3	1 hour	2
Soybeans	Yes	3	3½ hours or more	2

FIGURE 3-20 Beans. Top row: Navy beans, cannellini beans, black turtle beans, red kidney beans. Bottom row: Garbanzo beans, Anasazi beans, black-eyed peas, and pinto beans.

Photo by Peter Pioppo.

- Cooked chickpeas can be puréed, as in a hummus dip, spread, or sandwich filling. Cooked chickpeas can also be layered with grilled vegetables and roasted peppers, used as a filling for pasta or crêpes, or mixed with potatoes to create a twice-baked potato.

- Stone-ground chickpea flour can be cooked in water similar to polenta, poured into a half pan, cut to resemble French fries. then baked and serve with a variety of modified sauces such as Romesco (tomato, almonds, peppers and red wine vinegar) and Harrissa (a mixture of chili paste and spices) added to yogurt creating a creamy spiced dipping sauce.

- To enhance their flavor, cook soaked beans separately in vegetable stock flavored with herbs, shallots, onions, carrots, and celery.

- Presoaking dried beans helps remove some of the complex sugars that are hard to digest and can cause flatulence (gas). Soaking beans allows them to slowly absorb the moisture they need to cook evenly and be tender. When beans aren't soaked, it means longer cooking times. The longer the cooking times the more nutrients are lost.

- A number of dried beans are also available fresh: cannellini beans, cranberry beans, fava beans, Romano varieties, flageolets, lima beans, mung beans, and soybeans. They are seasonal, tend to be on the expensive side for beans, but are excellent products that can be prepared in a variety of ways to create a clean, fresh flavor with a colorful presentation.

- Use whole lentils, such as black or French green lentils, in ragouts and stew dishes or salads because whole lentils hold their shape better. Use split lentils, such as brown, red, or yellow lentils, in soups where they help thicken the liquid and shape is not as important.

FIGURE 3-21 Lentils and Split Peas. Clockwise from top: Golden lentils, black beluga lentils, red split lentils, green and yellow split peas.

Photo by Peter Pioppo.

LEGUMES: MENUING AND PRESENTATION

Legumes are plants that produce seedpods and seeds used for food. These include a variety of peas, lentils, and beans. They have played an important role in the global diet for thousands of years, and are especially important today given the increased interest in balance and wholesome ingredients. Legumes are also high in protein, making them an important element in vegetarian and vegan diets, but equally add flavor, texture, and additional nutritional benefits to many applications.

Breakfast

Researchers continue to prove that breakfast is the most important meal of the day providing fuel to get the body moving and the brain paying attention.

Culinarians are aggressively creating breakfast options using legumes, reinventing dishes taken from cultures around the world with interesting additions and modernist flare, using a more balanced approach in moderation of ingredients and cooking methods. Such items include traditional sweet breads, muffins, pancakes, crêpes, and pastries made with bean flours such as soy, black, garbanzo, pinto, fava, and white. These classics can be served with a variety of accompaniments such as pureed fruit spreads, fruit compotes, salsas, or relishes flavored with thai basil, peppermint, lemon balm, toasted spices or chilies, yogurt sauces, dried fruits, nuts, and fresh fruits. The end result adds balance, texture, and flavor to these classic favorites, while substituting for processed flours, spreads with excess fats, and sugars. Examples of these classics include the following:

- Ful Medames—an Egyptian breakfast dish of cooked mashed fava beans garnished with chopped tomato, parsley, and lemon juice served with a poached egg or scrambled egg whites.
- El Gallo Pinto—a dish from Costa Rica and Nicaragua that consists of rice and beans cooked together with a variety of garnishes, usually served for breakfast.
- Burritos—A Mexican tortilla rolled around a filling; this dish has gained increased popularity with a variety of heirloom bean stews and eggs wrapped and served with exotic fruit salads, or spicy greens tossed with nuts, orange supreme and grapes with a balanced vinaigrette.
- Huevos Rancheros—Corn tortillas topped with a Mexican stewed bean mixture with vegetables, poached eggs, or scrambled whites and Mexican string cheese served with a variety of spicy salsas.
- Natto—A Japanese fermented soybean stew, usually served over rice.
- Miso soup—A Japanese soup made with soybean paste, served with rice.
- Vietnamese style mung bean with steamed rice served with mango and coconut.

Other nontraditional breakfast items include some of the following:

- Bean and vegetable hash or chili.
- Seared tofu, beans, and vegetables.
- Bean cakes with poached eggs served with a variety of fresh salsas, relishes, and mojos.

These items are becoming increasingly popular with the acceptance of beans as a good source of fiber, protein, iron, and other important nutrients.

Lunch

For lunch, beans, lentils, and peas are great ingredients for salads (Figure 3-22), soups, purées, fillings, and sandwiches giving body and substance while adding nutritional value. The combinations are endless, the key is to use inviting comfort foods that are easy to visualize while still challenging the imagination. Example of these favorites include:

- White bean artichoke and grilled vegetable wrap
- Roasted balsamic eggplant, grilled chicken on a mozzarella panini
- Vegetable and bean chili with toasted whole wheat tortillas
- Six bean and corn salad with tomato, roasted peppers and fresh cilantro
- Black bean burgers with grilled heirloom tomatoes
- Falafel with cucumber yogurt sauce, grilled eggplant and roasted peppers

FIGURE 3-22 Lentil Salad.

Photo by Peter Pioppo.

- Red bean quesadillas with onions, mushrooms and peppers, avocado salad and salsa verte
- Smoked chicken and lentil soup

Dinner

For dinner, there are many dishes in which beans, lentils, and peas can be incorporated, such as stews, ragouts, soups, salads, appetizers, side dishes, dessert, and the main component of a dinner entrée. Legumes and pasta work well together, such as in Pasta e Fagioli—a rich Italian vegetable soup with pasta and beans; Pasta con Lenticche—a rich stew with pasta, lentils, and greens; or Spicy Mexican Black Bean and Pasta—with tomatoes, onions, fresh peppers, and dried chilics. There are many variations of this classic dish due to its nourishing value and spectacular taste. Many other cultures have similar traditional dishes associated with pasta and beans or rice and beans for their versatility of preparations, health benefits, cost factors, history, and utilization of a region's products. Other suggestions include the following:

- Sweet pea pancakes with smoked salmon and preserved lemon sauce
- White bean and vegetable Napoleon, with wilted spinach and spicy tomato sauce
- Stuffed portabellos with four bean, broccoli rabe, sautéed lemon asparagus, and oven-dried tomatoes
- Calypso, Romano, butter bean and escarole sauté with roasted halibut and tomato confit
- Beluga lentil stew with artichokes, chanterelles, and grilled shrimp
- Baked cranberry and anasazi beans with sage-crusted pork loin
- Adzuki bean and coconut pudding with banana chips

Well-crafted food, with proper methods, techniques, and flavoring is the key to successful menu items. Responsible cooking is moderating your ingredients while adding the right balance of fats, proteins, carbohydrates, and salt to create the same delicious result.

S U M M A R Y

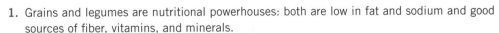

1. Grains and legumes are nutritional powerhouses: both are low in fat and sodium and good sources of fiber, vitamins, and minerals.
2. Tips are given for using grains and legumes on the menu.

CHECK-OUT QUIZ

1. Bread and fruits are good sources of carbohydrates.
   a. True
   b. False

2. Starch and fiber are only found in plant foods such as fruits, vegetables, and grains.
   a. True
   b. False

3. Glucose, fructose, and lactose are all made of a single sugar unit.
   a. True
   b. False

4. Sucrose is simply two sugars—glucose and galactose—linked together.
   a. True
   b. False

5. Starch is made of many chains of glucoses linked together.
   a. True
   b. False

6. When heated in liquid, sugars undergo a process called gelatinization.
   a. True
   b. False

7. Beef is a good source of fiber.
   a. True
   b. False

8. Functional fibers are fibers extracted from plants and then added to foods such as muffins to increase fiber content.
   a. True
   b. False

9. Refined grain foods contain more vitamins, minerals, fiber, and phytochemicals than whole grains.
   a. True
   b. False

10. A generous intake of fiber lowers blood cholesterol, helps to keep blood sugar at normal levels, and reduces your risk of obesity.
    a. True
    b. False

11. Before any carbohydrates can be absorbed into the small intestine, they must be broken down into monosaccharides.
    a. True
    b. False

12. Most fruits and vegetables are considered high glycemic load.
    a. True
    b. False

13. Meats and cheese can lead to cavities.
    a. True
    b. False

14. Increased consumption of added sugars is associated with a higher risk of obesity and developing type 2 diabetes.
    a. True
    b. False

15. Lactose-free milk and milk products, as well as yogurt and hard cheeses, can be used to manage lactose intolerance.
    a. True
    b. False

16. If a slice of apple pie has 32 grams of sugar, how many teaspoons of sugar does the slice contain?

17. Explain gelatinization.

18. List three foods rich in soluble fiber and three foods rich in insoluble fiber.

19. Describe three parts of a whole grain such as wheat.

20. What are the functions of carbohydrate?

21. What are the guidelines for consuming fiber and added sugar?

22. How can you tell if a loaf of bread is whole grain?

NUTRITION WEB EXPLORER

For a complete list of websites for the following activities, please visit the companion book page at www.wiley.com/college/drummond.

JOSLIN DIABETES CENTER

Joslin Diabetes Center is an excellent site to learn almost anything about diabetes. Click on "Diabetes Information," then click on "Diabetes & Nutrition." Read one of the articles under "Successful Eating with Diabetes" and summarize.

MEDLINE PLUS: HYPOGLYCEMIA

On this government website, click on "Hypoglycemia: Interactive Tutorial" to learn more about this topic. Write a short paragraph of what you learned.

NATIONAL DIGESTIVE DISEASES INFORMATION CLEARINGHOUSE: LACTOSE INTOLERANCE

Use this informative site to learn more about the causes and treatment of lactose intolerance. List 10 processed foods that have lactose added to them.

WHOLE GRAINS COUNCIL

Click on "Whole Grains 101" and then click on "Whole Grains A to Z." Read about a huge variety of whole grains. Write a description of three whole grains that are new to you.

MEDLINE PLUS: DENTAL HEALTH

Click on one of the articles listed under "Nutrition" and write a summary.

WHAT'S YOUR DISEASE RISK?

Click on "What's Your Diabetes Risk?" to take a questionnaire that assesses your risk to develop diabetes. What is your risk?

IN THE KITCHEN

Whole Grain and High Fiber Cooking

Each group of students will prepare two recipes as follows.

- Traditional Rice Pilaf and Cracked Wheat Pilaf
- Smoked Chicken Noodle Soup and Smoked Chicken Soup with Lentils
- Hearty Beef Chili and Hearty Turkey Chili with Beans
- Raisin Muffins and Raisin Bran Muffins
- Traditional Pasta with Marinara Sauce and Whole Wheat Pasta with Marinara Sauce
- Blueberry Buttermilk Pancakes and Whole Grain Pancakes with Blueberries

As you can see, the second recipe in each pair includes whole grains and good sources of fiber. Your instructor will hand you a worksheet at the end to compare the fiber content of all the dishes, as well as compare and evaluate each dish's taste and appearance.

HOT TOPIC

ALTERNATIVE SWEETENERS

The introduction of diet soda in the 1950s sparked the widespread use of alternative sweeteners, substitutes for sugar that provide no, or almost no, kcalories. If you drink diet soda, look at the food label and see which alternative sweeteners are present. The following alternative sweeteners are approved for use by the Food and Drug Administration: saccharin, aspartame, acesulfame potassium, sucralose, and neotame. The only one that contains kcalories is aspartame—but because so little is used, the kcalories are close to 0. Besides offering virtually no kcalories, alternative sweeteners are beneficial because they do not cause tooth decay or force insulin levels to rise as do added sugars such as high-fructose corn syrup.

Because they are considered food additives, the FDA requires that they be tested for safety before going on the market. The FDA uses the concept of an Acceptable Daily Intake (ADI) for many food additives, including alternative sweeteners. The ADI represents an intake level that, if maintained each day throughout a person's lifetime, would be considered safe by a wide margin. For example, the ADI for aspartame is 50 milligrams per kilogram of body weight per day. To take in the ADI for a 150-pound adult, someone would have to drink twenty 12-ounce cans of diet soft drinks daily. Most toxic effects associated with alternative sweeteners are at intakes 100 times greater than the ADI.

Approved Alternative Sweeteners

SACCHARIN

Saccharin, discovered in 1879, has been consumed by Americans for more than 100 years. Its use in foods increased slowly until the two World Wars, when its use increased dramatically due to sugar shortages. Saccharin is about 200 to 700 times sweeter than sucrose and is excreted unchanged directly into the urine. It is approved for use at specific maximum amounts in foods and beverages and as a tabletop sweetener. Known as Sweet'N Low or Sweet Twin, it is sold in liquid, tablet, packet, and bulk form. Because saccharin leaves some consumers with an aftertaste, it is frequently combined with other alternative sweeteners, such as aspartame.

ASPARTAME

In 1965, aspartame was discovered accidentally. Aspartame is made by joining two protein components, aspartic acid and phenylalanine, and a small amount of methanol. Aspartic acid and phenylalanine are building blocks of protein. Methanol is found naturally in the body and in many foods, such as fruit and vegetable juices. In the intestinal tract, aspartame is broken down into its three components, which are metabolized in the same way as if they had come from food. Aspartame contains 4 kcalories per gram, but so little of it is needed that the kcalorie content is not significant.

Aspartame is approximately 160 to 220 times sweeter than sucrose and has an acceptable flavor with no bitter aftertaste. It is marketed under the brand names NutraSweet and Equal. Aspartame is approved as a general-purpose sweetener and is found in diet sodas, cocoa mixes, pudding and gelatin mixes, fruit spreads and toppings, and other foods. If you drink diet soft drinks, chances are, they are sweetened with aspartame. Fountain-made diet soft drinks are more commonly sweetened with a blend of aspartame and saccharin, because saccharin helps provide increased stability.

Aspartame breaks down during prolonged heating and starts to lose its sweetness. For stovetop cooking, it is best to add aspartame at the end of cooking or after removing the food from the heat. For baking, it is best to use aspartame with regular sweeteners such as brown sugar in specially suggested and tested recipes. Aspartame is available in a granulated form that measures cup for cup just like regular white sugar.

The safety of aspartame provoked concerns in the past, and many studies have been done on it. Overall the consensus is that aspartame consumption is safe over the long term and is not associated with serious health effects.

The only individuals for whom aspartame is a known health hazard are those who have the disease phenylketonuria (PKU), because they are unable to metabolize phenylalanine.

For this reason, any product containing aspartame carries a warning label, "Phenylketonurics: Contains phenylalanine." Some other people may also be sensitive to aspartame and need to limit their intake.

ACESULFAME-POTASSIUM

Acesulfame-potassium is often abbreviated as Acesulfame-K, as K is the chemical symbol for potassium. Marketed under the brand names Sunett and Sweet One, it is about as sweet as aspartame but is more stable and can be used in baking and cooking. Coca-Cola mixes acesulfame-K with aspartame to sweeten one of its diet sodas. Its taste is clean and sweet, with no aftertaste in most products. Acesulfame-K passes through the digestive tract unchanged.

SUCRALOSE

Sucralose is the only alternative sweetener made from table sugar. Sucralose is 600 times sweeter than sugar and actually tastes similar to sugar. Sucralose cannot be digested, and so it adds no kcalories to food. Sucralose has exceptional stability and retains its sweetness over a wide range of conditions, including heat. Sucralose can be used in baking and cooking. It is even used in sweet microwave popcorn.

Sucralose is available as an ingredient for use in a broad range of foods and beverages under the name Splenda Brand Sweetener. Currently, a range of products sweetened with Splenda are on supermarket shelves, including diet soft drinks, low-calorie fruit drinks, maple syrup, and applesauce.

Like aspartame, sucralose is available in a granular form which pours and measures exactly like sugar. Maltodextrin, a starchy powder, is used to give it bulk, and the resulting product has no kcalories.

NEOTAME

In 2002, the Food and Drug Administration approved neotame for use as a general-purpose sweetener. Neotame is a high-intensity sweetener that is structurally similar to aspartame but is much sweeter and does not cause any problems for people with phenylketonuria. Once ingested, neotame is quickly metabolized and completely eliminated.

Depending on its food application, neotame is about 7,000 to 13,000 times sweeter than sugar. Its strength varies depending on the application and the amount of sweetness needed. While neotame can function as a stand-alone sweetener, it lends itself particularly well to blending with other sweeteners. It is a white crystalline powder that is heat-stable and can be used as a tabletop sweetener as well as in cooking. Neotame has a clean, sweet taste in foods.

Blending and Cooking with Alternative Sweeteners

Food scientists have discovered that blending certain alternative sweeteners with one another results in products that are much sweeter than expected. This is beneficial because you can reduce the amount of alternative sweeteners used and improve the taste profile. Some alternative sweeteners also work well with traditional caloric sweeteners, such as sucralose with high-fructose corn syrup. When combined with traditional sweeteners, sweeteners not only can improve taste but can also decrease kcalories.

Table 3-12 summarizes information on approved alternative sweeteners.

Stevia

Stevia is another alternative sweetener but it can be considered natural, not artificial. Stevia typically refers to a crude preparation of dried stevia leaves from the stevia bush found in South America. It contains a mixture of many substances, some of which are sweet. Some of the sweet compounds from the leaf are purified and approved for use in foods and beverages in the United States. It is now available as a tabletop sweetener with 0 kcalories under names such as Truvia™ or Pure

Via™. Some beverage companies are using stevia in drinks such as Sprite Green, Crystal Light, and vitamin waters.

Sugar Replacers

Sugar replacers, also called polyols, are a group of carbohydrates that are sweet and occur naturally in plants. Scientists call them sugar alcohols because part of their structure resembles sugar and part resembles alcohols; however, these sugar-free sweeteners are neither sugars nor alcohols. Many sugar replacers, such as xylitol, have been used for years in products such as sugar-free gums, candy, and fruit spreads. Table 3-13 lists the eight sugar replacers currently approved for foods in the United States and tells you how sweet each one is compared to sugar and also how many kcalories each one provides. Each sugar replacer has different characteristics.

Sugar replacers or polyols have the following benefits.

- They don't provide as many kcalories as sugar. The average kcalories per gram is 2, compared with 4 kcalories/gram from sugar.
- They don't promote tooth decay.
- They taste sweet, though not as sweet as sugar.
- They cause smaller increases in blood glucose and insulin levels than sugar.

Sugar replacers don't affect your blood glucose level the way sugar does because they are absorbed more slowly and incompletely from the small intestine. The portion that is absorbed is metabolized by processes that require almost no insulin.

The portion that is not absorbed is broken down by bacteria in the large intestine, which can sometimes cause abdominal gas, discomfort, and diarrhea in some individuals. These symptoms are more likely if large amounts of sugar replacers have been consumed. FDA regulations require food labels to use the following warning if reasonable consumption of the food could result in undesirable symptoms: "Excess consumption may have

TABLE 3-12 | Approved Alternative Sweeteners

ARTIFICIAL SWEETENER	BRAND NAME(S)	SAMPLE FOODS USED IN	KCALORIES/ GRAM	SWEETNESS COMPARED TO SUGAR
Saccharin	Sweet 'N Low Sugar Twin* Hermesetas Original (in Europe)	Fountain diet sodas	0	200 to 700 times as sweet

Description

- Not metabolized by body
- Some consumers report an aftertaste
- Available in liquid, tablet, packet, and bulk (granular) form. Sweet 'N Low Brown is a sugar substitute for brown sugar

ARTIFICIAL SWEETENER	BRAND NAME(S)	SAMPLE FOODS USED IN	KCALORIES/ GRAM	SWEETNESS COMPARED TO SUGAR
Aspartame	NutraSweet Equal Classic Canderel (in Europe)	Diet sodas, diet drink mixes, cocoa mixes, pudding and gelatin mixes, fruit spreads and toppings	4[†]	160 to 220 times as sweet

Description

- Made of two protein components and a small amount of methanol (found naturally in many foods).
- Foods containing aspartame must have a warning label because it contains phenylalanine, which a small number of people can't tolerate.
- Loses its sweet flavor with prolonged heating. Add at end of stovetop cooking. Can often be used successfully in baking.
- Available in tablets, packets, and bulk (granular) form. Also available: Equal Spoonful, which measures like sugar and has zero kcalories.

ARTIFICIAL SWEETENER	BRAND NAME(S)	SAMPLE FOODS USED IN	KCALORIES/ GRAM	SWEETNESS COMPARED TO SUGAR
Acesulfame Potassium (Acesulfame-K)	Sunett Sweet One	Diet sodas	0	200 times as sweet

Description

- Passes through the digestive system unchanged.
- Can be used in baking and cooking.
- Available in liquid, tablets, packets, and bulk (granular) form.

ARTIFICIAL SWEETENER	BRAND NAME(S)	SAMPLE FOODS USED IN	KCALORIES/ GRAM	SWEETNESS COMPARED TO SUGAR
Sucralose	Splenda Equal Sucralose	Diet sodas, low-calorie fruit drinks, maple syrup, applesauce	0	600 times as sweet

Description

- Cannot be digested.
- Stays sweet during cooking and baking.
- Available in packets and bulk (granular) form. Splenda Granulated measures like sugar and has 0 kcalories. Another product, called Splenda Sugar Blend for Baking, contains sugar and sucralose. One cup of sugar in a recipe can be replaced by ½ cup of Splenda Blend for Baking.

ARTIFICIAL SWEETENER	BRAND NAME(S)	SAMPLE FOODS USED IN	KCALORIES/ GRAM	SWEETNESS COMPARED TO SUGAR
Neotame	Not yet available	Products using neotame in the United States are still being developed	0	7,000 to 13,000 times as sweet

Description

- Structurally similar to aspartame.
- Can be used in cooking and baking.
- Can be blended with other sweeteners to increase sweetness and decrease kcalories.

*In Canada, Sugar Twin contains cyclamate, not saccharin.

[†]Because so little aspartame is used, the kcalorie content is very close to zero.

TABLE 3-13   Sugar Replacers (Polyols)

SUGAR REPLACER	KCALORIES/GRAM	USES	DESCRIPTION*
Mannitol	1.6	Chewing gum, powdered foods, chocolate coatings	50 to 70 percent as sweet as sugar May cause a laxative effect when 20 grams or more is consumed Does not absorb moisture, and so it works well as a dusting powder for chewing gum so that the gum doesn't stick to the wrapper.
Sorbitol	2.6	Candies, chewing gum, baked goods, frozen desserts	60 percent as sweet as sugar, gum, baked goods May cause a laxative effect when 50 grams or more is consumed
Xylitol	2.4	Chewing gum, candy	As sweet as sugar
Erythritol	0.2	Beverages, chewing gum, candy, baked goods	Pleasant taste Newest polyol Very heat-stable Much less of a laxative effect than other polyols Works well with other sweeteners to improve flavor and body
Isomalt	2.0	Candies, toffee, fudge, wafers	45 to 60 percent as sweet as sugar Used to add bulk and sweetness to foods Very heat-stable
Lactitol	2.0	Chocolate, candies, cookies and cakes, frozen dairy desserts	30 to 40 percent as sweet as sugar Mild sweetness with no aftertaste Used to add bulk and sweetness to foods
Maltitol	2.1	No-sugar-added ice cream, low-carb bagels, candy, chewing gum, chocolate, baked goods	90 percent as sweet as sugar Used to add bulk and sweetness to foods

*All sugar replacers have the following characteristics:

They occur naturally.

They don't provide as many kcalories as sugar. The average kcalories per gram is 2, compared with 4 kcalories per gram from sugar.

They don't promote tooth decay.

They cause smaller increases in blood glucose and insulin levels than sugar does.

a laxative effect." Most people who experience discomfort can eat a small amount of foods with sugar replacers and then slowly increase these foods in the diet. Not all sugar replacers are equally capable of causing discomfort. For example, sorbitol and mannitol are more likely to cause gas and diarrhea than maltitol is.

You find sugar replacers in a wide assortment of foods: chewing gums, chocolate, candies, frozen desserts such as ice cream, baked goods, salad dressings, beverages, and many foods designed to be lower in kcalories, carbohydrates, and/or fat.

Sugar replacers add bulk and texture and improve the mouthfeel of foods. They not only have been used successfully to replace sugar but can replace fat as well. Sugar replacers also enhance the flavor profile, retain moisture in foods, and provide a cooling effect or taste.

Information about sugar replacers is found in two places on the food label. First, the ingredient list must show the name of each sugar replacer the product contains.

Second, the Nutrition Facts panel shows the number of grams of total carbohydrates in a food, which includes the number of grams of any sugar replacers

1. Trans fats occur at low levels in meat and dairy. Most of the trans fats we eat are in hydrogenated fats found in shortening, margarine, and frying oils.
2. During hydrogenation, some unsaturated fatty acids are converted into trans fats, which behave like saturated fats in the body and raise blood cholesterol levels.
3. Trans fats can be found in some fried foods like French fries and doughnuts, pastries, pie crusts, frozen pizzas, biscuits, ready-made frosting, cookies, microwave popcorn, and some stick margarines and shortenings. Food labels must tell you how much trans fat is in food.

ESSENTIAL FATTY ACIDS: OMEGA-3 AND OMEGA-6 FATTY ACIDS

Essential fatty acids Two poly-unsaturated fatty acids (linoleic and alpha-linolenic acids) that the body can't make so they must be consumed in the diet; vital to growth and development, healthy cells, and the immune system.

Linoleic acid An omega-6 fatty acid found in vegetable oils such as soybean, sunflower, and corn oils.

Alpha-linolenic acid An omega-3 fatty acid found in canola, flaxseed, soybean, and walnut oils; ground flaxseed, walnuts, and soy products.

The body can make most of the fatty acids that it needs except for two, which are called essential fatty acids. These **essential fatty acids**, which must be consumed in the diet, are both polyunsaturated fatty acids:

- Linoleic acid (omega-6 fatty acid)
- **Alpha-linolenic acid (ALA)** (omega-3 fatty acid)

These essential fatty acids are vital to the health of all your body's cells and they help your immune system do its job. They are also needed for infants and children to grow and develop properly.

Linoleic acid is called an **omega-6 fatty acid** because its double bonds appear after the sixth carbon in the chain. The major omega-6 fatty acid in the diet is linoleic acid. Alpha-linolenic acid is the leading **omega-3 fatty acid** found in food, and its double bonds appear after the third carbon in the chain. Adequate Intakes have been set for linoleic and ALA (Table 4-5).

TABLE 4-5	Adequate Intake Values for Essential Fatty Acids			
	LINOLEIC ACID AL (GRAMS PER DAY)		**ALPHA-LINOLENIC ACID AL (GRAMS PER DAY)**	
Age Group	Male	Female	Male	Female
0–6 months	4.4 grams	4.4 grams	0.5 grams	0.5 grams
7–12 months	4.6	4.6	0.5	0.5
1–3 years	7	7	0.7	0.7
4–8 years	10	10	0.9	0.9
9–13 years	12	10	1.2	1.1
14–18 years	16	11	1.6	1.1
19–30 years	17	12	1.6	1.1
31–50 years	17	12	1.6	1.1
Over 50 years	14	11	1.6	1.1
Pregnancy		13		1.4
Lactation		13		1.3

Source: Adapted with permission from the Dietary References Intakes for Energy, Carbohydrates, Fiber, Fat, Protein, and Amino Acids (Macronutrients). © 2002 by the National Academy of Sciences. Courtesy of the National Academy Press, Washington, D.C.

Which foods contain linoleic acid or ALA?

- Linoleic acid (omega-6) is found in vegetable oils such as soybean, corn, and sunflower oils. Most Americans get plenty of linoleic acid from foods containing vegetable oils, including margarine, salad dressings, and mayonnaise. Whole grains and vegetables also supply some linoleic acid (and ALA).

- ALA (omega-3) is found in several oils, notably canola, flaxseed, and walnut oil or margarines made with these oils. Additional good sources of alpha-linolenic acid include ground flaxseed, walnuts, and soy products.

Table 4-6 shows food sources of omega-3 fatty acids.

Omega-6 fatty acids Fatty acids with double bonds after the sixth carbon in the chain; Americans get plenty of omega-6 fatty acids in diet.

Omega-3 fatty acids Fatty acid with double bonds after the third carbon in the chain; tend to be inadequate in many American diets.

TABLE 4-6	Food Sources of Omega-3 Fatty Acids	
FOOD PRODUCT	**ALPHA-LINOLENIC ACID (MG)** (ESSENTIAL OMEGA-3 FATTY ACID)	**EPA + DHA (MG)** (OTHER OMEGA-3 FATTY ACIDS)
Sources of ALA		
Walnuts (1 oz.)	250 mg	
Ground flaxseed (2 Tbsp)	280	
Flax cereal	500	
Canola oil (1 Tbsp)	130	
Walnut oil (1 Tbsp)	140	
Sources of EPA and DHA—Seafood (4 ounces cooked)		
Salmon: Atlantic, chinook, coho		1,200–2,400 mg
Herring and shad		2,300–2,400
Mackerel: Atlantic and Pacific (not King)		1,350–2,100
Tuna: Bluefin and albacore		1,700
Oysters: Pacific*		1,550
Trout: Freshwater		1,000–1,100
Tuna: White (albacore) canned		1,000
Salmon: Pink and sockeye		700–900
Pollock: Atlantic and walleye		600
Crab: Blue, king, snow, queen, and Dungeness		200–550
Tuna: Skipjack and yellowfin		150–350
Flounder and sole		350
Clams		200–300
Tuna: Light canned		150–300
Catfish		100–250
Cod: Atlantic and Pacific		200
Scallops: Bay and sea		200
Haddock		200
Shrimp		100
Omega-3 Fortified Foods**		
Fortified fruit juice blend, 1 cup		50 mg (DHA)
Fortified egg***, 1	125	75 (DHA)
Fortified soy milk		32 (DHA)
Fortified milk		32 (DHA + EPA)

*Eastern oysters have approximately 500–550 mg of EPA + DHA per 4 ounces.

**Omega-3 fatty acid content varies by brand.

***Regular eggs contain 25 mg ALA and 22 mg DHA + EPA.

Sources: US Department of Agriculture, Agricultural Research Service, 2011, USDA Nutrient Database for Standard Reference, Release 24, and manufacturers.

Two other important omega-3 fatty acids are:

- **Docosahexaenoic acid (DHA)**

- **Eicosapentaenoic acid (EPA)**

The best source of these omega-3 fatty acids are fatty fish such as salmon, mackerel, sardines, halibut, bluefish, trout, and tuna (Figure 4-9). Lean fish like haddock, cod, flounder, and sole contain only small amounts of DHA and EPA. DHA and EPA are not found in any plant foods.

DHA is found in high concentrations in the retina of the eye and in the brain, and is especially important for proper brain and eye development during pregnancy and infancy. DHA and EPA are excellent for heart health: they help reduce blood pressure, blood clots (which can start a heart attack), heart rate, and blood triglyceride levels.

Whereas Americans generally get plenty of omega-6 linoleic acid, that is not the case with the omega-3 fatty acids. The ratio of omega-6 to omega-3 fatty acids in the diet is important in regulating your blood pressure, blood clotting, and inflammation. Having a healthy ratio of omega-6 to omega-3 fatty acids seems to lower blood pressure, prevent blood clot formation, and reduce inflammation. Inflammation is the body's normal response to injury, and long-term inflammation seems to play a role in chronic diseases like heart disease.

To increase your intake of omega-3 fatty acids, the *2010 Dietary Guidelines for Americans* recommend you eat about 8 ounces per week of a variety of seafood, which provides an average of 250 mg per day of EPA and DHA. Taking in at least 250 mg per day of EPA and DHA is associated with preventing cardiovascular disease and deaths from cardiovascular disease.

However, nearly all fish and shellfish contain traces of mercury. In general, the health benefits from eating a variety of seafood in the amounts recommended outweigh the health risks associated with mercury. Yet, some fish and shellfish contain higher levels of mercury that may harm an unborn baby or young child's developing nervous system. The risks from mercury in fish and shellfish depend on the amount of fish and shellfish eaten and the levels of mercury in the fish and shellfish. Therefore, the Food and Drug Administration and the Environmental Protection Agency are advising women who may become pregnant, pregnant women, nursing mothers, and young children to avoid some types of fish and eat fish and shellfish that are lower in mercury.

By following three recommendations for selecting and eating fish or shellfish, women and young children will receive the benefits of eating fish and shellfish and be confident that they have reduced their exposure to the harmful effects of mercury:

1. Do not eat shark, swordfish, king mackerel, or tilefish because they contain high levels of mercury.

2. Eat up to 12 ounces (2 average meals) a week of a variety of fish and shellfish that are lower in mercury.

   - Five of the most commonly eaten fish that are low in mercury are shrimp, canned light tuna, salmon, pollock, and catfish.

   - White tuna (albacore) has more mercury than canned light tuna. So when choosing fish and shellfish, include only up to 6 ounces per week of white tuna.

3. Check local advisories about the safety of fish caught by family and friends in your local lakes, rivers, and coastal areas. If no advice is available, eat up to 6 ounces (one average meal) per week of fish from local waters, but don't eat any other fish that week.

FIGURE 4-9 Omega-3 Fats. Good sources of omega-3 fats include flaxseed oil, walnut oil, canola oil, walnuts, flaxseed, and fatty fish such as sardines, tuna, salmon, and blue fish.

Photo by Peter Pioppo.

Docosahexaenoic acid (DHA) and eicosapentaenoic acid (EPA) Two omega-3 fatty acids found mostly in fish that are especially important for heart health, growth, and proper brain and eye development during pregnancy and infancy.

SUMMARY

1. Linoleic acid (omega-6) and alpha-linolenic acid (omega-3) are both essential fatty acids. These essential fatty acids are vital to the health of all your body's cells, and they help your immune system do its job. They are also needed for infants and children to grow and develop properly.

2. Linoleic acid is found in vegetable oils such as soybean, corn, and sunflower oils. Alpha-linolenic acid (ALA) is found in ground flaxseed, walnuts, and soy products, as well as canola, flaxseed, and walnut oils.

3. DHA and EPA are two omega-3 fatty acids found mostly in fish that are especially important for heart health, growth, and proper brain and eye development during pregnancy and infancy. You should eat 8 ounces per week of a variety of seafood.

4. Americans get plenty of omega-6 fatty acids but not enough omega-3 fatty acids. Having a healthy ratio is important to lower blood pressure, prevent blood clots, and reduce inflammation in the body.

5. Recommendations are given for selecting seafood for women and young children to avoid mercury problems.

CHOLESTEROL

Cholesterol is also in the lipid family alongside triglycerides. Pure cholesterol is an odorless, white, waxy, powdery substance. You cannot taste it or see it in the foods you eat.

Your body needs cholesterol to function normally. It is part of every cell membrane and is present in every cell in your body, including the brain and nervous system, muscles, skin, liver, and skeleton. The body uses cholesterol to:

- Make bile acids, which allow us to digest fat
- Maintain cell membranes
- Make many **hormones** (hormones are the chemical messengers of the body; they enter the bloodstream and travel to a target organ to influence what the organ does), such as the sex hormones (estrogen and testosterone)
- Make vitamin D

Because the body makes it own cholesterol, it is not considered an essential nutrient.

So, which foods contain cholesterol? Cholesterol is found only in animal foods (Figure 4-10 and Table 4-7):

- Egg yolks (it's not in the whites)
- Meat
- Poultry
- Milk and milk products
- Fish

Egg yolks and organ meats (liver, kidney, sweetbreads, brain) contain the most cholesterol—one egg yolk (medium egg) contains 186 milligrams of cholesterol. About 4 ounces of meat, poultry, or fish (trimmed or untrimmed) contains 100 milligrams of cholesterol, with the exception of shrimp, which is higher in cholesterol. Eggs, meat, and whole milk provide most of the cholesterol we eat, and these sources are also rich in saturated fat. In milk products, cholesterol is mostly in the fat, and so lower-fat products contain less cholesterol. For example, 1 cup of whole milk contains 24 milligrams of cholesterol, whereas a cup of nonfat milk contains only 5 milligrams (Table 4-2).

Cholesterol A lipid; a soft, waxy substance found only in foods of animal origin; it is present in every cell of the body.

Hormones Chemical messengers in the body.

FIGURE 4-10 Cholesterol. Foods moderate to high in cholesterol include egg yolks, meat, poultry, seafood, milk, cheese, butter, and other dairy products.

Photo by Peter Pioppo.

When looking at fat in food, it is important to distinguish between two different concepts: the percentage of kcalories from fat and the percentage of fat by weight. To explain these two concepts, let's look at an example. In a supermarket, you find sliced turkey breast that is advertised as being "96 percent fat-free." What this means is that if you weighed a 3-ounce serving, 96 percent of the weight would be lean, or without fat. In other words, only 4 percent of its weight is actually fat. The statement "96 percent fat-free" does not tell you anything about how many kcalories come from fat.

Now, if you look at the Nutrition Facts on the label, you read that a 3-ounce serving of the turkey contains 3 grams of fat, 27 kcalories from fat, and 140 total calories. The label also states that the percentage of kcalories from fat in a serving is 19 percent. To find out the percentage of kcalories from fat in any serving of food, simply divide the number of calories from fat by the number of total kcalories and then multiply the answer by 100, as follows.

$$\frac{\text{Kcalories from fat}}{\text{Total kcalories}} \times 100 = \text{Percentage of kcalories from fat}$$

$$\frac{27 \text{ Kcalories from fat}}{140 \text{ kcalories}} \times 100 = 19 \text{ percent}$$

This percentage is useful when comparing foods.

SUMMARY

1. There is no DRI for total fat except for infants. The Acceptable Macronutrient Distribution Range for fat is 20 to 35 percent of kcalories for people over 18 years old.
2. The *Dietary Guidelines for Americans, 2010,* recommend consuming less than 10 percent of kcalories from saturated fatty acids by replacing them with monounsaturated and polyunsaturated fatty acids, consuming less than 300 mg of dietary cholesterol each day, keeping trans fatty acid consumption as low as possible, eating some protein foods that are low in saturated fat, and using oils to replace solid fats.
3. The percentage of kcalories from fat for a food is not the same as the percentage of fat by weight. The statement "96 percent fat-free" does not tell you anything about how many kcalories come from fat—only that 4 percent of its weight is actually fat.

FATS AND HEALTH

Most foods contain several different kinds of fat, and some are better for your health than others are. You don't need to eliminate fat from your diet, just choose the types of fat that are good for your health.

HEART DISEASE

Heart disease is the number-one killer of both men and women in the United States. It accounts for one out of every 2.9 deaths. The death rate from cardiovascular disease has declined recently, but the burden of the disease remains huge. In 2010, $450 billion was spent on treating heart disease. Many Americans have elevated blood cholesterol levels, one of the key risk factors for heart disease. Anyone can develop high blood cholesterol, regardless of age, gender, race, or ethnic background.

Too much circulating cholesterol can build up in the walls of the arteries, especially the heart's arteries (called the coronary arteries), which supply the heart with oxygen and nutrients to keep it pumping. This leads to accumulation of cholesterol-laden **plaque** in blood vessel linings, a condition called **atherosclerosis,** or "hardening of the arteries."

Plaque Deposits on arterial walls that contain cholesterol and other substances.

Atherosclerosis A disease in which arteries have plaque buildup along the arterial walls.

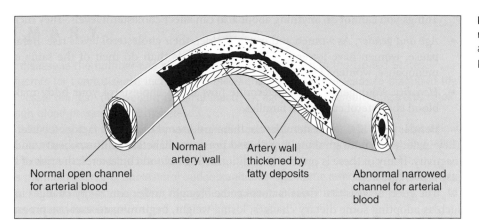

FIGURE 4-11 Cross-sectional representation of a coronary artery partially closed with plaque.

Normal open channel for arterial blood

Normal artery wall

Artery wall thickened by fatty deposits

Abnormal narrowed channel for arterial blood

If the coronary arteries become partially blocked by plaque (Figure 4-11), then the blood may not be able to bring enough oxygen and nutrients to the heart muscle itself. This can cause chest pain. A **heart attack** occurs if the flow of oxygen-rich blood to a section of heart muscle suddenly becomes blocked. If blood flow isn't restored quickly, the section of heart muscle begins to die. Healthy heart tissue is replaced with scar tissue. This heart damage may not be obvious, or it may cause severe or long-lasting problems.

If the blood supply to part of the brain is cut off, a **stroke** can occur, which can cause damage to brain cells. Like the heart, the brain must have a continuous supply of blood rich in oxygen and nutrients. If deprived of blood for more than a few minutes, brain cells die, and the functions these cells control—speech, muscle movement, and more—die with them.

LDL is the main source of cholesterol buildup and blockage in the arteries. The primary way in which LDL cholesterol levels become too high is through eating too much saturated fat, trans fat, and, to a lesser extent, cholesterol. Dietary factors that lower LDL cholesterol include:

Heart attack When flow of blood to a section of the heart muscle suddenly becomes blocked.

Stroke Damage to brain cells resulting from an interruption of blood flow to the brain.

- *Consume less than 10 percent of kcalories from saturated fats by replacing them with polyunsaturated and monounsaturated fats.* For example, when preparing foods, solid fats (such as butter) can be replaced with vegetable oils that are rich in monounsaturated and polyunsaturated fatty acids. Nuts, which are rich in monounsaturated and polyunsaturated fats, can be substituted for other foods. For example, have a peanut butter sandwich instead of roast beef and cheese. In addition, many of the major food sources of saturated fat can be purchased in lower fat forms, such as fat-free milk, low-fat cheese, or well-trimmed meats.

- *Limit your intake of trans fat to 1 percent of total kcalories.* Because meat and dairy contain small amounts of natural trans fats, you should avoid commercial foods with trans fats. Trans fats not only increase your LDL cholesterol but they also decrease your HDL cholesterol.

- *Increase intake of soluble fiber as found in oats.* Soluble fiber is also found in such foods as kidney beans, apples, pears, barley, and prunes. Soluble fiber can reduce the absorption of cholesterol into your bloodstream.

- *Foods are available that have been fortified with sterols or stanols*—substances found in plants that help block the absorption of cholesterol. Margarines, orange juice, and yogurt drinks with added plant sterols can help reduce LDL cholesterol by more than 10 percent.

A variety of other things can affect cholesterol levels. These are things you can do something about:

- *Weight.* Being overweight is a risk factor for heart disease. It also tends to increase your cholesterol. Losing weight can help lower your LDL and total cholesterol levels as well as raise your HDL and lower your triglyceride levels.

- *Physical activity.* Not being physically active is a risk factor for heart disease. Regular physical activity can help lower LDL cholesterol and raise HDL cholesterol levels. It also helps you lose weight. You should try to be physically active for 30 minutes on most, if not all, days.

cholesterol. Ounce for ounce, meat, poultry, and most cheeses have about the same amount of cholesterol. But cheeses tend to have much more saturated fat.

Determining which cheeses are high or low in saturated fat can be confusing, because there are so many different kinds on the market: part-skim, low-fat, processed, and so on. Not all reduced-fat or part-skim cheeses are low in fat; they are only lower in fat than similar natural cheeses. For instance, a reduced-fat cheddar gets 56 percent of its kcalories from fat—considerably less than the 71 percent of regular cheddar but not super-lean, either. The trick is to read the label. Table 4-9 is a guide to fat in cheeses.

Yogurt is made by adding two bacteria (*Streptococcus thermophilus* and *Lactobacillus bulgaricus*) into heated milk. Then the milk is held at 110°F until the milk is thickened and firm. With its slightly sour taste and creamy texture, yogurt can be used in many recipes, such as dips, dressings, sauces, smoothies, cold soups, curries, and desserts.

A very important difference among yogurts is whether they contain live bacteria. Although all yogurts must contain two types of bacteria, some yogurts are heated after they are made, thereby destroying the cultures. Also, some yogurts are initially formulated with a low level of cultures. You can be sure you are getting yogurt with significant levels of live and active cultures by looking for the National Yogurt Association *Live & Active Cultures* seal on the package. If a label says "heat-treated after culturing," you are not getting live and active cultures.

Yogurts also can differ tremendously in their nutritional content—especially carbohydrates, fats, and proteins. Plain, nonfat yogurt typically has about 11 grams of natural sugar in 6 ounces. Flavored yogurt varieties can contain a lot more sugar. The fat content of yogurt depends on what type of milk it was made from—which can range from nonfat to whole—so read the label.

Greek yogurt has become more popular in the United States. Greek yogurt is strained extensively to remove much of the liquid whey, lactose, and sugar. The result is a product that is thicker, creamier, and tangier. Nutritionally, is has less sugar and significantly more protein. However Greek yogurt made from whole milk is very high in saturated fat, so pick low-fat or nonfat varieties.

Eggs are very nutritious and full of high-quality protein, as well as varying amounts of many vitamins and minerals. The concern with overconsumption of eggs stems from the fact that they are very high in cholesterol—186 milligrams per egg (compare that to the suggested maximum of 300 milligrams daily). One egg also contributes 5 grams of fat, of which 2 grams are saturated fat.

PREPARATION

When cooking with milk, remember a very important rule: Use moderate heat and heat the milk slowly (but not too long) to avoid curdling—a grainy appearance with a lumpy texture. From a scientific point of view, milk curdles when the casein (protein in milk) becomes solid and separates out of the milk. Other preparation tips include the following:

- Add other food products to hot milk products slowly, stirring with either a spoon or a wire whisk, if preparing a sauce, to avoid lumps. Be especially careful when adding foods high in acid—milk has a tendency to curdle if not beaten quickly. Combining skim milk with stock for sauces and soups such as cheese sauce, chowder, white wine, and bisques will add a creamy texture when thickened without adding extra fats.

- There are several ways to use cheese in balanced cooking:
  - Use good-quality cheeses, both aged and fresh, yielding a better flavor and texture. You can ultimately use less cheese (less fat) with satisfying results.
  - Use cheese in the most critical part of the preparation, usually just before service, so richness of the cheese is the first flavor the customer tastes.
  - Experiment with low-fat cheese or a mixture of half whole cheese and half low-fat cheese.
  - Use low heat with cheese. It is best to use as low a heat as possible when cooking with cheese. Cheese has a tendency to toughen when subjected to high heat and long cooking

| | | TABLE 4-9 | Fat in Cheese[†] | |

LOW FAT	MEDIUM FAT	HIGH FAT	VERY HIGH FAT
0–3 g fat/oz.	4–5 g fat/oz.	6–8 g fat/oz.	9–10 g fat/oz.
Natural Cheeses			
*Cottage cheese (¼ cup) dry curd	*Mozzarella, part-skim *Ricotta (¼ cup), part-skim	Blue cheese *Brick	Cheddar Colby
Cottage cheese (¼ cup) low-fat 1%	String cheese, part-skim	Brie Camembert	*Cream cheese (1 oz. = 2 Tbsp.)
Cottage cheese (¼ cup) low-fat 2%		Edam Feta	Fontina *Gruyère
Cottage cheese (¼ cup) Creamed 4%		Gjetost Gouda	Longhorn *Monterey Jack
Sapsago		*Light cream cheese (1 oz. = 2 Tbsp.)	Muenster Roquefort
Look for special low-fat brands of mozzarella, ricotta, cheddar, and Monterey jack.	Look for reduced-fat brands of cheddar, Colby, Monterey jack, muenster, and Swiss.	Limburger Mozzarella, whole-milk Parmesan (1 oz. = 3 Tbsp.) *Port du Salut Provolone *Ricotta (¼ cup), whole-milk Romano (1 oz. = 3 Tbsp.) *Swiss Tilsit, whole milk	
Modified Cheeses			
Pasteurized process, imitation, and substitute cheeses with 3 g fat/oz or less	Pasteurized process, imitation, and substitute cheeses with 4–5 g fat/oz.	Pasteurized process Swiss cheese Pasteurized process Swiss cheese food Pasteurized process American cheese Pasteurized process American cheese food American cheese food cold pack Imitation and substitute cheeses with 6–8 g fat/oz.	Some pasteurized process cheeses are found in this category—check the labels.

[†]Check the labels for fat and sodium content. 1 serving = 1 oz. unless otherwise stated.

*These cheeses contain 160 mg or less of sodium per 1 oz.

Source: Reprinted by permission of the American Heart Association, Alameda County Chapter, 11200 Golf Links Road, Oakland, CA 94605.

1. Lipids include oils, sugars, cholesterol, and lecithin.
   a. True
   b. False

2. Triglycerides are the most common form of fats in food and in the body.
   a. True
   b. False

3. The concerns with fat in the diet have more to do with the type of fat you eat than the total quantity you eat.
   a. True
   b. False

4. When you leave the skin on chicken, it actually has more fat than beef.
   a. True
   b. False

5. Eggs are high in saturated fat and cholesterol.
   a. True
   b. False

6. Palm oil is a good source of monounsaturated and polyunsaturated fats.
   a. True
   b. False

7. Hydrogenation of vegetable oils to make margarine and shortening produces trans fatty acids.
   a. True
   b. False

8. The body makes both of the essential fatty acids.
   a. True
   b. False

9. Americans generally get plenty of omega-3 fatty acids in their diet.
   a. True
   b. False

10. DHA and EPA are omega-3 fatty acids found in fatty fish.
    a. True
    b. False

11. White tuna (albacore) has less mercury than canned light tuna.
    a. True
    b. False

12. Cholesterol is used to make some hormones and vitamin D.
    a. True
    b. False

13. Most of the cholesterol we eat comes from eggs, meat, and whole milk.
    a. True
    b. False

14. The American Heart Association recommends you eat 1% or less of your kcalories from trans fats.
    a. True
    b. False

15. HDL is called the "bad cholesterol" because it increases your risk for heart disease.
    a. True
    b. False

16. You should eat less than 10 percent of kcalories from saturated fatty acids and less than 1 percent from trans fatty acids to decrease your heart disease risk.
    a. True
    b. False

17. Being overweight and physically inactive also contributes to heart disease risk.
    a. True
    b. False

18. To decrease your cancer risk, you should increase intake of processed meats and red meats.
    a. True
    b. False

19. Which food groups contribute the most fat?

20. What are the biggest sources of saturated fat in the American diet? List five foods that may contain trans fats.

21. What foods can you eat to increase your intake of omega-3 fatty acids?

22. Name four fish that women and young children should avoid because they may contain high levels of mercury.

23. List two functions each of lipids, the essential fatty acids, and cholesterol.

24. What is the main function of chylomicrons, low-density lipoproteins, and high-density lipoproteins?

25. Give two dietary recommendations to prevent heart disease and two dietary recommendations to prevent cancer.

NUTRITION WEB EXPLORER

For a complete list of websites for the following activities, please visit the companion book page at www.wiley.com/college/drummond.

THE AMERICAN HEART ASSOCIATION

Visit the website of the American Heart Association and click on "Getting Healthy," then click on "Nutrition Center." Next click on "Healthy Cooking." Look around "Healthy Cooking" to find and write down three preparation methods, three seasonings, and three cooking substitutions that are heart-healthy.

GOLDEN VALLEY FLAX

Flaxseed is a wonderful source of omega-3 fatty acids. The body cannot break down whole seeds to access the omega-3 containing oil, so most recipes use ground seeds. Read the recipes at this website and try one out. Read the FAQs found on "Facts About Flax" page. What is the flavor of flaxseed like? Can you put it on cereal or in yogurt? Do you store whole and ground flaxseeds the same way?

CABOT CHEESE

Cabot makes some excellent cheeses that are lower in fat. Click on "Quality Products" and then click on "Light Cheddars." How many kcalories and fat does 1 ounce of 50 percent reduced fat cheddar have? Compare that to 1 ounce of Cabot Classic Cheddar.

DANNON YOGURT

Click on "Products," which shows you a variety of yogurts they make. Choose any three types of yogurt and compare the kcalories, fat, sugar, and protein each contains. Also click on "Recipes." What types of recipes can you use yogurt in?

AMERICAN CANCER SOCIETY

Click on "Stay Healthy" and then click on "Eat Healthy and Get Active." Click on "Eat Healthy" and read "Innovations in Home Cooking." Write a paragraph about what you learned.

IN THE KITCHEN

Using Different Types and Amounts of Fats and Oils in Cooking and Baking

You will be assigned to do one or more of the following.

1. Make 3 vinaigrette dressings using different amounts of oil.
2. Saute chicken using different oils.
3. Prepare traditional banana muffins, then replace half the oil with applesauce or grated zucchini.
4. Make an apple turnover using traditional pie dough and also with phyllo dough.

   Your instructor will hand you a worksheet at the end to examine and compare the fat content of all the dishes, as well as compare and evaluate each dish's taste and acceptability.

HOT TOPIC

CHOOSING FATS AND OILS

There is an ever-widening variety of oils, margarines, and buttery spreads on the market. They can differ markedly in their color, flavor, uses, and nutrient makeup (Figure 4-13).

When choosing vegetable oils, pick those high in monounsaturated fats, such as olive oil, canola oil, and peanut oil, or high in polyunsaturated fats, such as corn oil, sunflower oil, and soybean oil.

Olive oil contains from 73 to 77 percent monounsaturated fat, as compared to butter which contains about 60 percent saturated fat. The color of olive oil varies from pale yellow to dark green, and its flavor varies from subtle to a full, fruity taste. The color and flavor of olive oil depend on the olive variety, level of ripeness, and processing method. When buying olive oil, make sure you are buying the right product for your intended use.

- Extra virgin olive oil, the most expensive form, has a rich, fruity taste that is ideal for flavoring finished dishes and in salads, vegetable dishes, marinades, and sauces. It is not usually used for cooking because it loses some flavor. It is made by putting mechanical pressure on the olives, a more expensive process than using heat and chemicals.

- Olive oil, also called pure olive oil, is golden and has a mild, classic flavor. It is an ideal, all-purpose product that is great for sautéing, stir-frying, salad dressings, pasta sauces, and marinades.

- Light olive oil refers only to color or taste. These olive oils lack the color and much of the flavor found in the other products. Light olive oil is good for sautéing or stir-frying because the oil is used mainly to transfer heat rather than to enhance flavor.

Table 4-10 gives information on various oils. Be prepared to spend more for exotic oils such as almond, hazelnut, sesame, and walnut. Because these oils tend to be cold-pressed (meaning they are unrefined or processed without heat), they are not as stable as refined oils and they have a lower smoke point. They are strong in flavor, so you don't need too much of them.

Vegetable oils are also available in convenient sprays that can be used as nonstick spray coatings for cooking and baking with a minimal amount of fat. Vegetable-oil cooking sprays come in a variety of flavors (butter, olive, Italian, mesquite), and a quick, two-second spray adds about 1 gram of fat to the product. To use, first spray the cold pan away from any open flames (the spray is flammable), heat the pan, then add the food.

Margarine was first made in France in the late 1800s to provide an economical fat for Napoleon's army. It didn't become popular in the United States until World War II, when it was introduced as a low-cost replacement for butter. Margarine

FIGURE 4-13 Oils. From left to right: sesame oil, corn oil, roasted peanut oil, peanut oil, soybean oil, canola oil, toasted sesame oil, and olive oil.
Photo by Peter Pioppo.

TABLE 4-10 Oils*

OIL	CHARACTERISTICS	USES	SMOKE POINT
Almond oil	Light golden color Clean flavor	Sautéing, frying	420°F
Argan oil	Pressed from the nut of a Moroccan tree Nutty brown-butter flavor with a hint of musky herbaciousness	Sautéing Flavoring finished dishes such as grilled vegetables, meats and fish, and soups Good in salad dressings Used like olive oil in Morocco	410°F
Butternut squash seed oil	Hard-to-place taste—hints of apricot, peanuts	Good for flavoring finished dishes such as prepared vegetables, used to dress salads, or used as a dipping oil with hearty breads	—
Canola oil	Light yellow color Neutral flavor	Sautéing, salad dressings, baked goods	400°F
Corn oil	Golden color Bland flavor	Sautéing, baked goods Too heavy for salad dressings	450°F
Hazelnut oil	Dark amber color Nutty and smoky flavor	Good for flavoring finished dishes such as pasta, potatoes and beans, and used in salad dressings Sautéing	320–400°F
Olive oil	Color and flavor depend on olive variety, level of ripeness, and how oil was processed	Extra virgin or virgin—good for flavoring finished dishes and in salad dressings, strong olive taste	320–400°F
		Pure olive oil—can be used for sautéing and in salad dressings	400°F
		Light olive oil—the least flavorful, good for sautéing and stir-frying	400°F
Peanut oil	Pale yellow color Mild nutty flavor	Frying and sautéing Good for salad dressings	440°F
Pecan oil	Flavorful	Works well in salad dressings Sautéing and stir-frying	450°F
Pine nut oil	Subtle and mild	For flavoring finished dishes such as pesto, sauces, and soups	—
Safflower oil	Golden color Bland flavor	Good for sautéing and in baked goods Good oil for salad dressings	450°F
Sesame oil	Distinct strong flavor	Good for sautéing, flavoring dishes, and in salad dressings	410°F
Soybean oil	Light color Bland flavor	Good for sautéing, frying, and in baked goods Good oil for salad dressings	450°F
Sunflower oil	Somewhat neutral flavor	Good for sautéing, frying, and in baked goods Good oil for salad dressings	450°F
Walnut oil	Medium yellow to brown color Rich, nutty flavor	For flavoring finished dishes and in salad dressings Heating produces some bitterness	—

*All are refined oils, except those with no smoke points. Refining oils (taking out impurities) generally increases the smoke point.

must contain vegetable oil and water and/or milk or milk solids. Flavorings, coloring, salt, emulsifiers, preservatives, and vitamins are usually added. The mixture is heated and blended, then firmed by exposure to hydrogen gas at very high temperatures (see information about hydrogenation in the Trans Fats section of this chapter). The firmer the margarine, the greater the degree of hydrogenation and the longer the shelf life.

Standards set by the US Department of Agriculture and the Food and Drug Administration require margarine and butter to contain at least 80 percent fat by weight and to be fortified with vitamin A. One tablespoon of

- Define protein and explain the difference between essential and nonessential amino acids.

- Compare and contrast the nutrients in animal and plant sources of protein.

- Distinguish between complete and incomplete protein and give two examples of how to complement proteins.

- List five functions of protein in the body.

- Explain how protein is digested, absorbed, and metabolized.

- State the dietary recommendations for protein and explain the potential consequences of eating too much or too little protein.

- Explain the concept of denaturation, or what happens to protein when it is cooked.

- Identify six examples of meat, poultry, and fish that are moderate in fat and saturated fat, and describe three preparation techniques for balanced meat, poultry, and fish menu items.

- Give two examples of how to menu and/or present balanced meat, poultry, and fish items.

- List three benefits of vegetarian diets, use a vegetarian food guide to plan a balanced meal, and list nutrients (and their sources) that may be low in some vegetarian diets.

THE BASICS OF PROTEIN

Have you ever wondered why meat, poultry, and seafood are often considered entrées, or main dishes, whereas vegetables and potatoes are considered side dishes? The abundant protein found in meat, poultry, and seafood has long been considered by many to be the mainstay of a nutritious diet. With the arrival of MyPlate in 2011, protein has not lost its importance, but it does take up less space on the plate than fruits and vegetables.

Today, protein foods continue to be an important component of a nutritious diet; however, we are much more likely to see plant foods such as lentils and pasta, which provide protein, occupying the center of the plate. For adults who grew up when beef was king (and not nearly as expensive as it is today) and full-fat lunchmeat sandwiches filled many lunchboxes, making spaghetti without meatballs takes a little getting used to, but more and more meatless meals are being served.

So just what is **protein**? Proteins are nutrients found in all living cells in animals and plants that play a variety of important roles. The protein found in animals and plants is such an important substance that the term *protein* is derived from the Greek word meaning "first." You have protein in your skin, hair, nails, muscles, and blood, to name just a few places. Whereas carbohydrates and lipids are used primarily for energy, proteins function in a very broad sense to build and maintain your body.

Proteins are long chains of **amino acids** strung together the way different railroad cars make up a train. Amino acids are the building blocks of proteins. There are 20 different ones. Of these 20, 9 either cannot be made in the body or cannot be made in the quantities needed. You must get these amino acids from food in order for your body to function properly. This is why we call them **essential amino acids**. The remaining 11 can be made in the body, and so they are called **nonessential amino acids**. Essential amino acids become important when discussing vegetarian diets, addressed later in the chapter.

In the body, the instructions to make proteins reside in the core, or nucleus, of each of your cells. In the nucleus are molecules called **deoxyribonucleic acid**, or **DNA**, which contain your genetic information. Every human cell (with the exception of mature red blood cells, which have no nuclei) contains the same DNA. Segments of each DNA molecule are called **genes**. A gene carries a particular set of instructions that allows a cell to produce something—often a protein such as an enzyme. You have thousands of genes in your chromosomes (Figure 5-1).

Protein Major structural parts of the body's cells that are made of amino acids assembled in chains; performs many functions; especially rich in animal foods.

Amino acids The building blocks of protein.

Essential amino acids Amino acids that either cannot be made in the body or cannot be made in the quantities needed by the body; must be obtained in foods.

Nonessential amino acids Amino acids that can be made in the body.

DNA (Deoxyribonucleic acid) Molecules in the nucleus of cells that carry your genetic information.

Genes A tiny section of DNA that has a code to make proteins and other compounds.

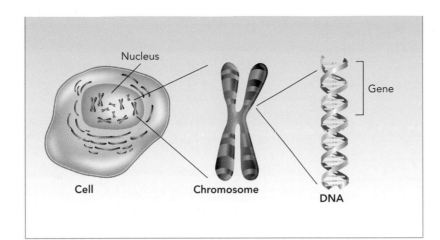

FIGURE 5-1 Genes in the cell nucleus.

S U M M A R Y

1. Proteins are long chains of amino acids. Of the 20 amino acids, 9 are essential amino acids and 11 are nonessential amino acids (can be made in the body).
2. Protein functions in a very broad sense to build and maintain your body. They are part of your skin, hair, and all other cells.
3. The instructions to make proteins are encoded in your DNA. Segments of each DNA molecule (genes) carry instructions to allow a cell to make specific products such as a protein.

PROTEIN IN FOOD AND PROTEIN QUALITY

Protein is found in animal and plant foods (Figure 5-2). Animal foods, such as beef, chicken, fish, and dairy products, have the most protein. Among the plant foods, legumes, grains, and nuts usually contribute more protein than do vegetables and fruits. Protein-rich animal foods are usually higher in fat and saturated fat and always higher in cholesterol than plant foods because plant foods have no cholesterol (Table 5-1). Animal proteins also tend to be the most expensive foods on the menu.

One ounce of meat, poultry, or fish is equal to any of the following:

- 1 egg
- 1 tablespoon peanut butter
- ½ ounce nuts (12 almonds, 7 walnut halves)
- ½ ounce of seeds (pumpkin, sunflower, or squash seeds, hulled, roasted)
- ¼ cup of cooked beans or peas
- ¼ cup tofu
- 1 ounce tempeh, cooked
- 2 tablespoons hummus

To understand the concept of protein quality, remember that 9 of the 20 amino acids either can't be made in the body or can't be made in sufficient quantities.

Food proteins that provide all the essential amino acids in the proportions needed by the body are called high-quality proteins, or **complete proteins**. Examples of complete proteins include the animal proteins, such as meats, poultry, fish, eggs, milk, and other dairy products.

Lower-quality proteins, or **incomplete proteins**, are low in one or more essential amino acids. Plant proteins, including dried beans and peas, grains, vegetables, nuts, and seeds, are incomplete. The essential amino acid in lowest concentration in a protein is referred to as a

FIGURE 5-2 Animal foods, such as beef, chicken, fish, eggs, and milk (on the left and center), have the most protein per serving. Plant foods also contain protein. Legumes (beans and peas), grains, and nuts are higher in protein than most fruits and vegetables.

Photo by Peter Pioppo.

TABLE 5-1

Protein, Fats, and Fiber in Animal and Plant Foods

	KCAL.	FAT (G)	SATURATED FAT (G)	CHOLESTEROL (MG)	PROTEIN (G)	FIBER (G)
ANIMAL FOODS						
Beef (all 4 oz. cooked)						
Beef outside bottom round steak, choice, grilled, 0" trim	216	9.5 g	3.5 g	88 mg	31 g	0 g
Beef flank steak, choice, broiled, 0" trim	228	10.5	4.5	92	31	0
Beef eye of round roast, choice, roasted, 1/8" trim	240	11	4.5	97	32	0
Beef top round steak, choice, broiled, 0" trim	228	8	3	101	36	0
Beef top sirloin steak, choice, broiled, 0" trim	235	10	4	99	34	0
Beef tenderloin steak, choice, broiled, 0" trim	231	12	5	90	28	0
Beef chuck eye steak, boneless, choice, grilled, 0" trim	320	23	10	97	28	0
Beef rib roast (large end), choice, roasted, 1/8" trim	427	35	14	96	25	0
Ground beef, 75/25, broiled	314	21	8	101	29	0
Ground beef, 90/10, broiled	245	13	5	96	29	0
Ground beef, 95/5, Broiled	193	7.5	3.5	86	30	0
Pork (all 4 oz. cooked)						
Tenderloin, broiled	227	9	3.5	106	34	0
Top loin chop, boneless, broiled	221	10.5	3.5	82	30	0
Loin rib chop, boneless, broiled	294	18	6.5	93	31	0
Fish (all 4 oz. cooked)						
Salmon, wild coho, cooked dry heat	157	5	1	62	26	0
Flounder, cooked dry heat	97	3	0.5	63	17	0
Crab, Alaska King, cooked moist heat	110	2	0	60	22	0
Poultry (all 4 oz. cooked)						
Chicken breast with skin, roasted	223	9	2.5	95	34	0
Chicken breast, no skin, roasted	186	4	1	96	35	0
Chicken drumstick, no skin, rotisserie	199	8	2	181	32	0
Milk and Eggs						
Whole milk, 1 cup	149	8	4.5	24	8	0
Nonfat milk, 1 cup	83	0	0	5	8	0
Egg, 1 large	72	5	1.5	186	6	0
PLANT FOODS						
Legumes						
Pinto beans, cooked, 1 cup	245	1	0	0	15.5	15
Kidney beans, cooked, 1 cup	225	1	0	0	15.5	13
Lentils, cooked, 1 cup	226	1	0	0	18	16
Tofu, raw, 1 cup	188	12	2	0	20	1
Veggie or soy burger, raw, 2.5 oz.	124	4.5	1	4	11	1
Breads and Cereals						
Whole wheat bread, 1 slice	76	1	0	0	4	2
Oatmeal, cooked, 1 cup	166	4	1	0	6	4
Nuts and Seeds						
Mixed nuts, dry roasted, 1 ounce	168	14.5	2	0	5	2.5

Peanut butter, smooth, 2 tablespoons	188	16	3	0	8	2
Sunflower seeds, dry roasted, 1 ounce	165	14	1.5	0	5.5	2.5
Vegetables						
Corn, whole kernel, sweet, boiled, 1 cup	130	2	0	0	4	3
Green beans, boiled, 1 cup	38	0	0	0	2	4

Source: U.S. Department of Agriculture, Agricultural Research Service, 2011, USDA Nutrient Database for Standard Reference, Release 24. Nutrient Data Laboratory Home Page: www.nal.usda.gov/fnic/foodcomp

limiting amino acid because it limits the protein's usefulness in the body unless another food in the diet contains it.

Although plant proteins are incomplete, it does not mean they are low-quality. When certain plant foods, such as peanut butter and whole-wheat bread, are eaten over the course of a day, the limiting amino acid in each of these proteins is supplied by the other food. Such combinations are called **complementary proteins**. This is the case when grains, such as whole-wheat bread, are consumed with legumes, such as peanut butter, or when rice is eaten with beans (Figure 5-3). Beans supply plenty of the essential amino acids lysine and isoleucine, which are both lacking in grains such as rice. Grains have plenty of methionine and tryptophan, which are lacking in beans and other legumes. Additional examples of legumes and grains include beans and corn or wheat tortillas, lentils and rice, pea soup and bread, and beans and pasta. Nuts/seeds and legumes (such as sesame paste with hummus, which contains chickpeas) also make a complete protein.

Grains, Nuts, or Seeds + Legumes = Complete protein

By having dairy with plant foods, you also get more complete protein (Table 5-2).

Some plant proteins, such as quinoa, and protein made from soybeans (isolated soy protein), are complete proteins. In adequate amounts and combinations, plant foods can supply the essential nutrients needed for growth and development and overall health. Many cultures around the world use plant proteins extensively.

Complete proteins Food proteins, generally animal proteins, that provide all the essential amino acids in the proportions needed by the body.

Incomplete proteins Food proteins that contain at least one limiting amino acid.

Limiting amino acid An essential amino acid in lowest concentration in a protein that limits the protein's usefulness unless another food in the diet contains it.

Complementary proteins The ability of two protein foods to make up for the lack of certain amino acids in each other when eaten over the course of a day.

FIGURE 5-3 An example of complementary protein: rice and bean burrito.

Photo by Peter Pioppo.

TABLE 5-2	Protein Pairings
COMPLEMENTARY PROTEINS	**EXAMPLES**
Legumes with grains	Refried beans on tortilla, cornbread and pinto beans, falafel with whole-wheat pita, peanut butter on whole-wheat bread, black bean quinoa burger
Legumes with nuts/seeds	Hummus (ground sesame seeds and chickpeas), lentil walnut burgers
Grains with dairy	Oatmeal and milk, whole-wheat macaroni and cheese, cheese sandwich on whole-grain bread
Legumes with dairy	Black bean and cheese burritos, chili with shredded cheese
Nuts/seeds with dairy	Yogurt dressing and toasted almonds on salad, feta/walnut pâté

S U M M A R Y

1. Animal foods contain more protein than plant foods. Animal foods are also higher in fat, saturated fat, and cholesterol than most plant foods.
2. One ounce of meat, poultry, or fish is equal to one egg, one tablespoon peanut butter, ½ ounce of nuts or seeds, or ½ cup of cooked bean, peas, or tofu.
3. Food proteins (such as meat, poultry, fish, eggs, dairy) that provide all the essential amino acids in the proportions needed by the body are high-quality or complete proteins. Plant proteins are incomplete (except quinoa and soy protein), because they are low in one or more essential amino acids (called the limiting amino acid).
4. When certain plant foods, such as peanut butter and whole-wheat bread, are eaten over the course of a day, the limiting amino acid in each of these proteins is supplied by the other food. Grains, nuts, or seeds, when eaten with legumes, make complete protein. Such combinations are called complementary proteins (Table 5-2).

FUNCTIONS OF PROTEIN

After reviewing all the jobs proteins perform, you will have a greater appreciation of this nutrient. In brief, protein is part of most body structures; builds and maintains the body; is a part of many enzymes, hormones, and antibodies; transports substances around the body; maintains fluid and acid-base balance; can provide energy for the body; and helps in blood clotting (Table 5-3). Now let's take a look at each function separately:

TABLE 5-3	Functions of Protein

- Acts as a structural component of the body
- Builds and maintains the body
- Found in many enzymes and hormones and all antibodies
- Transports iron, fats, minerals, and oxygen
- Maintains fluid and acid-base balance
- Provides energy as last resort
- Helps blood clot

Proteins function as part of the body's structure. For example, protein can be found in skin, bones, muscle, hair, fingernails, muscles, blood vessels, digestive tract, and blood. Proteins also give structure to individual cells. Collagen, the most abundant protein in your body, is the major protein in ligaments (which hold our bones together) and in tendons (which attach muscles to bones).

Proteins are used for building and maintaining body tissues. Worn-out cells are replaced throughout the body at regular intervals. For instance, your skin today will not be the same skin in a few months. A skin cell lives only about one month. Skin is constantly being broken down and rebuilt or remodeled, as are most body cells, including the protein within those cells. The cells that line the gastrointestinal tract are replaced every three to five days.

The greatest amount of protein is needed when the body is building new tissues rapidly, such as during pregnancy or infancy. A newborn baby requires 9 grams of protein each day for the first six months, which increases to 13.5 grams for the next 6 months up to his first birthday. By age 9 he needs about 34 grams of protein each day. Additional protein is also needed when body protein is lost or destroyed, as in burns, surgery, or infections.

Proteins are found in many enzymes, some hormones, and all antibodies. Thousands of enzymes have been identified. Almost all the reactions that occur in the body, such as food digestion, involve enzymes. **Enzymes** are catalysts, meaning that they increase the rate of these reactions, sometimes by more than a million times. They do this without being changed in the overall process. Enzymes contain a special pocket called the active site. You can think of the active site as a lock into which only the correct key will fit. Various substances fit into the pocket, undergo a chemical reaction, and then exit the enzyme in a new form, leaving the enzyme to perform its function again and again (Figure 5-4). Enzymes help do the following:

Enzymes Catalysts in the body that speed up reactions.

1. Break down substances (such as foods during digestion)
2. Build up substances (such as bone)
3. Change one substance into another (such as glucose into glycogen)

Hormones are chemical messengers secreted into the bloodstream by various organs, such as the liver, to travel to a target organ and influence what it does. Hormones regulate certain body activities so that a constant internal environment (called **homeostasis**) is maintained. For example, the hormone insulin is released from the pancreas when your blood sugar level goes up after you eat lunch. Insulin stimulates the transport of sugar from the blood into your cells, resulting in lower, more normal blood sugar levels. Amino acids are components of insulin, growth hormone, and some other hormones.

Hormones Chemical messengers secreted into the bloodstream by various organs that travel to a target organ and influence what it does.

Homeostasis A constant internal environment in the body.

Antibodies are proteins in the blood whose job is to bind with foreign bodies or invaders that do not belong in the body. The invaders could be viruses, bacteria, or toxins. Each antibody fights a specific invader. For example, many different viruses cause the common cold. An antibody that binds with a certain cold virus is of no use to you if you have a different strain of the virus. However, exposure to a cold virus results in increased amounts of the specific type of antibody that can attack it. Next time that particular cold virus comes around, your body remembers and makes the right antibodies. This time the virus is destroyed faster, and your body's response (called the **immune response**) is sufficient to combat the disease.

Antibodies Proteins in the blood that bind with foreign bodies or invaders.

Immune response The body's response to a foreign substance, such as a virus, in the body.

Proteins also act as taxicabs in the body, transporting iron and other minerals, some vitamins, fats, and oxygen through the blood. For example, hemoglobin, a protein in red blood cells, carries oxygen from the lungs to the cells of your body.

FIGURE 5-4 How enzymes work (The enzyme in this drawing is red.)

FIGURE 5-6 Endurance and power athletes benefit from additional protein in their diets.

Courtesy of James Peragine/Shutterstock.

Many people feel athletes need extra protein above the RDA. For many athletes, the normal RDA is sufficient. However, endurance athletes and power (strength or speed) athletes benefit from additional protein—from 1.2 to 1.7 grams per kilogram of body weight, depending on the situation (Figure 5-6). Many athletes can meet this increased need because they are eating more kcalories and therefore more protein anyway. Protein and amino acid supplements are rarely needed to meet protein needs of athletes or nonathletes. It is also a myth that a high-protein diet will promote muscle growth. Only strength training and exercise will change muscle.

The 2002 Dietary Reference Intake report established Acceptable Macronutrient Distribution Ranges (AMDR) for protein. Adults should get from 10 to 35 percent of total kcalories from protein. The AMDR for children from 1 to 3 years old is 5 to 20 percent of kcalories, and for children 4 to 18 years old it is 10 to 30 percent. Tolerable Upper Intake Levels for protein and individual amino acids could not be set due to inadequate or conflicting data.

SUMMARY

1. The RDA for protein is 0.8 gram per kilogram of body weight. For nonvegetarians, protein should come from plant and animal sources. Most Americans take in more than enough protein.
2. During periods of growth, the body needs more protein.
3. It is best to distribute your protein intake evenly among three meals to increase your satiety and to help preserve muscles as you get older.
4. Most athletes don't need more than 0.8 grams protein/kilogram. However, endurance and power athletes benefit from more protein.
5. The Acceptable Macronutrient Distribution Ranges for protein is 10 to 35 percent for adults.

HEALTH EFFECTS OF PROTEIN

In the United States, getting enough protein is rarely a problem. Most Americans eat more than the Recommended Dietary Allowance. Getting enough protein in the diet is more of a problem outside the United States. Before we discuss the problems associated with eating too little protein, let's take a look at eating too much.

Eating too much protein has no benefits. It will not result in bigger muscles, stronger bones, or increased immunity. In fact, eating more protein than you need may add kcalories beyond what you require. Extra protein is not stored as protein but is stored as fat if too many kcalories are taken in.

Diets high in protein can be a concern if you are eating a lot of high-fat animal proteins, such as hamburgers and cheese, and few vegetable proteins. Eating too many high-fat animal foods, which contain much saturated fat, raises your blood cholesterol levels. Higher blood cholesterol levels increase your risk of heart disease. Too many high-fat foods also increase the chances of eating too many kcalories and gaining weight.

Comparison of the fat and fiber content of animal and vegetable proteins (Table 5-1) shows that plant sources of protein contain less fat and more fiber (except for nuts, but they contain healthy monounsaturated fats). Plant foods also contain no cholesterol and are rich in vitamins and minerals. As a matter of fact, soy protein (about 25 grams a day) included in a diet low in saturated fat and cholesterol appears to reduce the risk of heart disease by lowering blood cholesterol levels.

Diets high in processed meats and/or red meats have been linked with a higher risk of colon cancer. The American Cancer Society advises you to limit consumption of both processed meats (such as hot dogs, pepperoni, and bacon) and red meats.

In contrast, eating too little protein can cause problems too, such as slowing down the protein rebuilding and repairing process and weakening the immune system. Developing countries have the most problems with protein-energy malnutrition (PEM), also called protein-kcalorie malnutrition. PEM refers to a broad spectrum of malnutrition, from mild to serious cases. PEM can occur in infants, children, adolescents, and adults, although it most often affects young children.

SUMMARY

1. Eating too much protein does not result in bigger muscles, stronger bones, or increased immunity. Extra protein is not stored as protein but is stored as fat if too many kcalories are taken in.
2. Diets high in protein with lots of high-fat animal protein can raise your blood cholesterol levels and increase your chances of eating too many kcalories and gaining weight.
3. Plant sources of protein contain less fat and more fiber (except for nuts).
4. Diets high in processed meats and/or red meats have been linked with a higher risk of colon cancer.
5. Eating too little protein can slow down repairing/maintaining of body tissues and weaken the immune system.

CULINARY FOCUS: MEAT, POULTRY, AND FISH

When you cook proteins, the protein structure changes in a process called **denaturation**. As you cook proteins, the proteins shrink, become firmer, and lose moisture. For example, when you fry an egg, the proteins in the egg white turn from clear to white and become firm. The yolk turns from liquid to a solid. As meat cooks, it loses moisture and shrinks in size. Proteins denature when they are exposed to high temperature, whipping, or addition of acid. Gluten, the protein in flour, denatures during baking to give bread and other baked goods their structure. When fresh vegetables are blanched quickly in boiling water, the heat kills the protein enzymes that make them spoil.

Denaturation The process in which protein structure changes due to high temperature, addition of acid, or whipping, resulting in the protein becoming firm, shrinking in size, and losing moisture.

PRODUCT

Table 5-1 compares the fat and cholesterol content of a variety of meat, poultry, and fish. To choose nutritious cuts of meat, poultry, or fish, use these guidelines:

- Most fish is lower in fat, saturated fat, and cholesterol than are meat and poultry (with the skin on).
- In comparison to red meats, skinless white-meat chicken and turkey are comparable in cholesterol but lower in total fat and saturated fat. The skin of chicken and turkey contains much of the bird's fat. Chicken breast and turkey breast (meaning white meat) without skin are low in fat—there is only about 4 grams of fat in 4 ounces of chicken or turkey. By comparison, white meat with skin and dark meat (such as thighs and drumsticks) are much higher in fat. Also, chicken wings may be considered white meat, but they are fattier than a drumstick.

- Poultry skin can be left on during cooking to keep in moisture, then removed before serving. Another alternative would be to use a dry or web rub—a seasoning mixture such as espresso or coriander crust, directly on the poultry and slow roast the poultry item to achieve a "skin," adding a crisp component to the dish.
- When buying ground turkey or chicken, make sure it is made from only skinless breast meat to get the least amount of fat. If the product includes skin and dark meat, it will be much higher in fat.
- Chicken and turkey are rich in protein, niacin, and vitamin B_6. They are also good sources of vitamin B_{12}, riboflavin, iron, zinc, and magnesium. Duck and goose are quite fatty in comparison, because they contain all dark meat.
- When choosing beef, you will get the least fat and saturated fat from: bottom round steak or roast, flank steak, eye round roast, top sirloin steak, tenderloin filet, top round roast or steak, 90/10 or 95/5 ground beef.
- Red meats are a good source of many important nutrients, including protein, iron, copper, zinc, and some of the B vitamins, such as B_6 and B_{12}. Meat is also a significant source of fat, saturated fat, and cholesterol. Pork is a good source of protein, thiamin, vitamins B_6 and B_{12}, riboflavin, niacin, and zinc.
- Lean veal cuts include loin or rib chops and top round.
- Leaner pork cuts include tenderloin and top loin chops. Stay away from very fatty country style ribs and spareribs.
- Niche pork includes products that are organically grown, heritage or heirloom varieties such as Berkshires and Tamworths, or locally produced. These breeds have superior meat quality, darker color, rich flavor and natural juiciness of the meat. They have slightly higher margins of fat then commercially raised pork, but when portion controlled, can add a distinct flavor profile.

Fish and shellfish are excellent sources of protein and are relatively low in kcalories. Most are also low to moderate in cholesterol. Certain fish (Table 5-4 and Figure 5-7) are fattier than others, such as salmon, mackerel, sardines and herring, but fatty fish are an important source of omega-3 fatty acids. There are additional health benefits to diets rich in omega-3 fatty acids, including protection against heart disease and strokes, and lowering of blood pressure.

TABLE 5-4	Fat Content of Fish	
LOW-FAT FISH (FAT CONTENT LESS THAN 2.5 PERCENT)	**MEDIUM-FAT FISH (FAT CONTENT 2.5–5 PERCENT)**	**HIGH-FAT FISH (FAT CONTENT OVER 5 PERCENT)**
Cod	Bluefish	Albacore tuna
Croaker	Swordfish	Bluefin tuna
Flounder	Yellowfin tuna	Herring
Grouper		Mackerel
Haddock		Sablefish
Pacific halibut		Salmon
Pollock		Sardines
Red snapper		Shad
Rockfish		Trout
Sea bass		Whitefish
Shark		
Sole		
Whiting		

Seafood

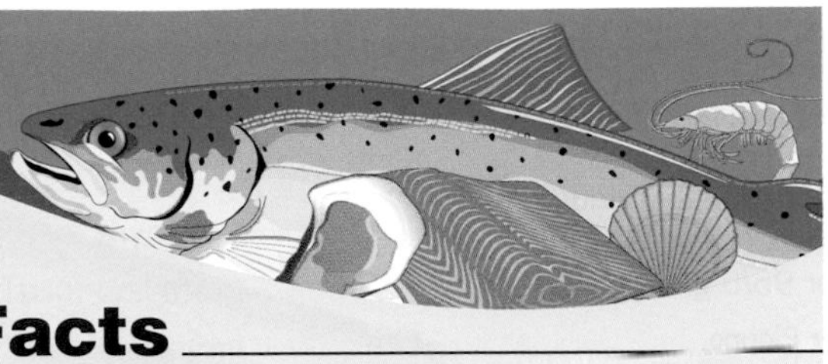

Nutrition Facts

Cooked (by moist or dry heat with no added ingredients), edible weight portion. Percent Daily Values (%DV) are based on a 2,000 calorie diet.

Seafood Serving Size (84 g/3 oz)	Calories	Calories from Fat	Total Fat g	Total Fat %DV	Saturated Fat g	Saturated Fat %DV	Cholesterol mg	Cholesterol %DV	Sodium mg	Sodium %DV	Potassium mg	Potassium %DV	Total Carbohydrate g	Total Carbohydrate %DV	Protein g	Vitamin A %DV	Vitamin C %DV	Calcium %DV	Iron %DV
Blue Crab	100	10	1	2	0	0	95	32	330	14	300	9	0	0	20g	0%	4%	10%	4%
Catfish	130	60	6	9	2	10	50	17	40	2	230	7	0	0	17g	0%	0%	0%	0%
Clams, about 12 small	110	15	1.5	2	0	0	80	27	95	4	470	13	6	2	17g	10%	0%	8%	30%
Cod	90	5	1	2	0	0	50	17	65	3	460	13	0	0	20g	0%	2%	2%	2%
Flounder/Sole	100	15	1.5	2	0	0	55	18	100	4	390	11	0	0	19g	0%	0%	2%	0%
Haddock	100	10	1	2	0	0	70	23	85	4	340	10	0	0	21g	2%	0%	2%	6%
Halibut	120	15	2	3	0	0	40	13	60	3	500	14	0	0	23g	4%	0%	2%	6%
Lobster	80	0	0.5	1	0	0	60	20	320	13	300	9	1	0	17g	2%	0%	6%	2%
Ocean Perch	110	20	2	3	0.5	3	45	15	95	4	290	8	0	0	21g	0%	2%	10%	4%
Orange Roughy	80	5	1	2	0	0	20	7	70	3	340	10	0	0	16g	2%	0%	4%	2%
Oysters, about 12 medium	100	35	4	6	1	5	80	27	300	13	220	6	6	2	10g	0%	6%	6%	45%
Pollock	90	10	1	2	0	0	80	27	110	5	370	11	0	0	20g	2%	0%	0%	2%
Rainbow Trout	140	50	6	9	2	10	55	18	35	1	370	11	0	0	20g	4%	4%	8%	2%
Rockfish	110	15	2	3	0	0	40	13	70	3	440	13	0	0	21g	4%	0%	2%	2%
Salmon, Atlantic/Coho/Sockeye/Chinook	200	90	10	15	2	10	70	23	55	2	430	12	0	0	24g	4%	4%	2%	2%
Salmon, Chum/Pink	130	40	4	6	1	5	70	23	65	3	420	12	0	0	22g	2%	0%	2%	4%
Scallops, about 6 large or 14 small	140	10	1	2	0	0	65	22	310	13	430	12	5	2	27g	2%	0%	4%	14%
Shrimp	100	10	1.5	2	0	0	170	57	240	10	220	6	0	0	21g	4%	4%	6%	10%
Swordfish	120	50	6	9	1.5	8	40	13	100	4	310	9	0	0	16g	2%	2%	0%	6%
Tilapia	110	20	2.5	4	1	5	75	25	30	1	360	10	0	0	22g	0%	2%	0%	2%
Tuna	130	15	1.5	2	0	0	50	17	40	2	480	14	0	0	26g	2%	2%	2%	4%

Seafood provides negligible amounts of *trans* fat, dietary fiber, and sugars.

U.S. Food and Drug Administration
(January 1, 2008)

FIGURE 5-7 Seafood nutrition chart.

Courtesy of the US Food and Drug Administration.

Fishing practices worldwide are depleting fish populations, destroying habitats, and polluting the water. Sustainable seafood comes from species of fish that are managed in a way that provides for today's needs without damaging the ability of the species to be available for future generations. Most fish and shellfish caught in federal waters of the United States are harvested under fishery management plans that must meet standards to ensure that fish stocks

FIGURE 5-9 Slice meat or poultry thin and fan around a base. Roast Pork with Cabbage Rolls.

Photo by Peter Pioppo.

placing the fish on top. Garnish with a fine-shaved lemon fennel salad, micro herbs or shaved radishes. The appearance when presented to your guest shouts flavor, style, appeal, and value.

- Cedar-planked fish is another way to infuse additional flavors with no additional calories. Soak an untreated cedar plank in water with a touch of vinegar or lemon and then add a few sprigs of thyme oregano or tarragon, place the lightly marinated fish on top, and bake in the oven. The cedar plank will impart a unique smoked flavor.

S U M M A R Y

1. Denaturation is the process in which protein structure changes due to high temperature or whipping, resulting in the protein becoming firm, shrinking in size, and losing moisture.
2. Animal proteins that are lower in fat and saturated fat include fish, skinless poultry, and pork tenderloin or top loin chops. For beef, choose bottom round steak (outside), flank steak, eye of round roast, top round steak or roast, and 90/10 or 95/5 ground beef.
3. Consult Monterey Bay Aquarium Seafood Watch to pick sustainable seafood.
4. To have flavorful proteins, use rubs, marinades, and cooking methods that will produce a moist product that adds little or no fat (such as roasting, grilling, etc.). Smoking can be used to complement the taste of meat, poultry, or fish.

CHECK-OUT QUIZ

1. Proteins contain amino acids.
   a. True
   b. False

2. Essential amino acids cannot be made in the body or made in sufficient amounts.
   a. True
   b. False

3. There is about the same amount of protein in one ounce of meat as there is in two tablespoons of peanut butter.
   a. True
   b. False

4. Plant proteins are low in one or more essential amino acids.
   a. True
   b. False

5. When nuts are eaten with seeds, it makes a complete protein.
   a. True
   b. False

6. Collagen, the most abundant protein in the body, transports oxygen to the body's cells.
   a. True
   b. False

7. Hormones are catalysts in the body that speed up reactions.
   a. True
   b. False

8. If the body is making a protein and can't find an essential amino acid for it, the protein can't be completed.
   a. True
   b. False

9. During periods of growth, a person needs less protein.
   a. True
   b. False

10. If you eat more protein than you need, it will be used to build muscle.
    a. True
    b. False

11. Americans tend to eat too little protein.
    a. True
    b. False

12. Most fish is lower in fat, saturated fat, and cholesterol than are meat and poultry (skin on).
    a. True
    b. False

13. During cooking, proteins lose their shape and emulsify.
    a. True
    b. False

14. A leaner cut of beef is eye of round.
    a. True
    b. False

15. Rubs and marinades can add flavor to proteins.
    a. True
    b. False

16. Where are the body's instructions for making protein?

17. Using MyPlate, 1 ounce of meat, poultry, or fish is equal to how many/how much eggs, peanut butter, nuts, seeds, cooked beans or peas, tofu, tempeh, and hummus?

18. Explain the concept of complementary proteins and give examples.

19. List five functions of proteins.

20. Calculate your daily protein requirement.

21. List three situations in which someone's protein needs would increase.

22. Name five animal proteins that are tend to be lower in fat and saturated fat than most animal proteins.

NUTRITION WEB EXPLORER

For a complete list of websites for the following activities, please visit the companion book page at www.wiley.com/college/drummond.

BASS ON HOOK—FISH COOKING TECHNIQUES

Read "Fish Cooking Techniques." How long does it take to cook fresh fish that is 1 inch thick? Frozen fish that is 1 inch thick? What is the difference between blackening and bronzing?

EATING WELL MAGAZINE—EGG BUYER'S GUIDE

Use this website to define the following terms: cage-free and free-range eggs, certified organic eggs, certified humane eggs, and high in omega-3's eggs.

SPORTS, CARDIOVASCULAR AND WELLNESS NUTRITION (A PRACTICE GROUP OF THE ACADEMY OF NUTRITION AND DIETETICS)

At this website, click on "Areas of Expertise" and then "Sports Nutrition." Next click on "Access Fact Sheets. Read the Sports Nutrition Fact Sheet on "Gaining Weight, Building Muscle" and summarize what you learned.

SOY CONNECTION—

Click on "Soyfoods Guide" under the "Soyfoods" tab. Starting on page 11, learn about edamame, tofu, tempeh, miso, and soy cheese, and how to use them to prepare menu items.

MORNINGSTAR FARMS—BURGERS

Soy protein is used in many of Morningstar Farms products. Select five Morningstar Farms burgers and compare the ingredients and nutrition in each—using kcalories, protein, fat, and fiber. Which of the burgers is most appealing to you? Which burger would be most appealing to offer as a vegetarian option on a menu?

VEGETARIAN NUTRITION (A PRACTICE GROUP OF THE ACADEMY OF NUTRITION AND DIETETICS)

Under "Resources," click on "Frequently Asked Questions." Read the response to one of the Frequently Asked Questions of interest to you, and write a summary of what you learned.

IN THE KITCHEN

Complementary Protein

You will prepare one of the following vegetarian recipes.

Creamy Pumpkin Curry Soup

Hummus with Pita

Fiesta Wrap

Mediterranean Quinoa Salad

Butternut Squash Lasagna

Smokin' Powerhouse Chili

You will be given a worksheet to identify the complementary protein combinations in each recipe, determine which recipes would be appropriate for vegans, and evaluate each dish for taste and acceptability.

HOT TOPIC

VEGETARIAN EATING

The number of vegetarians in the United States has been increasing. Instead of eating the meat entrées that have traditionally been the major source of protein in the American diet, vegetarians dine on main dishes emphasizing legumes (dried beans and peas), grains, and vegetables. Vegetarian entrées such as red beans and rice can supply adequate protein with less fat and cholesterol and more fiber than their meat counterparts.

Whereas vegetarians do not eat meat, poultry, or fish, the largest group of vegetarians, referred to as *lacto-ovo vegetarians*, do consume animal products in the form of eggs (ovo-) and milk and milk products (lacto-). Another group of vegetarians, *lacto vegetarians*, consume milk and milk products but forgo eggs. Most vegetarians are either lacto-ovo vegetarians or lacto vegetarians.

Vegans, a third group of vegetarians, do not eat eggs or dairy products and therefore rely exclusively on plant foods to meet protein and other nutrient needs. Vegans are a small group; it is estimated that only a little over 10 percent of vegetarians are vegans.

Among the numbers of nonvegetarians in the United States, growing numbers of adults frequently or occasionally buy meat or dairy alternatives. Others simply try to reduce the amount of meat they eat by choosing occasional meatless meals.

One reason people give for being vegetarian is health benefits. Vegetarians tend to be leaner and to keep their body weight and blood lipid levels closer to desirable levels than nonvegetarians do. Vegetarians tend to have a lower risk of the following diseases: hypertension, heart disease, cancer, and type 2 diabetes. Being vegetarian does not mean that you automatically get these benefits. It is possible to be a lacto-ovo vegetarian and still eat too much fat, saturated fat, and cholesterol. It's probably the exception rather than the rule, but it is still possible. Being vegetarian does not guarantee that your diet will meet current dietary recommendations.

Some other reasons for becoming vegetarian include the following:

1. **Ecology.** For ecological reasons, vegetarians choose plant protein because livestock and poultry require much land, energy, water, and plant food (such as soybeans), which they consider wasteful. According to the North American Vegetarian Society, the grains and soybeans fed to US livestock could feed 1.3 billion people. Livestock also waste loads of water—it takes 2,500 gallons of water to produce a pound of meat but only 25 gallons to produce a pound of wheat.

2. **Economics.** A vegetarian diet is more economical—in other words, less expensive. This can be easily demonstrated by the fact that in a typical foodservice operation, the largest component of food purchases is for meats, poultry, and fish.

3. **Ethics.** Vegetarians do not eat meat for ethical reasons; they believe that animals should not suffer or be killed unnecessarily. They feel that animals suffer real pain in crowded feed lots and cages and that both their transportation to market and their slaughter are traumatic.

4. **Religious beliefs.** Some vegetarians, such as the Seventh-Day Adventists, practice vegetarianism as a part of their religion, which also encourages exercise and discourages drinking alcohol and smoking.

Vegetarian diets can be nutritionally adequate when appropriately planned, varied, and adequate in kcalories. In some cases, supplements or fortified foods are necessary to provide certain nutrients. Vegetarian diets, including vegan diets, can be used for individuals during all state of the life cycle, including pregnancy, lactation, infancy, childhood, and adolescence. Most vegetarians get enough protein, and their diets are typically lower in fat, saturated fat, and cholesterol (as long as they avoid whole-milk dairy, high-fat snacks, and fried foods).

Nutrients that may be of concern, especially for vegans, include vitamin B_{12}, vitamin D, calcium, iron, and zinc (see Table 5-6). Vitamin B_{12} is only in animal foods so vegans need to use supplements and/or vitamin B_{12}-fortified foods. Vitamin D and calcium are mainly in dairy products, but can be obtained through fortified foods such as soymilk, supplements, or sun exposure (vitamin D only). The bioavailability of iron and zinc from vegetarian diets is lower than from nonvegetarian diets. Therefore, iron and zinc may be of concern for some vegetarians.

When plant proteins are eaten with other foods, the food combinations usually result in complete protein. Eating complementary proteins during

TABLE 5-6	Nutrients of Concern for Vegetarians

NUTRIENT | **SOURCES**

Vitamin B₁₂

Vitamin B_{12} is found only in animal foods. Lacto-ovo vegetarians usually get enough of this vitamin unless they limit their intake of dairy products and eggs. Vegans definitely need either a supplement or vitamin B_{12}-fortified foods such as most ready-to-eat cereals, most meat analogs, and some soy beverages.

Dairy products

Eggs

Fortified cereals

Fortified meat analogs

Vitamin D

Milk is fortified with vitamin D, and vitamin D can be made in the skin with sunlight. Generally, only vegans without enough exposure to sunlight need a supplementary source of vitamin D. Some ready-to-eat breakfast cereals and some soy beverages are fortified with vitamin D.

Fortified milk and soy milk

Egg yolks

Fortified cereals

Sunlight

Calcium

Lacto vegetarians and lacto-ovo vegetarians generally don't have a problem here, but vegans sometimes do if they don't eat enough calcium-rich foods. Good choices include calcium-fortified soymilk or orange juice and tofu made with calcium sulfate. Some green leafy vegetables (such as spinach and Swiss chard) are rich in calcium, but they also contain a binder (called oxalic acid) that prevents the calcium from being absorbed. Without calcium-fortified drinks or calcium supplements, it can be difficult to consume enough calcium.

Milk and milk products

Calcium-fortified juice, soy milk, tofu, or other food

Almonds

Sesame seeds

Broccoli

Kale

Some seaweeds

Iron

Because iron from plants is not absorbed as well as iron from animal sources, and because the vegetarian diet contains substances that interfere with iron absorption, vegetarians need 1.8 times the RDA. Iron is widely distributed in plant foods, and its absorption is greatly enhanced by vitamin C–containing fruits and vegetables that are consumed at the same time.

Soy foods such as soybeans and tofu

Legumes

Nuts and seeds

Enriched and whole grain breads and cereals

Dried fruit

Vegetables (such as potatoes, mushrooms, and broccoli)

Zinc.

Zinc is highest in animal foods, but is found in whole grains and legumes if eaten frequently.

Milk and cheese

Legumes

Nuts and seeds

Peanut butter

Whole grains

---

the day generally ensures a balance of dietary amino acids. Some vegetable proteins, such as those found in amaranth, quinoa, and soybeans, are complete proteins.

Meat alternatives (such as veggie burgers) usually use soy- or vegetable-based protein (Figure 5-10). Soy or vegetable proteins are used in many vegetarian products because they offer a meat-like texture and low cost. Soy protein is used most often in processed vegetarian foods, with wheat protein and vegetable protein second and third. At one time, vegetarian burgers were tasteless and tough. Now, with improved

and more varied ingredients and flavorings, as well as new technology, products are much tastier and actually have a more meat-like texture. Also, some meat alternatives are fortified with important nutrients for vegetarians. These foods can add substantially to vegetarians' intakes of vitamins B_{12} and D, calcium, iron, zinc, and omega-3 fatty acids. See Figure 5-11 and Table 5-7 for more on other soy products such as tofu.

Figure 5-12 shows a vegetarian food pyramid for lacto vegetarians and vegans.

Use these guidelines when planning vegetarian meals, and don't forget that variety is key:

1. Use a variety of plant protein sources at each meal: legumes, grain products (preferably whole grains), nuts and seeds, and/or vegetables. Vegetarian entrees commonly use cereal grains such as rice and bulgur (precooked and dried whole wheat) in combination with legumes and/or vegetables. Use small amounts of nuts and seeds in dishes.
2. Use a wide variety of vegetables. Steaming, stir-frying, or microwaving vegetables retains flavor, nutrients, and color.
3. Choose low-fat and fat-free varieties of milk and products and limit

FIGURE 5-10 Meat alternatives or analogs. Top row: vegetarian frankfurters, burger, chicken strips, ground meat. Bottom row: vegetarian breaded chicken, sausage patties, barbecue ribs.

Photo by Peter Pioppo.

FIGURE 5-11 Soyfood products (clockwise from top): soy milk, tofu, edamame, soy flour, and soy nuts.

Courtesy of The Soyfoods Council. Photo by Karry Hasford.

TABLE 5-7	Soyfoods
Tofu	Tofu is made in a process similar to the one used in making cheese. Soybeans are crushed to produce soymilk, which is then coagulated, causing solid curds (the tofu) and liquid whey to form. Tofu is white in color, soft in consistency, and bland in taste. It readily picks up other flavors, making it a great choice for mixed dishes such as lasagna. Tofu is available shaped in cakes of varying textures and packed in water, which must be changed daily to keep it fresh. *Firm tofu* is compressed into blocks and holds its shape during preparation and cooking. Firm tofu can be used for stir-frying, grilling, or marinating. Soft tofu contains much more water and is more delicate. *Soft tofu* is good to use in blenderized recipes to make dips, sauces, salad dressings, spreads, puddings, cream pies, pasta filling, and cream soups. *Silken tofu* is even softer and more delicate and works well in creamy desserts. Tofu should be kept refrigerated and used within one week.
Tempeh	Tempeh is a white cake made from fermented soybeans with a smoky or nutty flavor. It is a high-protein food that can be cooked quickly to make dishes such as barbecued or fried tempeh or cut into pieces to add to soups. Tempeh is cultured like cheese and yogurt and therefore must be used when fresh or it will spoil.
Miso	Miso is similar to soy sauce but is pasty in consistency. It is made by fermenting soybeans with or without rice or other grains. A number of varieties are available, from light-colored and sweet to dark and robust. It is used in soups and gravies, as a marinade for tofu, as a seasoning, and as a spread on sandwiches and fried tofu.
Meat alternatives or meat analogs	Meat alternatives contain soy protein or tofu and other ingredients to simulate various kinds of meat. They are offered in forms such as hamburgers, hot dogs, bacon, ham, and chicken patties and nuggets. They contain little or no fat and no cholesterol but may be high in sodium. Some are fortified with vitamin B_{12} and iron.
Green vegetable soybeans (edamame)	Edamame are soybeans that have been harvested when still green and sweet. They are boiled and then can be served as a vegetable dish, appetizer, or snack. They are high in protein and fiber.
Natto	Natto is fermented, cooked soybeans with a sticky coating and a cheesy texture. In traditional preparations, natto is used as a topping for rice, in miso soups, and with vegetables.
Soymilk	Soymilk is made from soaked soybeans that have been ground finely and strained. It is an excellent source of protein and B vitamins.
Soy cheese	Made from soymilk, soy cheese is creamy and can be substituted for sour cream or cream cheese.
Soy nuts	Soy nuts are roasted and available in a number of flavors.
Soy sauce	Soy sauce is made from fermented soybeans. There are several types of soy sauce. Shoyu is made from soybeans and wheat. The wheat is roasted first and contributes both the soy sauce's brown color and its sharp, distinctive flavor. Tamari is made only from soybeans. Teriyaki sauce includes other ingredients, such as sugar, vinegar, and spices and can be thicker than other types of soy sauce.

the use of eggs. This is important to prevent a high intake of saturated fat, which is found in whole milk, low-fat milk, regular cheeses, and eggs.

4. Offer dishes made with soybean-based products, such as tofu and tempeh. Soybeans are unique in that they contain the only plant protein that is nutritionally equivalent to animal protein.

5. For menu ideas, don't forget to look at the cuisine of other countries (Table 5-8).

6. For vegan diets, avoid honey or gelatin.

7. Since vegans do not use eggs, here are possible substitutes for 1 egg.

- ¼ cup (2 ounces) soft tofu blended with the liquid ingredients of the recipe
- ¼ cup mashed banana
- ¼ cup applesauce
- 1-½ teaspoons Ener-G Egg Replacer mixed with 2 tablespoons water

The following substitutions can be made for dairy products:

- Soy milk, rice milk, potato milk, nut milk, or water (in some recipes) may be used.
- Buttermilk can be replaced with soured soy or rice milk. For each cup of buttermilk, use 1 cup soy milk plus 1 tablespoon of vinegar.
- Soy cheese is good if it does not contain casein (from milk).
- Crumbled tofu can be substituted for cottage cheese or ricotta cheese in lasagna and similar dishes.
- Several brands of nondairy cream cheese are available.

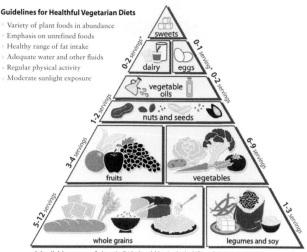

Guidelines for Healthful Vegetarian Diets
- Variety of plant foods in abundance
- Emphasis on unrefined foods
- Healthy range of fat intake
- Adequate water and other fluids
- Regular physical activity
- Moderate sunlight exposure

* A reliable source of vitamin B12 should be included if no dairy or eggs are consumed.

Other Lifestyle Recommendations · Daily Exercise · Water—eight, 8 oz. glasses per day · Sunlight—10 minutes a day to activate vitamin D

Calories/day ▶	1600kcal/day	2000kcal/day	2500kcal/day	1600kcal/day	2000kcal/day	2500kcal/day
Food Groups	vegan servings/day			lacto-ovo servings/day		
Whole Grains	5	7	12	5	6	9
Legumes and Soy	3	3	3	3	3	3
Vegetables	6	8	9	6	8	9
Fruits	3	4	4	3	4	4
Nuts and Seeds	2	2	2	1	1	2
Vegetable Oils	1	2	2	1	2	2
Dairy Products	0	0	0	2	2	2
Eggs	0	0	0	1/2 egg	1/2 egg	1/2 egg
Sweets	Optional					

FIGURE 5-12 Vegetarian food pyramid.

©2008 Loma Linda University, School of Public Health, Department of Nutrition.

TABLE 5-8	**International Vegetarian Dishes**		
Africa	Couscous Spicy vegetable stew Vegetable tajine Peanut soup	Italy	Pasta e fagioli (pasta and bean soup) Minestrone soup Risotto Polenta
Asian (Chinese, Japanese, Thai, and Vietnamese)	Vegetable, rice, and noodle dishes Meatless stir-fries Tofu and vegetables Vegetable moo shu wraps Vegetable tempura Vegetarian spring rolls		Marinara sauce Vegetable lasagna Pasta and vegetable dishes
		Mexican	Vegetarian quesadillas, tacos, burritos, tamales, and enchiladas Gazpacho Refried beans (without lard) Spanish rice
Caribbean	Black beans and rice Spinach and potato croquettes		
France	Ratatouille (vegetable casserole based on roasted eggplant) Onion soup	Middle East	Falafel Tabbouleh Hummus Couscous
Indian	Dal (lentil stew) Vegetable curries Vegetable samosas Chutney Raita		

CHAPTER 6

- State four general characteristics of vitamins, and explain how water-soluble and fat-soluble vitamins are different.

- Identify which vitamin is deficient in the American diet and two vitamins that are toxic when taken in excess.

- Identify functions and food sources of each vitamin presented.

- List benefits of eating fruits and vegetables.

- Discuss the use of fruits and vegetables on the menu, and describe ways to conserve vitamins when handling and cooking fruits and vegetables.

- Define phytochemicals and give examples of foods in which they are found.

BASICS OF VITAMINS

In the early 1900s, scientists thought they had found the compounds needed to prevent scurvy and pellagra, two diseases caused by vitamin deficiencies. These compounds originally were believed to belong to a class of chemical compounds called amines and were named from the Latin *vita*, or life, plus *amine*—vitamine. Later, the e was dropped when it was found that not all the substances were amines. At first, no one knew what they were chemically, and so vitamins were identified by letters. Later, what was thought to be one vitamin turned out to be many, and numbers were added, such as vitamin B$_6$. Later on, some vitamins were found to be unnecessary for human needs and were removed from the list; this accounts for some of the numbering gaps. For example, vitamin B$_8$, adenylic acid, was later found not to be a vitamin.

Vitamins are organic substances that carry out processes in the body that are vital to health. Let's start with some basic facts about vitamins:

Vitamins Noncaloric, organic nutrients found in a wide variety of foods that are essential in small quantities to regulate body processes, maintain the body, and allow for growth and reproduction.

1. Very small amounts of vitamins are needed by the human body, and very small amounts are present in foods. Some vitamins are measured in IUs (international units), a measure of biological activity; others are measured by weight, in micrograms or milligrams. To illustrate how small these amounts are, remember that 1 ounce is 28.3 grams. A milligram is 1/1000 of a gram, and a microgram is 1/1000 of a milligram.

2. Although vitamins are needed in small quantities, the roles they play in the body are enormously important, as you will see in a moment.

3. Most vitamins are obtained through food. Some are also produced by bacteria in the intestine (and are absorbed into the body), and one (vitamin D) can be produced when the skin is exposed to sunlight.

4. There is no perfect food that contains all the vitamins you need in just the right amounts. The best way to ensure an adequate vitamin intake is to eat a varied and balanced diet of plant and animal foods, although it is possible to get the nutrients you need in a vegetarian diet (supplements may be needed).

5. Vitamins do not contain kcalories, and so they do not directly provide energy to the body. Vitamins provide energy indirectly because they are involved in extracting energy from carbohydrate, protein, and fat.

Precursors Forms of vitamins that the body converts to active vitamin forms.

Fat-soluble vitamins A group of vitamins that generally occur in foods containing fats; include vitamins A, D, E, and K.

Water-soluble vitamins A group of vitamins that are soluble in water and are not stored appreciably in the body; include vitamin C, thiamin, riboflavin, niacin, vitamin B$_6$, folate, vitamin B$_{12}$, pantothenic acid, and biotin.

6. Some vitamins in foods are not the actual vitamins but are **precursors**. The body chemically changes the precursor to the active form of the vitamin.

Vitamins are classified according to how soluble they are in fat or water (Table 6-1). **Fat-soluble vitamins** (A, D, E, and K) generally occur in foods containing fat, and they are not readily excreted (except for vitamin K) from the body. **Water-soluble vitamins** (vitamin C and the B-complex vitamins) are readily excreted from the body (except vitamins B$_6$ and B$_{12}$) and therefore don't often reach toxic levels.

Fat-soluble vitamins include vitamins A, D, E, and K. They generally occur in foods containing fats and are stored in the body either in the liver or in adipose (fatty) tissue until they

TABLE 6-1	Fat-Soluble and Water-Soluble Vitamins
FAT-SOLUBLE VITAMINS	**WATER-SOLUBLE VITAMINS**
Vitamin A	Vitamin C
Vitamin D	Thiamin
Vitamin E	Riboflavin
Vitamin K	Niacin
	Vitamin B_6
	Folate
	Vitamin B_{12}
	Pantothenic Acid
	Biotin

are needed. Fat-soluble vitamins are absorbed and transported around the body in the same way as other fats. If anything interferes with normal fat digestion and absorption, these vitamins may not be absorbed. Although it is convenient to be able to store these vitamins so that you can survive periods of poor intake, excessive vitamin intake (higher than the Tolerable Upper Intake Level) causes large amounts of vitamins A and D to be stored and may lead to undesirable symptoms such as weakened bones.

Water-soluble vitamins include vitamin C and the B-complex vitamins. The B vitamins include thiamin, riboflavin, niacin, folate, vitamin B_6, Vitamin B_{12}, biotin, and pantothenic acid. The B vitamins work in every body cell, where they function as part of **coenzymes**. A coenzyme combines with an enzyme which then allows the enzyme to do its job. The body stores only limited amounts of water-soluble vitamins (except vitamins B_6 and B_{12}). Due to their limited storage, these vitamins need to be taken in daily.

Coenzyme A molecule that combines with an enzyme to make the enzyme active.

Excesses of water-soluble vitamins are excreted in the urine. Even though the excesses are excreted, excessive supplementation of certain water-soluble vitamins can cause toxic side effects.

A majority of adults and many children in the United States do not meet the RDA for vitamin D. In addition, some elderly are deficient in vitamin B_{12} and women capable of becoming pregnant are often deficient in folate. In terms of toxicity due to high doses, vitamin D is the most likely vitamin to be toxic when consumed in excess. Additionally, high doses of vitamins A, C, and B_6 can also create concerns. The use of vitamin supplements is discussed in the Hot Topic for Chapter 7.

SUMMARY

1. Although vitamins are needed in small quantities, the roles they play in the body are enormously important.
2. Most vitamins are obtained through food, although there is no perfect food that contains all the vitamins in just the right amounts. Some vitamins in foods are not the actual vitamins but are precursors.
3. Vitamins do not contain kcalories but they provide energy indirectly to the body because they are involved in extracting energy from carbohydrate, protein, and fat.
4. Fat-soluble vitamins (A, D, E, and K) generally occur in foods containing fat, and they are not readily excreted (except for vitamin K) from the body. Fat-soluble vitamins are absorbed and transported in the same way as other fats.
5. Water-soluble vitamins (vitamin C and the B-complex vitamins) are readily excreted from the body (except vitamins B_6 and B_{12}) and therefore don't often reach toxic levels. The B vitamins work in every body cell where they function as part of coenzymes, which combine with an enzyme to make it active.
6. A majority of adults and many children in the United States do not meet the RDA for vitamin D. In addition, some elderly are deficient in vitamin B_{12} and women capable of becoming pregnant are often deficient in folate. Vitamin D is also toxic in excess.

VITAMIN A

During World War I, many children in Denmark developed eye problems. Their eyes became dry and their eyelids swollen, and eventually blindness resulted. A Danish physician read that an American scientist had given milkfat to laboratory animals to cure similar eye problems in animals. At that time Danish children were drinking skim milk, because all the milkfat was being made into butter and sold to England. When the Danish doctor gave whole milk and butter to the children, they got better. The Danish government later restricted the amount of exported dairy foods. Dr. E. V. McCollum, an American scientist, eventually found vitamin A (the first vitamin to be discovered) in milkfat to be the substance that cured the children.

WHAT VITAMIN A DOES

Vitamin A has the following functions:

1. Maintains the health of the eye and vision
2. Promotes healthy skin and surface linings of the lungs, gastrointestinal tract, and urinary tract
3. Supports reproduction, growth, and development

Vitamin A is important to prevent night blindness, an inability to see well at night or in poor light.

Red, yellow, or orange plant foods, such as carrots, contain colored pigments called **carotenoids**. Certain carotenoids, such as **beta carotene**, are converted to vitamin A in the body. Beta carotene also acts as an antioxidant in the body. **Antioxidants** combine with oxygen so that the oxygen is not available to oxidize, or destroy, important substances in the cell. Antioxidants prevent the oxidation of unsaturated fatty acids in the cell membrane, DNA (the genetic code), and other cell parts that substances called **free radicals** destroy. Free radicals are highly reactive compounds that normally result from cell metabolism and the functioning of the immune system.

In the absence of antioxidants, free radicals destroy cells (possibly accelerating the aging process) and alter DNA (possibly increasing the risk for cancerous cells to develop). Free radicals may also contribute to the development of cardiovascular disease. In the process of functioning as an antioxidant, beta-carotene is itself oxidized or destroyed.

WHERE TO GET VITAMIN A

Animal foods, such as eggs, provide compounds that are absorbed as retinol. **Retinol** is often called preformed vitamin A and is one of the active, or usable, forms of vitamin A in the body. You can find retinol in foods including animal products such as:

- Liver (a very rich source)
- Milk and milk products
- Eggs

Whole milk naturally contains vitamin A. However, because vitamin A is fat soluble, the vitamin is lost when fat is removed to make lower-in-fat milks. Therefore, lower-in-fat milks are all fortified with vitamin A. Butter and margarine are also usually fortified with vitamin A. Most ready-to-eat and instant cereals you'll find at a grocery store are also fortified with at least 25 percent of the Daily Value for vitamin A.

Many red, orange, or yellow fruits and vegetables contain carotenoids and are therefore sources of vitamin A (Figure 6-1). Beta-carotene is the orange pigment you see in carrots and sweet potatoes that is split in the body to make retinol. Not all red, orange, or yellow fruits and vegetables are good sources of vitamin A. Sweet corn is yellow but it doesn't contain carotenoids. In some vitamin A-rich fruits and vegetables, beta carotene's color is masked by

Carotenoids A class of pigments that contribute a red, orange, or yellow color to fruits and vegetables; some can be converted to vitamin A in the body.

Beta-carotene A precursor of vitamin A that functions as an antioxidant in the body; the most abundant carotenoid.

Antioxidant A compound that combines with oxygen to prevent oxygen from oxidizing or destroying important substances; antioxidants prevent the oxidation of unsaturated fatty acids in the cell membrane, DNA, and other cell parts that substances called free radicals try to destroy.

Free radicals An unstable compound resulting from cell metabolism and the functioning of the immune system that reacts quickly with other molecules in the body.

Retinol An animal form of vitamin A; one of the active forms of vitamin A in the body.

<figure>
FIGURE 6-1 Many red, orange, or yellow fruits and vegetables contain carotenoids and are therefore sources of vitamin A. Some green vegetables, such as spinach and romaine, also contain vitamin A.

Photo by Peter Pioppo.
</figure>

the dark green chlorophyll found in vegetables such as broccoli and spinach. Some good plant sources of vitamin A are:

- Dark leafy greens such as broccoli and spinach
- Deep orange vegetables such as carrots and sweet potatoes
- Deep orange fruits such as apricots and cantaloupe

When eating vitamin A rich vegetables in a salad, more carotenoids are absorbed when you use a regular salad dressing than if you use a low-fat or no-fat dressing. Table 6-2 lists the vitamin A content of selected foods.

TABLE 6-2	Food Sources of Vitamin A	
	VITAMIN A (MICROGRAMS RAE)	**PERCENT ADULT RDA**
Sweet potato, baked, 1 medium	1096 micrograms RAE	122%
Pumpkin, cooked from fresh or canned, ½ cup	306–953	34–106%
Carrots, frozen, boiled, ½ cup	618	69%
Spinach, frozen, chopped, boiled, ½ cup	573	64%
Carrots, raw, ½ cup	509	57%
Fortified ready-to-eat cereals (various)	177–307	20–34%
Winter squash, cooked, ½ cup	268	30%
Romaine lettuce, 1 cup shredded	205	23%
Cantaloupe, raw, ½ cup	150	17%
Milk, nonfat, fortified, 1 cup	149	17%
Cheese, ricotta, whole milk, ½ cup	149	17%
Milk, 1%, fortified, 1 cup	142	16%
Milk, 2%, fortified, 1 cup	134	15%
Margarine, stick, 1 tablespoon	115	13%
Milk, regular, 1 cup	112	12%
Butter, 1 tablespoon	97	11%
Apricots, canned, ½ cup	80	9%
Ice cream, vanilla, ½ cup	78	9%
Cheddar cheese, 1 ounce	75	8%
Broccoli, frozen spears, cooked, ½ cup	47	5%

Source: US Department of Agriculture, Agricultural Research Service, 2011, USDA Nutrient Database for Standard Reference, Release 24. Nutrient Data Laboratory Home Page: www.nal.usda.gov/fnic/foodcomp.

AM I GETTING ENOUGH?

Most Americans are unlikely to be deficient in vitamin A.

CAN VITAMIN A BE HARMFUL?

Prolonged use of high doses of preformed vitamin A (the Tolerable Upper Intake Level is 3,000 micrograms/day) damages cells and weakens the bones. High doses are particularly dangerous for pregnant women, as they can cause birth defects. Supplemental doses of vitamin A also do not get rid of acne in adolescents. Overconsumption of beta-carotene supplements can be quite harmful as well, causing joint pain and other problems.

SUMMARY

VITAMIN A		
FUNCTIONS	**SOURCES**	**RDA**
1. Maintains the health of the eye and normal vision.	Milk/milk products	Men: 900 micrograms RAE
2. Promotes healthy skin and surface linings of the lungs, etc.	Eggs	Women: 700 micrograms RAE
3. Supports reproduction, growth, and development.	Fortified butter, margarine, cereals	
4. Anti-oxidant (beta carotene)	Deep orange fruits and vegetables	
	Dark leafy greens	

VITAMIN D

WHAT VITAMIN D DOES

The major function of vitamin D is to maintain normal blood levels of calcium and phosphorus mainly by promoting calcium and phosphorus absorption. By doing so, vitamin D makes sure that calcium and phosphorus are available to make and maintain strong bones. Vitamin D works with a number of other vitamins, minerals, and hormones to promote bone mineralization. Without vitamin D, bones can become thin, brittle, or misshapen.

WHERE TO GET VITAMIN D

Vitamin D differs from most other vitamins in that it can be made in the body. When ultraviolet rays shine on your skin, a cholesterol-like compound is converted into a precursor of vitamin D and absorbed into the blood (Figure 6-2). Over a few days, the precursor is converted into the active form of vitamin D. The body gets most of its vitamin D from the skin, not from food.

Of course, if you are not in the sun much or if the ultraviolet rays are cut off by heavy clothing, clouds, smog, fog, sunscreen (SPF of 8 or higher), or window glass, less vitamin D will be produced. On the positive side, a light-skinned person needs only about 15 minutes of sun on the face, hands, and arms two to three times per week to make enough vitamin D. A dark-skinned person needs more time in the sun because melanin (dark brown to black pigments in the skin) acts like a sunscreen. Several months' supply of vitamin D can be stored in the body; this is helpful during winter months when the sun is not as strong in northern climates and you need to wear more clothing. That's why you need to get an adequate amount

FIGURE 6-2 When ultraviolet rays shine on your skin, you make a precursor of vitamin D.

Courtesy of Maridav/Shutterstock.

TABLE 6-3

Food Sources of Vitamin D

	IUs PER SERVING	PERCENT ADULT RDA (19–70 YEARS OLD)
Cod liver oil, 1 tablespoon	1,360	226%
Salmon (sockeye), cooked, 3 ounces	447	75%
Mackerel, cooked, 3 ounces	388	97%
Tuna fish, canned in water, drained, 3 ounces	154	26%
Orange juice fortified with vitamin D, 1 cup (check product labels, as amount of vitamin D varies)	137	23%
Milk, all types, vitamin D-fortified, 1 cup	115–124	19–21%
Yogurt, fortified with vitamin D, 6 ounces (more heavily fortified yogurts provide more vitamin D)	88	15%
Margarine, fortified, 1 tablespoon	60	10%
Liver, beef, cooked, 3.5 ounces	49	8%
Sardines, canned in oil, drained, 2 sardines	46	8%
Egg, 1 large (vitamin D is found in yolk)	41	7%
Ready-to-eat cereal, fortified 0.75–1 cup	40	7%
Cheese, Swiss, 1 ounce	6	1%

Source: US Department of Agriculture, Agricultural Research Service, 2011, USDA Nutrient Database for Standard Reference, Release 24. Nutrient Data Laboratory Home Page: www.nal.usda.gov/fnic/foodcomp

of exposure to sunlight in the spring, summer, and fall to get you through the winter. As you get older, your body makes less active vitamin D.

Very few foods in nature contain vitamin D except for fatty fish such as salmon and mackerel. Other significant food sources of vitamin D are foods fortified with the vitamin (Table 6-3 and Figure 6-3).

- Vitamin D fortified milk (all U.S. milk is fortified with at least 100 IU/cup)
- Vitamin D fortified orange juice
- Vitamin D fortified cereals
- Vitamin D fortified butter and margarine

Most cereals are fortified with vitamin D; however, that is not always the case with orange juice, butter, and margarine. If you drink 2 cups of milk each day, you will get close to half the RDA of vitamin D (the rest comes from sun exposure and other foods). Other dairy products made from milk, such as cheese and ice cream, are generally not fortified, but more yogurt companies are now adding vitamin D to yogurt. Egg yolks also contain some vitamin D.

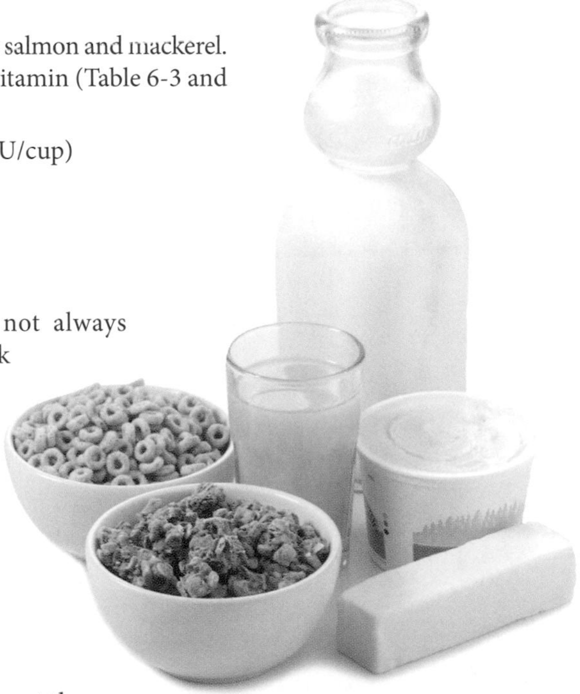

AM I GETTING ENOUGH?

A majority of adults and many children in the United States do not meet the RDA for vitamin D. All children, adults, and the elderly are encouraged to meet the RDA for vitamin D by consuming vitamin D-rich foods in both naturally occurring and fortified forms. If necessary, individuals should consider vitamin D supplements if they have difficulty meeting the RDA through foods, get little sun exposure, or are elderly. As you get older, the amount of vitamin D your body makes from sun exposure decreases by more than 50 percent.

FIGURE 6-3 All milk in the United States is fortified with vitamin D. Vitamin D is also often added to breakfast cereals, butter, and margarine. Some orange juice is also vitamin D-fortified.

Photo by Peter Pioppo.

Rickets A childhood disease in which bones do not grow normally, resulting in bowed legs and knock knees; it is generally caused by a vitamin D deficiency.

Osteomalacia Softening of the bones in adults due to a lack of vitamin D.

Osteoporosis The most common bone disease, characterized by loss of bone density and strength, it is associated with debilitating fractures (such as hip and wrist), especially in people 50 years and older.

Vitamin D deficiency in children causes **rickets**, a disease in which bones do not grow normally, resulting in soft bones and bowed legs. The fortification of milk with vitamin D beginning in the 1930s has made rickets a rare disease in the United States. Adolescents who are deficient in vitamin D may not build the strongest bones they can, which can cause problems later in life.

In adults, vitamin D deficiency causes **osteomalacia**, a disease in which bones become soft and painful. Older adults are at increased risk of developing vitamin D insufficiency: as they age, skin cannot synthesize vitamin D as efficiently, the body doesn't activate vitamin D as well, and they are likely to spend more time indoors and have inadequate intakes of the vitamin from food.

Vitamin D deficiency can contribute to and worsen the bone disease called **osteoporosis**, typically seen in older adults—especially women. Osteoporosis is characterized by fragile bones with low bone mass, which are more likely to break. Having normal storage levels of vitamin D in your body is one of several steps needed to keep your bones strong. Osteoporosis is discussed in more detail in the next chapter.

CAN VITAMIN D BE HARMFUL?

Vitamin D, when taken in excess of the RDA, is the most likely vitamin to be toxic when consumed in excess. Excess vitamin D can lead to calcium deposits in the heart, blood vessels, and kidneys that can cause severe health problems and even death. Young children and infants are especially susceptible to the toxic effects of too much vitamin D.

SUMMARY

VITAMIN D		
FUNCTIONS	**SOURCES**	**RDA**
1. Maintain normal blood levels of calcium and phosphorus mainly by increasing their absorption from the gastrointestinal tract—therefore, calcium and phosphorus can be used to build/rebuild strong bones	Fatty fish Vitamin D fortified milk Vitamin D fortified orange juice Vitamin D fortified cereals Vitamin D fortified butter and margarine Vitamin D fortified yogurt Also made in the skin when exposed to sunlight	19–70 years old: 600 I.U. 71+ years old: 800 I.U.

VITAMIN E

WHAT VITAMIN E DOES

Alpha-tocopherol The active vitamin E compound; powerful anti-oxidant.

The active vitamin E compound in the body is named **alpha-tocopherol**. Vitamin E has an important function in the body as an antioxidant in the cell membrane and other parts of the cell. Antioxidants such as vitamin E act to protect your cells against the effects of free radicals. Free radicals can damage cells and may contribute to the development of cardiovascular disease and cancer. Studies are underway to determine whether vitamin E, through its ability to limit production of free radicals, might help prevent or delay the development of those chronic diseases.

TABLE 6-4   Food Sources of Vitamin E

	MILLIGRAMS/SERVING	PERCENT ADULT RDA
Wheat germ oil, 1 tablespoon	20.3 mg	135%
Sunflower seeds, dry roasted, 1 ounce	7.4	49%
Almonds, dry roasted, 1 ounce	6.8	45%
Rice Krispies, 1 cup	5.7	38%
Sunflower oil, 1 tablespoon	5.6	37%
Safflower oil, 1 tablespoon	4.6	31%
Hazelnuts, dry roasted, 1 ounce	4.3	29%
Mixed nuts, dry roasted, 1 ounce	3.1	21%
Peanut butter, 2 tablespoons	2.9	19%
Canola oil, 1 tablespoon	2.4	16%
Peanuts, dry roasted, 1 ounce	2.2	15%
Oat squares, fortified with vitamin E, 1 cup	2.1	14%
Corn oil, 1 tablespoon	1.9	13%
Spinach, boiled, ½ cup	1.9	13%
Broccoli, chopped, boiled, ½ cup	1.2	8%
Soybean oil, 1 tablespoon	1.1	7%
Kiwifruit, 1 medium	1.1	7%
Mango, sliced, ½ cup	0.7	5%
Tomato, raw, 1 medium	0.7	5%
Spinach, raw, 1 cup	0.6	4%

Source: US Department of Agriculture, Agricultural Research Service, 2011, USDA Nutrient Database for Standard Reference, Release 24. Nutrient Data Laboratory Home Page: www.nal.usda.gov/fnic/foodcomp

WHERE TO GET VITAMIN E

Vitamin E is widely distributed in plant foods (Table 6-4 and Figure 6-4). Rich sources include vegetable oils, margarine and shortening made from vegetable oils, and salad dressings made from vegetable oils. You can also find vitamin E in the following foods:

- Nuts
- Seeds
- Leafy green vegetables
- Whole grains
- Fortified ready-to-eat cereals

While many breakfast cereals are fortified with vitamins A and D, not as many are fortified with vitamin E. Just check the Nutrition Facts panel to see if your breakfast cereal is a good source of vitamin E.

In oils, vitamin E acts like an antioxidant, thereby preventing the oil from going rancid or bad. Unfortunately, vitamin E is easily destroyed by heat and oxygen, so lightly processed foods are better sources.

AM I GETTING ENOUGH?

Vitamin E deficiency due to poor dietary intake is rare.

CAN VITAMIN E BE HARMFUL?

Toxicity of vitamin E is rare. However, very high doses of alpha-tocopherol supplements may cause hemorrhages.

FIGURE 6-4 Good sources of vitamin E include vegetable oils (and salad dressings and margarines made with oil), nuts, peanut butter, seeds, and leafy green vegetables such as broccoli and spinach. Vitamin E is also added to many breakfast cereals.

Photo by Peter Pioppo.

VITAMIN E		
FUNCTIONS	SOURCES	RDA
1. Antioxidant	Widely distributed in plant foods	15 mg alpha-tocopherol
	Vegetable oils and margarine made from oil	
	Salad dressings made with oil	
	Nuts and seeds	
	Leafy green vegetables	
	Whole grains	
	Fortified cereals	

VITAMIN C

Scurvy A vitamin C deficiency disease marked by bleeding gums, weakness, loose teeth, and broken capillaries under the skin.

Scurvy, the name for vitamin C deficiency disease, has been known since biblical times. It was most common on ships, where sailors developed bleeding gums, weakness, loose teeth, and broken capillaries (small blood vessels) under the skin and eventually died. Because sailors' diets included fresh fruits and vegetables only for the first part of a voyage, longer voyages resulted in more cases of scurvy. Once it was discovered that citrus fruits prevent scurvy, British sailors were given daily portions of lemon juice. In those days, lemons were called limes; hence, British sailors got the nickname "limeys."

WHAT VITAMIN C DOES

Vitamin C (its chemical name is ascorbic acid, meaning "no-scurvy acid") has many functions in the body:

Collagen The most abundant protein in the body; a fibrous protein that is a component of skin, bone, teeth, ligaments, tendons, and other connective structures.

1. Vitamin C is important in forming **collagen**, a protein substance that provides strength and support to bones, teeth, skin, cartilage, and blood vessels, as well as healing wounds. You can think of vitamin C as acting like cement, holding together cells and tissues.
2. Vitamin C is also important in helping to make some hormones, such as thyroxin, which regulates the metabolic rate.
3. Vitamin C strengthens resistance to infection.
4. Like vitamin E, vitamin C is an important antioxidant in the body. For example, it prevents the oxidation of iron in the intestine so that the iron can be absorbed.

The use of vitamin C supplements might shorten the duration of the common cold and improve symptom severity in the general population, possibly due to the antihistamine effect of high-dose vitamin C. However, taking vitamin C after the onset of cold symptoms does not appear to be beneficial.

WHERE TO GET VITAMIN C

Foods rich in vitamin C include fruits and vegetables (Figure 6-5):

- Citrus fruits (oranges, grapefruits, limes, and lemons)
- Bell peppers
- Kiwi fruit
- Strawberries

FIGURE 6-5 Fruits and vegetables rich in vitamin C include citrus fruits, bell peppers, strawberries, broccoli, potatoes, and tomatoes.

Photo by Peter Pioppo.

- Broccoli
- Potatoes
- Tomatoes

Only foods from the fruit and vegetable groups contribute vitamin C (Table 6-5). There is little or no vitamin C in meats (except in liver), grains, legumes, and dairy. Some juices are fortified with vitamin C (if not already rich in vitamin C), as are most ready-to-eat cereals. Many people meet their needs for vitamin C simply by drinking orange juice.

The vitamin C content of food may be reduced by prolonged storage and by cooking because it is water-soluble and is destroyed by heat. Steaming or microwaving may lessen

TABLE 6-5	Food Sources of Vitamin C	
FOOD	**MILLIGRAMS/SERVING**	**PERCENT ADULT RDA**
Red pepper, raw, ½ cup	95 mg	105%
Orange juice, ¾ cup	93	103%
Orange, 1 medium	70	78%
Grapefruit juice, ¾ cup	70	78%
Kiwifruit, 1 medium	64	71%
Fortified ready-to-eat cereals (various), about 1 ounce	60–61	67–68%
Green pepper, raw, ½ cup	60	67%
Broccoli, cooked, ½ cup	51	57%
Strawberries, fresh, ½ cup	49	54%
Brussels sprouts, cooked, ½ cup	48	53%
Grapefruit, ½ medium	39	43%
Broccoli, raw, ½ cup	39	43%
Tomato juice, ¾ cup	33	37%
Cantaloupe, ½ cup	29	32%
Cabbage, cooked, ½ cup	28	31%
Cauliflower, raw, ½ cup	26	29%
Potato, baked, 1 medium	20	22%
Tomato, raw, 1 medium	17	19%
Spinach, cooked, ½ cup	9	10%
Green peas, cooked, ½ cup	8	9%

Source: US Department of Agriculture, Agricultural Research Service, 2011, USDA Nutrient Database for Standard Reference, Release 24. Nutrient Data Laboratory Home Page: www.nal.usda.gov/fnic/foodcomp

cooking losses. Fortunately, many of the best food sources of vitamin C, such as fruits and vegetables, are usually consumed raw.

AM I GETTING ENOUGH?

Low intakes of vitamin C tend to reflect low intakes of fruits and vegetables. Deficiencies resulting in scurvy are rare. Scurvy may occur in people with poor diets, especially if coupled with alcoholism, drug abuse, or diabetes.

Certain situations require additional vitamin C. They include pregnancy and breastfeeding, growth, fevers and infections, and cigarette smoking. Smoking produces oxidants, which deplete vitamin C. The RDA for smokers is 35 milligrams of vitamin C daily in addition to the normal RDA (75 mg for women, 90 mg for men).

CAN VITAMIN C BE HARMFUL?

The Tolerable Upper Intake Level for vitamin C is 2,000 mg, or 2 grams per day. Over 2 grams per day can cause gastrointestinal symptoms such as stomach cramps and diarrhea. High doses can interfere with certain clinical laboratory tests.

SUMMARY

VITAMIN C		
FUNCTIONS	**SOURCES**	**RDA**
1. Forms collagen—cement which holds together cells and tissues	Citrus fruits	Men: 90 mg
2. Makes some hormones such as thyroxin	Bell peppers	Women: 75 mg
3. Strengthens resistance to infection	Kiwi fruit	More vitamin C needed for smokers
4. Antioxidant	Strawberries	
	Broccoli	
	Potatoes	
	Tomatoes	

THIAMIN, RIBOFLAVIN, AND NIACIN

WHAT THIAMIN, RIBOFLAVIN, AND NIACIN DO

Thiamin, riboflavin, and niacin all play key roles as part of coenzymes in energy metabolism, meaning that they are essential in getting energy from carbohydrates, fats, and proteins. They are also needed for normal growth.

WHERE TO GET THIAMIN, RIBOFLAVIN, AND NIACIN

Thiamin is widely distributed in foods, but mostly in moderate amounts. Pork is an excellent source of thiamin. Other sources include dry beans, whole-grain and enriched/fortified breads and cereals, watermelon, and acorn squash.

Milk and milk products are the major source of riboflavin in the American diet. Other sources include organ meats such as liver (very high in riboflavin), whole-grain and enriched/fortified breads and cereals, eggs, and some meats.

The main sources of niacin are meat, poultry, and fish. Organ meats are quite high in niacin. Whole-grain and enriched/fortified breads and cereals, as well as peanut butter, are also important sources of niacin. All foods containing complete protein, such as those just mentioned and also milk and eggs, are good sources of the precursor of niacin, tryptophan. Tryptophan, an amino acid present in some of these foods, can be converted to niacin in the body. This is why the RDA for niacin is stated in **niacin equivalents**. One niacin equivalent is equal to 1 milligram of niacin or 60 milligrams of tryptophan. Less than half the niacin we use is made from tryptophan.

AM I GETTING ENOUGH?

Deficiencies in thiamin, riboflavin, and niacin are rare in the United States, in large part because breads, cereals, and flour are enriched with all three nutrients (Figure 6-6). General symptoms for B-vitamin deficiencies include fatigue, decreased appetite, and depression. Alcoholism can create deficiencies in these three vitamins, due in part to limited food intake.

CAN THIAMIN, RIBOFLAVIN, OR NIACIN BE HARMFUL?

Toxicity is not a problem except in the case of niacin. Nicotinic acid, a form of niacin, is often prescribed by physicians to lower elevated blood cholesterol levels. Unfortunately, it has some undesirable side effects. Starting at doses of 100 milligrams, typical symptoms include flushing, tingling, itching, rashes, hives, nausea, diarrhea, and abdominal discomfort.

FIGURE 6-6 Breads, cereals, and flours are enriched with thiamin, riboflavin, and niacin.

Courtesy of the US Department of Agriculture.

Niacin equivalents The unit for measuring niacin. One niacin equivalent is equal to 1 milligram of niacin or 60 milligrams of tryptophan.

SUMMARY

THIAMIN, RIBOFLAVIN, AND NIACIN		
FUNCTIONS	**SOURCES**	**RDA**
1. Part of coenzymes needed to get energy from carbohydrates, fats, and proteins	Thiamin	Thiamin
	Pork	Men: 1.2 mg
	Beans	Women: 1.1 mg
2. Needed for normal growth	Whole-grain and enriched breads and cereals	
	Watermelon	
	Acorn squash	
	Riboflavin	Riboflavin
	Milk and milk products	Men: 1.3 mg
	Whole-grain and enriched breads and cereals	Women 1.1 mg
	Eggs	
	Niacin	Niacin
	Meat, poultry, fish	Men: 16 mg niacin equivalent
	Whole-grain and enriched breads and cereals	Women: 14 mg niacin equivalent
	Peanut butter	
	Milk and eggs	

VITAMIN B₆

WHAT VITAMIN B₆ DOES

Vitamin B₆ has three primary roles:

1. Vitamin B₆ is part of a coenzyme involved in carbohydrate, fat, and protein metabolism, and is particularly crucial to protein metabolism.

2. Your body needs vitamin B₆ to make hemoglobin. Hemoglobin is found in red blood cells and carries oxygen to the tissues.

3. Vitamin B₆ is important to the immune system.

FIGURE 6-7 Most protein-rich foods, such as this Chicken alla Cacciatora, provide ample vitamin B₆, and some fruits and vegetables are good sources, too.

Photo by Frank Pronesti.

WHERE TO GET VITAMIN B₆

Good sources for vitamin B₆ include meat, poultry, fish, and fortified, ready-to-eat cereals (Figure 6-7). Vitamin B₆ also appears in plant foods; however, it does not seem to be as well absorbed from these sources. Good plant sources include potatoes, some fruits (such as bananas and watermelon), and some leafy green vegetables (such as broccoli and spinach). Table 6-6 lists the amount of vitamin B₆ in many foods.

AM I GETTING ENOUGH?

B₆ deficiency is uncommon. Inadequate vitamin B₆ status is usually associated with low concentrations of other B-complex vitamins, such as vitamin B₁₂ and folic acid.

CAN VITAMIN B₆ BE HARMFUL?

Excessive use of vitamin B₆ above the Upper Level can cause irreversible nerve damage. The problem with B₆ is that, unlike other water-soluble vitamins, it is not largely excreted, but is stored in the muscles. Too much vitamin B₆ can cause pain and numbness in your legs and arms.

SUMMARY

VITAMIN B₆		
FUNCTIONS	**SOURCES**	**RDA**
1. Part of a coenzyme involved in carbohydrate, fat, and protein metabolism	Meat, poultry, fish	19–50 years old 1.3 mg
2. Needed to make hemoglobin	Fortified cereals	Men 51+ 1.7 mg
3. Important to immune system	Potatoes	Women 51+ 1.5 mg
	Some fruits (bananas and watermelon)	
	Broccoli and spinach	

TABLE 6-6   Food Sources of Vitamin B$_6$

FOOD	MILLIGRAMS PER SERVING	PERCENT ADULT RDA (19–50 YEARS OLD)
Chickpeas, canned, 1 cup	1.1 mg	85%
Beef liver, pan fried, 3.5 ounces	1.0	77%
Yellowfin tuna, cooked, 3 ounces	0.9	69%
Salmon, cooked, 3.5 ounces	0.7	54%
Chicken breast, roasted, 3.5 ounces	0.6	46%
Breakfast cereals, fortified with 25% of the DV for vitamin B$_6$	0.5	38%
Potatoes, boiled, 1 cup	0.5	38%
Turkey, meat only, roasted, 3.5 ounces	0.5	38%
Banana, 1 medium	0.4	31%
Ground beef, 15% fat, pan-browned, 3 ounces	0.4	31%
Marinara (spaghetti) sauce, ready to serve, 1 cup	0.4	31%
Waffles, plain, ready to heat, toasted, 1 waffle	0.3	23%
Bulgur, cooked, 1 cup	0.2	15%
Cottage cheese, 1% milkfat, 1 cup	0.2	15%
Onions, chopped, 1 cup	0.2	15%
Squash, winter, baked, ½ cup	0.2	15%
White rice, long-grain, enriched, parboiled, 1 cup	0.2	15%
Nuts, mixed, dry-roasted, 1 ounce	0.1	8%
Raisins, seedless, ½ cup	0.1	8%
Spinach, boiled, ½ cup	0.1	8%
Broccoli, frozen, chopped, boiled, ½ cup	0.1	8%
Tofu, raw, firm, prepared with calcium sulfate, ½ cup	0.1	8%
Watermelon, raw, 1 wedge	0.1	8%

Source: US Department of Agriculture, Agricultural Research Service, 2011, USDA Nutrient Database for Standard Reference, Release 24.
Nutrient Data Laboratory Home Page: www.nal.usda.gov/fnic/foodcomp

FOLATE

WHAT FOLATE DOES

Folate is a component of coenzymes required to form DNA, the genetic material contained in every body cell. Folate is therefore needed to make all new cells. Much folate is used to produce adequate numbers of red blood cells and digestive tract cells, since these cells are replaced frequently.

WHERE TO GET FOLATE

Folate gets its name from the Latin word for leaf, *folium*. Excellent sources of folate include the following (Figure 6-8 and Table 6-7).

- Green leafy vegetables
- Legumes
- Fortified breads and cereals
- Orange juice

FIGURE 6-8 Folate is found in green leafy vegetables, legumes, and orange juice. Breads and cereals are fortified with folate.

Photo by Peter Pioppo.

TABLE 6-7	Food Sources of Folate	
FOOD	**MICROGRAMS (μG)/SERVING**	**% ADULT RDA**
*Breakfast cereals fortified with 100% of the DV, ¾ cup	400 μg	100%
Asparagus, boiled, ½ cup	121	30%
Spinach, frozen, boiled, ½ cup	115	29%
*Egg noodles, cooked, enriched, ½ cup	110	28%
Cowpeas (blackeyes), immature, cooked, boiled, ½ cup	105	26%
*Breakfast cereals, fortified with 25% of the DV, ¾ cup	100	25%
Great northern beans, boiled, ½ cup	90	23%
*Rice, white, long-grain, parboiled, enriched, cooked, ½ cup	64	16%
Avocado, raw, all varieties, sliced, ½ cup sliced	59	15%
Spinach, raw, 1 cup	58	15%
Broccoli, chopped, frozen, cooked, ½ cup	51	13%
*Flour tortilla, 6" diameter	50	13%
Green peas, frozen, boiled, ½ cup	47	12%
Peanuts, all types, dry roasted, 1 ounce	41	10%
Broccoli, raw, 2 spears (each 5 inches long)	40	10%
Orange, all commercial varieties, fresh, 1	39	10%
Tomato juice, canned, 6 ounces	37	9%
Orange juice, chilled, includes concentrate, ¾ cup	35	9%
Turnip greens, frozen, cooked, boiled, ½ cup	32	8%
Lettuce, Romaine, shredded, ½ cup	32	8%
Vegetarian baked beans, canned, 1 cup	30	8%
Papaya, raw, ½ cup cubes	26	7%
*Bread, white, 1 slice	25	6%
*Bread, whole wheat, 1 slice	25	6%
Egg, whole, raw, fresh, 1 large	24	6%
Banana, raw, 1 medium	24	6%
Cantaloupe, raw, ½ cup cubes	17	4%

*Items marked with an asterisk (*) are fortified with folic acid as part of the Folate Fortification Program.

Source: US Department of Agriculture, Agricultural Research Service, 2011, USDA Nutrient Database for Standard Reference, Release 24.
Nutrient Data Laboratory Home Page: www.nal.usda.gov/fnic/foodcomp

Meats and dairy products contain little folate. Much folate is lost during food preparation and cooking, and so fresh and lightly cooked foods are likely to contain more folate.

A synthetic form of folate is used in supplements and fortified foods such as breads and cereals. The synthetic form of folate is absorbed at 1.7 times the rate of naturally occurring folate in foods such as leafy green vegetables.

AM I GETTING ENOUGH?

Current intakes in the United States indicate that the majority of the population is taking in adequate amounts of folate. However, pregnant women are at risk for folate deficiency. The need for folate is critical during the earliest weeks of pregnancy, when most women don't know they are pregnant. A folate deficiency may cause **neural tube defects**, which are malformations of the brain, spinal cord, or both of the fetus during pregnancy that can result in death or lifelong disability. Despite folic acid fortification of grains, many women capable of becoming pregnant still do not meet the recommended intake for folate. All women capable of becoming pregnant are advised to consume 400 mcg of synthetic folic acid daily from fortified foods and/or supplements in addition to natural folate in foods. Women who are pregnant are advised to consume 600 mcg of folate from all sources.

Neural tube defects Diseases in which the brain and/or spinal cord form improperly in early pregnancy.

CAN FOLATE BE HARMFUL?

Your intake of supplemental folate should not exceed the RDA in order to avoid triggering symptoms of vitamin B_{12} deficiency, which is discussed next.

SUMMARY

FOLATE		
FUNCTIONS	**SOURCES**	**RDA**
1. Part of coenzymes required to make DNA—needed to make all new cells	Green leafy vegetables Legumes Fortified breads and cereals Orange juice	400 micrograms dietary folate equivalent

VITAMIN B_{12}

WHAT VITAMIN B_{12} DOES

Vitamin B_{12} is called cobalamin because it contains the metal cobalt. It has three very important functions:

1. Vitamin B_{12} functions as part of a coenzyme necessary to make new cells and DNA.

2. It activates the folate coenzyme so that folate can make new cells and DNA.

3. Vitamin B_{12} is needed to maintain the protective cover around nerve fibers and ensure the normal functioning of the nervous system.

TABLE 6-8

Food Sources of Vitamin B₁₂

FOOD	MICROGRAMS (μG)/SERVING	% ADULT RDA
Breakfast cereals, fortified with 100% of the DV for vitamin B$_{12}$, 1 serving	6.0 μg	250%
Trout, rainbow, wild, cooked, 3 ounces	5.4	225%
Salmon, sockeye, cooked, 3 ounces	4.8	200%
Trout, rainbow, farmed, cooked, 3 ounces	3.5	146%
Soymilk, fortified 1 cup	2.7	113%
Cheeseburger, double patty and bun, 1 sandwich	2.1	88%
Haddock, cooked, 3 ounces	1.8	75%
Breakfast cereals, fortified, 1 serving	1.5	63%
Yogurt, plain, 1 cup	1.4	58%
Beef, top sirloin, broiled, 3 ounces	1.4	58%
Milk, low-fat, 1 cup	1.3	54%
Clams, breaded and fried, ¾ cup	1.1	46%
Tuna, white, 3 ounces	1.0	42%
Cheese, Swiss, 1 ounce	0.9	40%
Beef taco, 1 taco	0.8	33%
Ham, cured, roasted, 3 ounces	0.6	25%
Egg, large, 1 whole	0.6	25%
Chicken, roasted, ½ breast	0.3	13%

Source: US Department of Agriculture, Agricultural Research Service, 2011, USDA Nutrient Database for Standard Reference, Release 24.
Nutrient Data Laboratory Home Page: www.nal.usda.gov/fnic/foodcomp

WHERE TO GET VITAMIN B₁₂

Vitamin B$_{12}$ differs from other vitamins in that it is found only in animal foods such as meat, poultry, fish, shellfish, eggs, milk, and milk products. Plant foods do not naturally contain any vitamin B$_{12}$. Many ready-to-eat cereals are fortified with vitamin B$_{12}$. Table 6-8 lists the vitamin B$_{12}$ content of foods. Unlike the other B vitamins, vitamin B$_{12}$ is easily destroyed when foods containing it are microwaved.

AM I GETTING ENOUGH?

A vitamin B$_{12}$ deficiency in the body is usually due not to poor intake but to a problem with absorption. Two steps are required for your body to absorb vitamin B$_{12}$ from food. First, hydrochloric acid in the stomach separates vitamin B$_{12}$ from the protein to which vitamin B$_{12}$ is attached. After this, vitamin B$_{12}$ combines with a different protein made by the stomach called an **intrinsic factor** and is absorbed by the body.

On average, Americans aged 50 years or older consume adequate vitamin B$_{12}$. Nonetheless, a substantial proportion of individuals aged 50 years and older may have a reduced ability to absorb naturally occurring vitamin B$_{12}$. These individuals are encouraged to eat foods fortified with vitamin B$_{12}$, such as fortified cereals, or take dietary supplements. **Crystalline vitamin B$_{12}$**, the type of vitamin B$_{12}$ used in supplements and in fortified foods, is much more easily absorbed because it is not attached to protein, unlike naturally occurring vitamin B$_{12}$.

Vitamin B$_{12}$ deficiency is characterized by fatigue, weakness, constipation, loss of appetite, and weight loss. Neurological changes, such as numbness and tingling in the

Intrinsic factor A protein secreted by stomach cells that is necessary for the absorption of vitamin B$_{12}$.

Crystalline vitamin B$_{12}$ The type of vitamin B$_{12}$ used in supplements and fortified foods; much more easily absorbed than natural vitamin B$_{12}$ in foods.

hands and feet, can also occur. Additional symptoms of vitamin B_{12} deficiency include difficulty maintaining balance, depression, confusion, dementia, poor memory, and soreness of the mouth or tongue. Typically, vitamin B_{12} deficiency is treated with vitamin B_{12} injections, since this method bypasses potential barriers to absorption.

Strict vegetarians (called vegans) who do not eat meats, fish, eggs, milk or milk products, or vitamin B_{12}-fortified foods, consume no vitamin B_{12} and are at a high risk of developing a deficiency. When adults adopt a vegetarian diet, deficiency symptoms can be slow to appear because it usually takes years to deplete normal body stores. However, severe symptoms of B_{12} deficiency, most often featuring poor neurological development, can show up quickly in children and breast-fed infants of women who follow a strict vegetarian diet. Vitamin B_{12}-fortified cereals, meat analogs, and soy beverages are important dietary sources of B_{12} for vegetarians who consume no eggs, milk, or milk products (Figure 6-9). Vegetarian adults who do not eat vitamin B_{12}-fortified foods need to consider taking a supplement. Likewise, vegetarian mothers should consult with a pediatrician regarding appropriate vitamin B_{12} supplementation for their infants and children.

FIGURE 6-9 Because vitamin B_{12} is only found in animal foods, vegans need to eat foods that are sometimes fortified with vitamin B_{12} such as soymilk, cold cereals, tofu, and meat alternatives such as vegetarian chicken patties, sausage, and chicken strips.

Photo by Peter Pioppo.

CAN VITAMIN B_{12} BE HARMFUL?

No Upper Level is set for vitamin B_{12} and harmful effects for excess vitamin B_{12} have not been reported.

S U M M A R Y

VITAMIN B_{12}		
FUNCTIONS	**SOURCES**	**RDA**
1. Part of coenzymes required to make DNA—needed to make all new cells	Only in animal foods and vitamin B_{12} fortified foods such as cereals	2.4 micrograms
2. Activates the folate coenzyme		
3. Maintains the nervous system		

CULINARY FOCUS: FRUITS AND VEGETABLES

PRODUCT

Three reasons support the recommendation for Americans to eat more vegetables and fruits. First, most vegetables and fruits are major contributors of a number of nutrients that are underconsumed in the United States, including folate, magnesium, potassium, dietary fiber,

TABLE 6-9 — Nutrient Contributions of Fruits and Vegetables

HIGH IN VITAMIN A*	HIGH IN VITAMIN C*	HIGH IN FIBER OR GOOD SOURCE OF FIBER*	HIGH IN FOLATE	HIGH IN VITAMIN K
Apricots	Apricots	Apple	Asparagus	Broccoli
Cantaloupe	Broccoli	Banana	Banana	Cabbage
Carrots	Brussels sprouts	Blackberries	Broccoli	Collards
Kale, collards	Cabbage	Blueberries	Green peas	Kale
Leaf lettuce	Cantaloupe	Brussels sprouts	Oranges—orange juice	Lettuce—butterhead and iceberg
Mango	Cauliflower	Carrots	Romaine lettuce	Mustard greens
Mustard greens	Chili peppers	Cherries	Spinach	Prunes
Pumpkin	Collards	Cooked beans and peas (kidney, navy, lima, and pinto beans, lentils, black-eyed peas)	Tomato juice	Turnip Greens
Romaine lettuce	Grapefruit	Dates		
Spinach	Honeydew melon	Figs		
Sweet potato	Kiwi fruit	Grapefruit		
Winter squash (acorn, Hubbard)	Mango	Kiwi fruit		
	Mustard greens	Orange		
	Orange	Pear		
	Orange juice	Prunes		
	Pineapple	Raspberries		
	Plum	Spinach		
	Potato with skin	Strawberries		
	Spinach	Sweet potato		
	Strawberries			
	Bell peppers			
	Tangerine			
	Tomatoes			
	Watermelon			

*Based on FDA's food labeling regulations.

Source: National Cancer Institute.

and vitamins A, C, and K (Table 6-9). Several of these are of public health concern for the general public (e.g., dietary fiber and potassium) or for a specific group (e.g., folic acid for women who are capable of becoming pregnant).

Second, consumption of vegetables and fruits is associated with reduced risk of many chronic diseases. Specifically, moderate evidence indicates that intake of at least 2½ cups of vegetables and fruits per day is associated with a reduced risk of cardiovascular disease, including heart attack and stroke. Some vegetables and fruits may be protective against certain types of cancer.

Third, most vegetables and fruits, when prepared without added fats or sugars, are relatively low in calories. Eating them instead of higher calorie foods can help adults and children achieve and maintain a healthy weight.

To get a healthy variety of fruits and vegetables, think color. Eating fruits and vegetables of different colors gives your body a wide range of valuable nutrients, like folate, potassium, vitamin A, and vitamin C (Figures 6-10 and 6-11). Some examples

Fruits

Nutrition Facts

Raw, edible weight portion.
Percent Daily Values (%DV) are
based on a 2,000 calorie diet.

Fruits Serving Size (gram weight/ounce weight)	Calories	Calories from Fat	Total Fat g / %DV	Sodium mg / %DV	Potassium mg / %DV	Total Carbohydrate g / %DV	Dietary Fiber g / %DV	Sugars g	Protein g	Vitamin A %DV	Vitamin C %DV	Calcium %DV	Iron %DV
Apple 1 large (242 g/8 oz)	130	0	0 / 0	0 / 0	260 / 7	34 / 11	5 / 20	25g	1g	2%	8%	2%	2%
Avocado California, 1/5 medium (30 g/1.1 oz)	50	35	4.5 / 7	0 / 0	140 / 4	3 / 1	1 / 4	0g	1g	0%	4%	0%	2%
Banana 1 medium (126 g/4.5 oz)	110	0	0 / 0	0 / 0	450 / 13	30 / 10	3 / 12	19g	1g	2%	15%	0%	2%
Cantaloupe 1/4 medium (134 g/4.8 oz)	50	0	0 / 0	20 / 1	240 / 7	12 / 4	1 / 4	11g	1g	120%	80%	2%	2%
Grapefruit 1/2 medium (154 g/5.5 oz)	60	0	0 / 0	0 / 0	160 / 5	15 / 5	2 / 8	11g	1g	35%	100%	4%	0%
Grapes 3/4 cup (126 g/4.5 oz)	90	0	0 / 0	15 / 1	240 / 7	23 / 8	1 / 4	20g	0g	0%	2%	2%	0%
Honeydew Melon 1/10 medium melon (134 g/4.8 oz)	50	0	0 / 0	30 / 1	210 / 6	12 / 4	1 / 4	11g	1g	2%	45%	2%	2%
Kiwifruit 2 medium (148 g/5.3 oz)	90	10	1 / 2	0 / 0	450 / 13	20 / 7	4 / 16	13g	1g	2%	240%	4%	2%
Lemon 1 medium (58 g/2.1 oz)	15	0	0 / 0	0 / 0	75 / 2	5 / 2	2 / 8	2g	0g	0%	40%	2%	0%
Lime 1 medium (67 g/2.4 oz)	20	0	0 / 0	0 / 0	75 / 2	7 / 2	2 / 8	0g	0g	0%	35%	0%	0%
Nectarine 1 medium (140 g/5.0 oz)	60	5	0.5 / 1	0 / 0	250 / 7	15 / 5	2 / 8	11g	1g	8%	15%	0%	2%
Orange 1 medium (154 g/5.5 oz)	80	0	0 / 0	0 / 0	250 / 7	19 / 6	3 / 12	14g	1g	2%	130%	6%	0%
Peach 1 medium (147 g/5.3 oz)	60	0	0.5 / 1	0 / 0	230 / 7	15 / 5	2 / 8	13g	1g	6%	15%	0%	2%
Pear 1 medium (166 g/5.9 oz)	100	0	0 / 0	0 / 0	190 / 5	26 / 9	6 / 24	16g	1g	0%	10%	2%	0%
Pineapple 2 slices, 3" diameter, 3/4" thick (112 g/4 oz)	50	0	0 / 0	10 / 0	120 / 3	13 / 4	1 / 4	10g	1g	2%	50%	2%	2%
Plums 2 medium (151 g/5.4 oz)	70	0	0 / 0	0 / 0	230 / 7	19 / 6	2 / 8	16g	1g	8%	10%	0%	2%
Strawberries 8 medium (147g/5.3 oz)	50	0	0 / 0	0 / 0	170 / 5	11 / 4	2 / 8	8g	1g	0%	160%	2%	2%
Sweet Cherries 21 cherries; 1 cup (140 g/5.0 oz)	100	0	0 / 0	0 / 0	350 / 10	26 / 9	1 / 4	16g	1g	2%	15%	2%	2%
Tangerine 1 medium (109 g/3.9 oz)	50	0	0 / 0	0 / 0	160 / 5	13 / 4	2 / 8	9g	1g	6%	45%	4%	0%
Watermelon 1/18 medium melon; 2 cups diced pieces (280 g/10.0 oz)	80	0	0 / 0	0 / 0	270 / 8	21 / 7	1 / 4	20g	1g	30%	25%	2%	4%

Most fruits provide negligible amounts of
saturated fat, *trans* fat, and cholesterol; avocados
provide 0.5 g of saturated fat per ounce.

U.S. Food and Drug Administration
(January 1, 2008)

FIGURE 6-10 Fruits nutrition guide.

Courtesy of the US Department of Agriculture.

Vegetables
Nutrition Facts

Raw, edible weight portion. Percent Daily Values (%DV) are based on a 2,000 calorie diet.

Vegetables Serving Size (gram weight/ounce weight)	Calories	Calories from Fat	Total Fat (g)	Total Fat %DV	Sodium (mg)	Sodium %DV	Potassium (mg)	Potassium %DV	Total Carbohydrate (g)	Total Carbohydrate %DV	Dietary Fiber (g)	Dietary Fiber %DV	Sugars (g)	Protein (g)	Vitamin A %DV	Vitamin C %DV	Calcium %DV	Iron %DV
Asparagus 5 spears (93 g/3.3 oz)	20	0	0	0	0	0	230	7	4	1	2	8	2g	2g	10%	15%	2%	2%
Bell Pepper 1 medium (148 g/5.3 oz)	25	0	0	0	40	2	220	6	6	2	2	8	4g	1g	4%	190%	2%	4%
Broccoli 1 medium stalk (148 g/5.3 oz)	45	0	0.5	1	80	3	460	13	8	3	3	12	2g	4g	6%	220%	6%	6%
Carrot 1 carrot, 7" long, 1 1/4" diameter (78 g/2.8 oz)	30	0	0	0	60	3	250	7	7	2	2	8	5g	1g	110%	10%	2%	2%
Cauliflower 1/6 medium head (99 g/3.5 oz)	25	0	0	0	30	1	270	8	5	2	2	8	2g	2g	0%	100%	2%	2%
Celery 2 medium stalks (110 g/3.9 oz)	15	0	0	0	115	5	260	7	4	1	2	8	2g	0g	10%	15%	4%	2%
Cucumber 1/3 medium (99 g/3.5 oz)	10	0	0	0	0	0	140	4	2	1	1	4	1g	1g	4%	10%	2%	2%
Green (Snap) Beans 3/4 cup cut (83 g/3.0 oz)	20	0	0	0	0	0	200	6	5	2	3	12	2g	1g	4%	10%	4%	2%
Green Cabbage 1/12 medium head (84 g/3.0 oz)	25	0	0	0	20	1	190	5	5	2	2	8	3g	1g	0%	70%	4%	2%
Green Onion 1/4 cup chopped (25 g/0.9 oz)	10	0	0	0	10	0	70	2	2	1	1	4	1g	0g	2%	8%	2%	2%
Iceberg Lettuce 1/6 medium head (89 g/3.2 oz)	10	0	0	0	10	0	125	4	2	1	1	4	2g	1g	6%	6%	2%	2%
Leaf Lettuce 1 1/2 cups shredded (85 g/3.0 oz)	15	0	0	0	35	1	170	5	2	1	1	4	1g	1g	130%	6%	2%	4%
Mushrooms 5 medium (84 g/3.0 oz)	20	0	0	0	15	0	300	9	3	1	1	4	0g	3g	0%	2%	0%	2%
Onion 1 medium (148 g/5.3 oz)	45	0	0	0	5	0	190	5	11	4	3	12	9g	1g	0%	20%	4%	4%
Potato 1 medium (148 g/5.3 oz)	110	0	0	0	0	0	620	18	26	9	2	8	1g	3g	0%	45%	2%	6%
Radishes 7 radishes (85 g/3.0 oz)	10	0	0	0	55	2	190	5	3	1	1	4	2g	0g	0%	30%	2%	2%
Summer Squash 1/2 medium (98 g/3.5 oz)	20	0	0	0	0	0	260	7	4	1	2	8	2g	1g	6%	30%	2%	2%
Sweet Corn kernels from 1 medium ear (90 g/3.2 oz)	90	20	2.5	4	0	0	250	7	18	6	2	8	5g	4g	2%	10%	0%	2%
Sweet Potato 1 medium, 5" long, 2"diameter (130 g/4.6 oz)	100	0	0	0	70	3	440	13	23	8	4	16	7g	2g	120%	30%	4%	4%
Tomato 1 medium (148 g/5.3 oz)	25	0	0	0	20	1	340	10	5	2	1	4	3g	1g	20%	40%	2%	4%

Most vegetables provide negligible amounts of saturated fat, *trans* fat, and cholesterol.

U.S. Food and Drug Administration
(January 1, 2008)

FIGURE 6-11 Vegetables nutrition guide.

Courtesy of the US Department of Agriculture.

include green spinach, orange sweet potatoes, black beans, yellow corn, purple plums, red watermelon, and white onions. Many fruits and vegetables are like superfoods—nutritional powerhouses loaded with nutrients and phytochemicals crucial to a healthy, long life.

Green leafy vegetables are readily available and have a wide range of applications and preparations. However, most people do not eat enough of them. Examples of nutritious greens include collards, escarole, kale, sorrel, turnip greens, beet greens, spinach, dandelion greens, arugula, watercress, and chicory. Once thoroughly cleaned, there are many exciting ways to prepare these greens, adding texture, color, and variety to your dishes. They lend themselves to flavorings such as smoked lean meats, smoked vegetables, or smoked spices with the addition of garlic, stock, vinegar or citrus juice, and toasted chilies, all of which add depth to your preparations.

Fruits and vegetables present different challenges to a chef. Fruits are naturally sweet; they are like the candy of the plant world. Unlike vegetables, fruits soften as they become ripe and change more dramatically in color, taste, and aroma. You have to be careful to prepare them at just the right time to get perfect sweetness and a soft, yet not mushy, texture. It is important when purchasing these items to monitor them in storage. Constant sorting of berries, apples, oranges, and grapes to eliminate any spreading of mold is important. Rotate fruits such as melons, stone fruits, and bananas to ensure use at optimum time.

Vegetables require the same care to ensure their quality, freshness and integrity before preparation. Many vegetables require cooking prior to serving so a good culinary foundation is important to preserve their nutrients, color, texture and flavor. The flavors of vegetables range from mild to strong—from mustard greens to garlic to sweet potatoes to eggplant. You will need to be equipped with the proper cooking method for each of these types. Blanching, stewing, steaming, grilling, sautéing, baking, boiling, and braising are all methods used in vegetable cookery, providing a wholesome and balanced end product.

Once fruits and vegetables have been picked, they don't have access to a water source. Water makes up most of their weight—on average, about 70 percent. The plant cells in picked fruits and vegetables continue to function and take in oxygen for breathing (called respiration in plants). Respiration at this point causes the fruit or vegetable to deteriorate quickly and its quality will go straight downhill. To slow down respiration, you need to keep it cold, keep it moist (this also helps the plant retain its water), and sometimes wrap the produce to prevent it from taking in lots of oxygen.

PREPARATION

Fruits and vegetables each require different cooking times. Sweet and white potatoes can be baked, roasted, or boiled. Broccoli, green beans, and asparagus can be blanched, sautéed, or roasted. With the application of heat, plant cells die, lose water, and soften (except for some of the fiber in and around the cell wall). The amount of fiber varies in vegetables. For example, parsnips have more fiber then asparagus.

Besides controlling texture changes as you cook, you also need to monitor flavor changes. The longer a vegetable is cooked, the more flavor is lost into the cooking liquid and into the air through evaporation. To decrease flavor loss, it is best to cook in just enough water to cover and cook for as short a time as possible. The best way to cook vegetables is to steam them; this method cooks fast and helps retain both flavor and nutrients.

Not only does overcooking produce flavor loss, it often creates flavor change. Some vegetables take on new, undesirable flavors when overcooked. Members of the cabbage family—cabbage, turnips, cauliflower, Brussels sprouts, and broccoli—will develop a strong acrid or bitter taste and unpleasant smell if overcooked, owing to chemical changes. These vegetables will taste best when properly cooked using the appropriate method, with the cover off the pot to allow evaporation of the strong-flavored substances.

Table 6-10 shows methods to reduce loss of nutrients in fruits and vegetables when preparing and cooking.

- Buy fresh, high-quality fruits and vegetables.
- Most frozen fruits and vegetables today are flash-frozen, meaning they were frozen at extremely low temperatures at harvest. Flash freezing stops nutrient loss completely until the fruit or vegetable is thawed out or cooked.
- Examine fresh fruits and vegetables thoroughly for appropriate color, size, and shape.
- Store fresh fruits and vegetables in the refrigerator (except green bananas, avocados, potatoes, and onions) to inhibit enzymes that make fruits and vegetables age and lose nutrients. The enzymes are more active at warm temperatures. Refrigerated goods should be maintained at a temperature of 40°F or lower, freezer goods at 0°F or lower. Thermometers should be kept in the refrigerator and the freezer to monitor temperatures.
- Store canned goods in a cool place.
- Foods should not be stored for too long, as nutrient loss occurs during storage.
- When storing fresh fruits and vegetables, close up the wrapping tightly to decrease exposure to the air, which pulls out water and decreases the nutrient content.
- For best results when cooking vegetables:
  - Avoid peeling vegetables when possible. Potatoes and other vegetables that are cooked without being peeled retain many more nutrients than do peeled and cut vegetables.
  - Prepare small amounts. Avoid long exposure to heat. Fresh or frozen vegetables can be cooked by several different methods. You can steam, bake, or sauté them. Regardless of the cooking method you choose, it is better to prepare small amounts than to cook single large batches. Nutritive value is lost and quality is lowered with long exposure to heat. Microwaving is great to prepare several portions.
  - To retain nutrients and bright colors, cook "just until tender." Steaming is a good way to cook vegetables.
  - Use carefully timed "batch cooking" to avoid having vegetables held too long before serving. A good rule of thumb: The quantity you cook should not exceed the amount you can serve in 15 minutes. This applies both to vegetables served alone and to vegetables used in recipes such as beef and chicken stir-fry.

MENUING AND PRESENTATION

- Fruits work especially well as an accompaniment to proteins, bean cakes, or brunch items in relishes, compotes, mojos, salsas, coulis and chutney (Figure 6-12). For example, serve potato cakes with a pear compote made with fresh or dried pears cooked in their juices and flavored with toasted spices and fresh herbs, or a glazed grapefruit with vanilla bean and honey as a great start to the morning. Dessert options that are appealing and satisfying include baked apple wrapped with phyllo dough and whipped spiced yogurt, or a peach and cottage cheese chimichanga with kiwi salsa.
- Fruits are a natural in salads with vegetables: combine pineapple and raisins with carrots, or apples and grapes with celery. Feature colorful fruits and vegetables such as orange, frisee, fennel, or mango (Figure 6-13). Or in a mixed salad try avocado, onion, tomato, and mango. Cooked fruits and vegetables have many possibilities as well: sweet potatoes and cranberries; butternut squash and apples; green bean, shallots and roasted pears; roasted parsnips and dried apricots or lemon-glazed baby carrots.
- Roasted fruits with shallots are a wonderful base on which to place proteins. For example, place a chicken paillard (a cutlet pounded and grilled or sautéed) on a bed of roasted peaches and mango with jicama or sautéed bok choy. Another combination is monkfish on a bed of roasted pears, endive, and fennel.
- Fruits have always been a natural for dessert: fresh, roasted, or baked into a cobbler with an oatmeal almond crust or stuffed into phyllo as a strudel, or pureed in a soufflé batter.
- Fruits such as pineapple, kiwi, orange, mango, and papaya hold up well in salsas, mojos, and relishes.
- Berries, such as blueberries and strawberries, are wonderful when you want vibrant colors in sauces, purees, toppings, and garnishes.

FIGURE 6-12 Berry Compote. Fruits work especially as an accompaniment to entrées.

Photo by Peter Pioppo.

- Vegetables allow you to serve what appears to be an overgenerous portion with reduced calories and no fat or cholesterol. Additionally, when in season vegetables are reasonably priced, aiding profitability.

- When using vegetables, be mindful of the season for maximum flavor, visual appeal, and customer value. Coffee-crusted roast chicken breast with butternut squash, Brussels sprouts and cauliflower puree is a heartier winter style dish. Summer flounder works well with marinated baby beets, haricots verts, grilled zucchini, summer squash, or eggplant with a tomato relish, lending itself to lighter warmer weather.

- Also think variety. Not only does a selection of vegetables create a better constructed and interesting plate but it will also open up new techniques and methods that can become part of your culinary repertoire. Examples such as fresh artichokes, crones, celery root, salsify, white asparagus, beets, and heirloom squashes have unique flavors, challenging preparations and meaningful results.

- Most of a meal's eye appeal comes from vegetables—use them to your advantage. The length of asparagus and green beans is a great way to bring a single entrée plate or buffet platter together. Be sure to use not only vegetables' colors but also their ability to be cut into many different classic shapes that can accentuate your food (Figure 6-14).

- Pickled vegetables have gained increased popularity from side-dish accompaniments and garnishes to flavoring components in salads, relishes, and salsa.

- Olives add exquisite flavor with even just a small amount. They are included in many of today's eating styles. Here are some great varieties:

  - Picholine: French green, salt brine-cured, delicate flavor; great for salads and relishes.
  - Kalamata: Greek black, harvested fully ripened, deep purple with great flavor; great for vegetable stews, component salads, relishes, and compotes.
  - Nicoise: French black, small, rich nutty flavor; great for roasting with fish, vegetables, or poultry; wonderful in relishes, salads, and ragouts.
  - Manzanilla: Spanish green, available pitted and stuffed; adds zest to a salad or a vegetable dish.
  - Linguiria: Black Italian, vibrant flavor; good with roasted vegetables and meats, sautéed vegetables, relishes with fish, and composition salads.
  - Gaeta: Black Italian, wrinkled appearance, usually marinated with rosemary; makes a great stuffing addition as well as a flavoring for stews, ragouts, and sautés.

FIGURE 6-13 Citrus Fennel Salad. Fruits are a natural in salads with vegetables.

Photo by Peter Pioppo.

FIGURE 6-14 Most of a meal's eye appeal comes from vegetables—use them to your advantage.

Photo by Peter Pioppo.

SUMMARY

1. Eating fruits and vegetables is important to get adequate vitamins and minerals, and most are low in kcalories. Moderate evidence indicates that intake of at least 2½ cups of vegetables and fruits per day is associated with a reduced risk of cardiovascular disease, including heart attack and stroke. Some vegetables and fruits may be protective against certain types of cancer.

2. Fruits and vegetables are mostly carbohydrates and are fairly easy to cook. They cook evenly and become soft and tender, although you have to be careful about getting the right texture. With heat, plant cells die, lose water, and soften (except for some of the fiber in and around the cell wall). The amount of fiber varies in vegetables.

3. Overcooking produces flavor loss and flavor change. The best way to cook vegetables is to steam them; this method cooks fast and helps retain both flavor and nutrients.

4. Ideas are given for menuing and presenting.

CHECK-OUT QUIZ

1. Vitamin E deficiency in adults causes osteomalacia.
   a. True
   b. False

2. Water-soluble vitamins are never toxic when taken in excess of the RDA because they are fully excreted.
   a. True
   b. False

3. Vitamin E is a fat-soluble vitamin.
   a. True
   b. False

4. Vitamins are needed in very small amounts.
   a. True
   b. False

5. Vitamins contain small amounts of kcalories.
   a. True
   b. False

6. Beta-carotene is the orange pigment you see in carrots and sweet potatoes that is split in the body to make vitamin A.
   a. True
   b. False

7. Water-soluble vitamins are more likely than fat-soluble vitamins to be needed in the diet on a daily basis.
   a. True
   b. False

8. Vitamin B_6 requires intrinsic factor to be absorbed.
   a. True
   b. False

9. Thiamin, riboflavin, and niacin all play key roles as coenzymes to extract energy from carbohydrates, fats, and protein.
   a. True
   b. False

10. Good sources of folate include green leafy vegetables, legumes, and orange juice.
    a. True
    b. False

11. Name the vitamins described in the following.
    a. Which vitamin is present only in animal foods?
    b. Which vitamin is found in high amounts in pork and ham?
    c. Which vitamin is found mostly in fruits and vegetables?
    d. Which vitamin needs a compound made in the stomach to be absorbed?
    e. Which vitamin, when deficient, causes osteomalacia?
    f. Which vitamin is made from tryptophan?
    g. Which vitamin is most likely deficient in the American diet?
    h. Which vitamin is known for forming a cellular cement?
    i. Which vitamin has a precursor called beta-carotene?
    j. Which vitamin is made in the skin?
    k. Which vitamin(s) is an antioxidant?
    l. Which vitamin(s) is needed for bone growth and maintenance?
    m. Which vitamin, when deficient, causes night blindness?
    n. Which vitamin is purposely put into milk because there are no other good sources of it available?

NUTRITION WEB EXPLORER

For a complete list of websites for the following activities, please visit the companion book page at www.wiley.com/college/drummond.

CENTERS FOR DISEASE CONTROL AND PREVENTION: NUTRITION FOR EVERYONE: FRUITS AND VEGETABLES

On this home page for the Fruits and Vegetables campaign, do the exercise "How Many Fruits and Vegetables Do You Need?" Then click on "What Counts as a Cup?" How many grapes, baby carrots, raisins, or grapefruit count as 1 cup?

NATIONAL CENTER FOR COMPLEMENTARY AND ALTERNATIVE MEDICINE

Click on "Herbs at a Glance," a series of fact sheets that provides basic information about specific herbs or botanicals. Pick one, such as garlic, and read about how garlic is used, cautions, and what evidence there is of any possible benefits. Write up a summary.

NATIONAL INSTITUTES OF HEALTH, OFFICE OF DIETARY SUPPLEMENTS

Click on "Dietary Supplement Fact Sheets." Then select a vitamin from the list under "V," and then click on "Quick Facts." Read about the vitamin and write up a summary of the section titled "What are Some Effects of Vitamin X on Health?"

PURDUE UNIVERSITY: VITAMIN QUIZ

Take the vitamin quiz on this website. You should know the answers for at least 13 of the 15 questions.

FOOD AND DRUG ADMINISTRATION: DIETARY SUPPLEMENTS—WHAT YOU NEED TO KNOW

Read "What You Need to Know" at the FDA website. Summarize what you learned including the benefits and risks of taking supplements and who is responsible for the safety of dietary supplements.

IN THE KITCHEN

Vitamins in Salads

You will prepare one of the following vitamin-rich salads.

1. Garden Salad with Sprouts
2. Greek Salad
3. Arugula, Citrus, and Fennel Salad
4. Tricolor Spiral Pasta with Fresh Garden Vegetables
5. Waldorf Salad

You will be given a worksheet to identify the vitamins in some ingredients from each recipe, and then will evaluate each salad based on appearance and taste.

HOT TOPIC

FUNCTIONAL FOODS AND PHYTOCHEMICALS

In contrast to dietary supplements, functional foods are foods you normally eat that contain something that is real health promoting. For example, lycopene, which is found in tomato products, may provide health benefits beyond basic nutrition by helping prevent certain cancers. The Food and Nutrition Board has defined a functional food as "any food or food ingredient that may provide a health benefit beyond the traditional nutrients it contains." Many of the health-promoting ingredients in functional foods are phytochemicals. Phytochemicals are bioactive compounds found in plants that are linked to decreased risk of chronic diseases. Following are some examples of what you can think of as superfoods because they naturally contain lots of phytochemicals and have health benefits.

- *Beans.* Beans are an inexpensive way to get lots of nutrients, such as fiber, potassium, and many B vitamins. In addition, beans contain several phytochemicals known as saponins, which are naturally occurring compounds widely distributed in the cells of beans and peas. Studies suggest that saponins may stimulate the immune response and help protect the body from developing cancer.

- *Nuts.* A growing number of clinical studies indicate that the beneficial effect of tree nuts may be due not only to the fact that they contain healthy types of fats—monounsaturated and polyunsaturated fats—but that they contain phytochemicals that may be heart healthy.

- *Cocoa.* Cocoa, which is used to make chocolate, is made from cacao beans. Cocoa is a rich source of antioxidants that may help protect your blood vessels and heart. Dark chocolate contains more cocoa than milk chocolate and therefore contains more antioxidants.

- *Tea.* In recent years, scientists have investigated the potential benefits of green and black tea because tea is a rich source of polyphenols, which act as antioxidants in the body. Green tea goes through a fermentation process in order to make black tea. Black tea also contains polyphenols, but does not contain quite as much as green tea. Polyphenols in tea appear to be heart healthy and may have a cancer-fighting role.

- *Spinach.* Spinach contains lutein, a phytochemical that seems to help protect the eyes from cataracts and muscular degeneration, a progressive condition affecting the central part of the retina that leads to the loss of sharpness in vision. Spinach is a powerhouse of antioxidants.

- *Plant stanols and sterols.* Plant stanols and sterols occur naturally in small amounts in many plants. They are added to certain margarines and some other foods, and help block the absorption of cholesterol from the digestive tract, which helps to lower LDL.

Additional superfoods could include berries, cabbage, broccoli, Brussels sprouts, citrus fruits, pumpkin, soy, sweet potatoes, tomatoes, and whole grains.

In addition to superfoods, which naturally contain phytochemicals, food manufacturers also add vitamins, minerals, phytochemicals, and/or herbs into food and beverages. This results in functional foods and beverages, such as bottled water enhanced with vitamins and minerals, margarine with plant stanols to lower cholesterol, and energy drinks with taurine. Beverages have become a popular way for people to consume healthy ingredients.

There are many questions about functional foods. Although it may be possible to isolate specific components of food that may reduce the risk of diseases such as cancer, it is unclear whether phytochemicals added to foods have the same health benefits as whole foods because compounds in foods may act synergistically to impart health benefits. Like dietary supplements, functional foods will not compensate for a poor diet. Whole foods still contain the right amount and balance of nutrients and phytochemicals to promote health.

As you have seen, phytochemicals are rich in plants, especially fruits and vegetables. Other foods known for phytochemicals include the following:

- Red wine, which contains reversatrol, seems to influence heart health in a positive way.
- Tomatoes, which contain lycopene, may have anticancer properties.
- Berries such as blueberries and raspberries contain flavonoids which act as antioxidants.
- Green tea, which contains polyphenols, may have anticancer properties.

The way phytochemicals work and optimum amounts for consuming are still being investigated. Some phytochemicals act as antioxidants, such as lycopene in tomatoes and phenols in tea, or they may act in other ways.

- Broccoli contains the chemical sulforaphane, which seems to lower risk of both the incidence of prostate cancer and of developing aggressive prostate cancer.
- Isoflavones, found mostly in soy foods, are known as phytoestrogens. They may help to reduce risk of estrogen-sensitive cancers such as cancer of the breast and ovaries.
- Members of the cabbage family (cabbage, broccoli, cauliflower, mustard greens, kale), also called cruciferous vegetables, contain phytochemicals such as indoles and dithiolthiones. They may activate enzymes that block the damage of carcinogens. The consumption of cruciferous vegetables has been associated with a modest reduced risk of lung cancer, colorectal cancer, and prostate cancer.
- Flavonoids, which are found in citrus fruits, onions, apples, grapes, wine, and tea, are thought to be helpful in preventing heart disease and cancer.
- Garlic is the edible bulb from a plant in the lily family. It has been used as both a medicine and a spice for thousands of years. Garlic contains allicin, a chemical that acts as an antioxidant similar to vitamins A, C, and E, and may help protect the body from free radicals. Some evidence indicates that taking garlic may slightly lower blood cholesterol levels and blood pressure.

Phytochemicals are usually related to the color of fruits and vegetables: green, yellow-orange, red, blue-purple, and white. Table 6-11 lists phytochemicals according to color. You can benefit from all of them by eating five to nine servings of colorful fruits and vegetables every day.

TABLE 6-11	Phytochemicals in Foods	
PHYTOCHEMICALS	**FOODS**	**HEALTH EFFECTS**
Flavonoids	Citrus fruits, berries, purple grapes	Antioxidant
	Onions	Decreases inflammation
	Tea (black and green)	May protect against heart disease, stroke, and cancer
	Cocoa and chocolate	
	Legumes, soybeans, and soy products	
	Whole grain	
Anthocyanins	Red, Purple, and Blue Vegetables:	Antioxidant
	Red cabbage	Decreases inflammation
	Strawberries	May help combat cancer cells
	Radishes	
	Plums	
	Blackberries	
	Eggplant	
	Blueberries	
Quercetin	Yellow Fruits and Vegetables:	Antioxidant
	Apples and pears	Decreases inflammation
	Citrus fruits	May help ptrotect against heart disease, stroke, and cancer
	Onions	
	Also: Some berries	
	Black and green tea	
Carotenoids	Red, orange, and yellow fruits and vegetables, and	Antioxidants
	Some green vegetables	May help protect against heart disease and cancer
Lycopene	Tomatoes	May help protect against heart disease and cancer
	Watermelon	
	Pink grapefruit	
Beta carotene	Carrots	May help protect against heart disease
	Yellow bell pepper	
	Pumpkin	
Lutein	Green leafy vegetables such as kale, collards, spinach, romaine, and broccoli	Maintain healthy vision and protects eyes from cataracts and macular degeneration
	Yellow corn	May help protect against heart disease
Phenolic compounds	Olive oil	Antioxidant
	Red wine	May help protect against heart disease and cancer
	Black tea	
	Coffee	
	Some fruits and vegetables	
Resveratrol	Red and purple grapes	May improve blood flow to brain and decrease chances of stroke
	Red wine	
	Purple grape juice	May help protect against heart disease and cancer
	Cocoa and dark chocolate	
	Peanuts	

WATER AND MINERALS

CHAPTER 7

- Identify the percentage of body weight made up of water, list the functions of water in the body, and discuss the Adequate Intake for total water.

- Identify possible causes of dehydration and symptoms.

- Distinguish between different types of bottled waters.

- Distinguish between different types of functional beverages, and list three considerations in choosing a functional beverage.

- Explain why drinking alcohol with energy drinks is dangerous.

- Discuss what caffeine does, where it is found, and its side effects.

- State the general characteristics of minerals, and identify which minerals are most likely to be deficient in the American diet.

- Identify functions and food sources of each mineral presented.

- Discuss the nutrient content, preparation, and use of nuts and seeds on the menu.

- Explain how dietary supplements are regulated and labeled, and identify instances when supplements may be necessary.

 WATER

Without food, you can live for weeks. But death comes quickly, in a matter of a few days, if you are deprived of water. Nothing survives without water, and virtually nothing takes place in the body without water playing a vital role.

The cells in your body are full of water. Because water dissolves so many substances, your cells use valuable nutrients, minerals, and chemicals in everyday processes to give you energy and maintain your body's systems. Water is also sticky and elastic and tends to clump together in drops rather than spread out in a thin film. This allows water (and its dissolved substances) to move through the tiny blood vessels in your body.

The average adult's body weight is generally 60 percent water—enough, if the water were bottled, to fill 40 to 50 quarts. For example, in a 150-pound man, water accounts for about 90 pounds and fat for about 30 pounds, with protein, carbohydrates, vitamins, and minerals making up the balance. Men generally have proportionally more water than women because they have more muscle, and a lean person has proportionally more water than an obese person. Some parts of the body have more water than others. Human blood is about 92 percent water, muscle and brain tissue about 75 percent, and bone 22 percent.

FUNCTIONS

Almost all body cells need and depend on water to perform their functions. Water serves as the medium for many metabolic activities and also participates in some metabolic reactions. Water carries nutrients to the cells and carries away waste materials to the kidneys and out of the body in urine.

Water is needed in each step of the process of converting the food you eat into energy and tissue. Water in the digestive secretions softens, dilutes, and liquefies the food to make digestion easier. It also helps move food along the gastrointestinal tract. Differences in the fluid concentration on the two sides of the intestinal wall improve absorption of nutrients.

Over 90 percent of your blood is water, so water maintains blood volume in your body. Water in blood also helps maintain normal temperatures. For example, when you exercise, your muscles work extremely hard and create energy and heat. Your body needs to get rid of the heat, and so your blood circulates to the muscles, picks up the heat, and circulates to your skin. Sweating takes place, and you lose some of the water and heat. The end result is that you are cooled down.

Water serves as an important part of body lubricants, helping to cushion the joints and internal organs; keeping tissues in the eyes, lungs, and air passages moist; and surrounding and protecting the fetus during pregnancy.

HOW MUCH WATER DO YOU NEED?

In the United States, women who are adequately hydrated consume about 2.7 liters (about 91 ounces) of total water daily from all beverages and foods. Men consume about 3.7 liters (about 125 fluid ounces) from all sources. The amount actually needed varies depending on factors such as your size, diet, and activity, as well as the temperature and humidity of your environment.

About 80 percent of your total water intake comes from beverages, including water, coffee, tea, and cola. Contrary to popular belief, there is no convincing evidence that caffeine in coffee, colas, and other beverages is dehydrating. The remaining 20 percent of your total water intake comes from moisture found in foods. Nearly all foods have some water. Milk, for example, is about 87 percent water, eggs about 75 percent, meat between 40 and 75 percent, vegetables from 70 to 95 percent, cereals from 8 to 20 percent, and bread around 35 percent.

The Adequate Intake for total water is based on the average water consumption of people who are adequately hydrated. The Adequate Intakes for total water for adults and the elderly are as follows. Don't forget that these numbers include all fluids *and* the water in food.

Adequate Intake for Total Water

Men: 3.7 liters/day (about 15½ cups)

Women: 2.7 liters/day (about 11½ cups)

The AI for water for women is based on a woman who is 5'4" and weighs 126 pounds. The AI for men is based on a man who is 5'10" and weighs 154 pounds.

The Food and Nutrition Board does not offer any rule of thumb for how many glasses of water you should drink each day. This is because our hydration needs can be met through drinking and eating juice, milk, coffee, tea, soda, fruits, vegetables, and other foods and beverages.

When healthy, the body maintains water at a constant level. A number of mechanisms, including the sensation of thirst, keep your body water within narrow limits. If you engage in strenuous or prolonged physical activity or are exposed to hot temperatures, you need to consume more water to replace that lost in sweat. There are, of course, conditions in which the various body mechanisms for regulating fluid balance do not work, such as severe vomiting, diarrhea, excessive bleeding, high fever, burns, and excessive perspiration. In these situations, large amounts of fluids and minerals are lost. These conditions are medical problems that should be managed by a physician.

With aging, thirst declines, as does the ability of the kidneys to conserve water (Figure 7-1). Dehydration is more common among the elderly and is one of the most common causes of hospitalization for people over 65. Symptoms of dehydration include dry mouth, fatigue, irritability, and headaches.

The body gets rid of the water it doesn't need through the kidneys and skin and, to a lesser degree, from the lungs and gastrointestinal tract. Water is excreted as urine by the kidneys, along with waste materials carried from the cells. About 4 to 6 cups a day is excreted as urine. The amount of urine reflects, to some extent, the amount of fluid you take in. However, the kidneys always excrete a certain amount each day (about 2 cups) to eliminate waste products generated by the body's metabolic actions. In addition to urine, air released from the lungs contains some water, and evaporation that occurs on the skin (whether one is sweating or not sweating) contains water as well.

FIGURE 7-1 With aging, thirst declines and dehydration is more common among the elderly.

Courtesy of Goodluz/Shutterstock.

SUMMARY

1. The average adult body is 60 percent water by weight. Men generally have proportionally more water than women.
2. Water serves as the medium for many metabolic activities, such as digestion and absorption, and also participates in some metabolic reactions.
3. Water maintains blood volume and carries nutrients to the cells and carries wastes away to the kidneys. Water in the blood also helps you sweat so that your body maintains normal temperatures.
4. Water serves as an important part of body lubricants—cushioning joints and internal organs and keeping tissues moist.
5. The AI for total water for women is 2.7 liters, for men is 3.7 liters. It is based on the average water and food consumption of people who are adequately hydrated. About 80 percent of total water intake comes from beverages, with 20 percent from moisture in foods.
6. When healthy, the body maintains water at a constant level. Thirst is one way your body tries to keep its water content within narrow limits. With aging, thirst declines. Dehydration is more common among the elderly.

BOTTLED WATER

Bottled water Water that is intended for human consumption that is tested for safety and sealed in bottles with no added ingredients except that it may optionally contain safe and suitable antimicrobial agents. May be used as an ingredient in beverages such as flavored bottled waters.

People buy bottled water for what it does *not* have: kcalories, sugar, caffeine, additives, preservatives, and, in most cases, higher sodium. The Food and Drug Administration (FDA) has established standards of quality for bottled drinking water. **Bottled water** is defined as water that is intended for human consumption that is sealed in bottles or other containers with no added ingredients except that it may optionally contain safe and suitable antimicrobial agents. The FDA has established maximum allowable levels for contaminants in the standard recommendations for bottled water quality. Similar standards are set for tap water.

Bottled water is different from tap water in that it has more consistent taste. The taste of water has to do with the way it is treated and the quality of its source, including its natural mineral content. One of the key taste differences between tap water and bottled water is due to how the water is disinfected. Tap water is often disinfected with chlorine or chloramines while bottled water is often disinfected with ozone. Ozone is preferred by bottlers because it does not leave a taste.

While bottled water may originate from protected sources such as underground springs and aquifers, tap water comes mostly from lakes and rivers. Bottled water is regulated by the FDA as a food product, and tap water is regulated by the US Environmental Protection Agency (EPA).

The FDA has published standard definitions for different types of bottled water to promote honesty and fair dealing in the marketplace.

- *Artesian well water* is water from a well that taps an aquifer—layers of porous rock, sand, and earth that contain water—which is under pressure from surrounding upper layers of rock or clay. When tapped, the pressure in the aquifer, commonly called artesian pressure, pushes the water above the level of the aquifer, sometimes to the surface. Other means may be used to bring the water to the surface. According to the EPA, water from artesian aquifers often is more pure because the confining layers of rock and clay impede the movement of contamination. However, despite the claims of some bottlers, there is no guarantee that artesian waters are any cleaner than groundwater from an unconfined aquifer.

- *Mineral water* is water from an underground source that contains at least 250 parts per million of total dissolved solids. Minerals and trace elements must come from the source of the underground water. They cannot be added later.

- *Spring water* is derived from an underground formation from which water flows naturally to the earth's surface. Spring water must be collected only at the spring or through a borehole that taps the underground formation feeding the spring. If some external force is used to collect the water, the water must have the same composition and quality as the water that naturally flows to the surface.
- *Well water* is water from a hole drilled into the ground that taps into an aquifer.

Bottled water may be used as an ingredient in beverages (e.g., diluted juices, flavored bottled waters).

Some bottled water also comes from municipal sources—in other words, the tap. Municipal water is usually treated before it is bottled. Bottled water that has been treated by one of the following methods can be labeled as *purified water*. Examples of water treatments include:

- Reverse osmosis: Water is forced through membranes to remove minerals in the water.
- Absolute 1 micron filtration: Water flows through filters that remove particles larger than 1 micron in size.
- Ozonation: Bottlers of all types of waters typically use ozone gas, an antimicrobial agent, to disinfect the water instead of chlorine, since chlorine can leave a residual taste and odor in the water.

Like all other foods regulated by the FDA, bottled water must be processed, packaged, shipped, and stored in a safe and sanitary manner and be truthfully and accurately labeled.

Beverages labeled as containing "sparkling water," "seltzer water," "soda water," "tonic water," or "club soda" are not included as bottled water under the FDA's regulations, because these beverages have historically been considered soft drinks.

- *Sparkling water* is any carbonated water.
- *Seltzer* is filtered, artificially carbonated tap water that generally has no added mineral salts. It is available with assorted flavor essences, such as black cherry and orange. If seltzer contains sweeteners (and therefore calories), it must be called a flavored soda.
- *Club soda*, sometimes called soda water or plain soda, is filtered, artificially carbonated tap water to which mineral salts are added to give it a unique taste. Most average 30 to 70 milligrams of sodium per 8 ounces.
- *Tonic water* is not really water or low in calories. It contains 84 calories per 8 ounces. Diet tonic water uses sugar substitutes.

Different mineral and carbonation levels of waters make them appeal to different customers and appropriate for different eating situations. For instance, a heavily carbonated sparkling water such as Perrier is excellent as an aperitif, yet some customers may prefer the lighter sparkle of San Pellegrino. Still (noncarbonated) waters such as Evian are generally more popular and appropriate to have on the table during the meal.

Although some people prefer bottled water, it is important to know that the public water supply is regulated by the U.S. Environmental Protection Agency (EPA). All municipal water systems are tested regularly for up to 118 chemicals and bacteria specified by the Safe Drinking Water Act (SDWA). Individual states may require additional testing. Everyone who gets their tap water from a public system is therefore assured of regular testing and certain standards.

Besides being cheaper than bottled water, tap water also doesn't need a plastic bottle. Plastic water bottles create huge waste problems. They are also produced from petroleum and require energy to manufacture. Be sure to recycle all plastic bottles.

S U M M A R Y

1. Bottled water is regulated by the FDA, which has published definitions for types of bottled water such as spring water, purified water, artesian well water, mineral water, and well water. Bottled water may be used as an ingredient in beverages such as flavored bottled waters.
2. Sparkling water, seltzer water, soda water, tonic water, or club soda are not included as bottled water because they are considered soft drinks.
3. Tap water is tested regularly and is much cheaper than bottled water. It also doesn't need a plastic bottle, which creates waste problems.

FUNCTIONAL BEVERAGES

Functional beverages Drinks that have been enhanced with ingredients added to provide specific health benefits beyond general nutrition.

Sports drinks Drinks that contain a dilute mixture of carbohydrate and electrolytes that are designed to be used during exercise lasting 60 minutes or more.

Fitness waters Drinks that contain fewer kcalories, carbohydrates, and electrolytes than sports drinks but taste better than plain water due to some sugar and/or no kcalorie sweeteners.

Energy drinks Drinks that contain varying amounts of caffeine as well as other plant-based stimulants such as ginseng.

Enhanced drinks Drinks such as fruit drinks, teas, dairy drinks, and waters that contain added vitamins, minerals, phytochemicals, and/or herbs.

Functional beverages are drinks that have been enhanced with ingredients added to provide specific health benefits beyond general nutrition. These beverages have become very popular and include sports drinks, fitness drinks, energy drinks, and enhanced drinks.

Sports drinks contain a diluted mixture of carbohydrate and electrolytes such as sodium and potassium (Figure 7-2). Most contain about 50 kcalories per cup, with 3–4 teaspoons of carbohydrate and small amounts of sodium and potassium (which are lost in sweat). Sports drinks are purposely made to be weak solutions so that they can empty faster from the stomach, and the nutrients they contain are therefore available to the body more quickly. They are designed mostly for exercise lasting 60 minutes or more to help replace water and electrolytes and provide some carbohydrates for energy.

Carbohydrates taken during exercise can help you maintain a normal blood-sugar level and enhance (as well as lengthen) performance. Athletes often consume ½ to 1 cup of sports fluids every 15 to 20 minutes during exercise.

Most sports drinks contain about 14 grams of carbohydrate/cup from several carbohydrate sources, such as glucose, fructose, and sucrose. Sports drinks that contain several different carbohydrates may improve the amount of carbohydrates that gets to the muscles as fuel.

Fitness waters fall somewhere between the sports drinks and plain water. They contain fewer kcalories, carbohydrates, and electrolytes than sports drinks, but offer more taste than plain water (they usually contain a small amount of sugar and/or no kcalorie sweeteners). Fitness waters often have added vitamins and/or minerals.

You have probably seen **energy drinks**, such as Red Bull, or energy shots. Energy drinks are not at all like sports drinks. Energy drinks contain varying amounts of caffeine—from 80 to over 500 mg—as well as other plant-based stimulants like ginseng and guarana. Red Bull, for example, contains 80 mg of caffeine, which is just a little less than that found in a cup of coffee. Energy drinks are marketed to people under 30, especially to college students, who are looking for a stimulus to get through studying or other activities. Many energy drinks also contain various B vitamins.

When energy drinks are consumed with alcohol, safety issues arise. The caffeine and other stimulants in energy drinks work against the sedative effect of alcohol. Sedation, meaning that you feel very tired, is a cue for you to stop drinking at the end of an evening. But mixing alcohol with an energy drink may lead you to drink more than you normally would because you feel more awake. It may also lead you to make other poor decisions, like driving while intoxicated, because you aren't really aware how intoxicated you are. Research has shown that drinkers who consume alcohol mixed with energy drinks are three times more likely to binge drink than drinkers who don't mix alcohol and energy drinks.

Unfortunately, energy drinks are risky for children, especially those with diabetes or attention deficit disorder. Caffeine can affect children and adolescents more than adults because they may not have developed a tolerance for it and their bodies are smaller. Among adolescents, caffeine consumption has been associated with sleep disturbances, nervousness, and elevated blood pressure. Caffeine overdose can occur and results in dehydration, vomiting, irregular heartbeat, and convulsions. Too much caffeine is also hazardous to people with heart problems, as well as pregnant women (can cause miscarriages).

Enhanced drinks include fruit drinks, teas, water, and other beverages that contain added vitamins, minerals, phytochemicals, herbs, and/or other substances. Most people do not benefit from the typically low levels of vitamins and minerals found in enhanced drinks. In addition, some enhanced drinks contain quite a bit of sugar, and some contain ingredients that have not been studied enough to know if certain amounts will have healthy effects.

FIGURE 7-2 Sports drinks contain about 50 kcalories and about 3–4 teaspoons of carbohydrate per cup.

Photo by Peter Pioppo.

When choosing functional beverages, consider the following:

1. Kcalories: Many functional beverages are now available in low- or no-kcalorie versions.

2. Sweeteners: Unless the product uses low- or no-kcalorie sweeteners such as stevia or sucralose, look for what is used to sweeten the drink. Quite a number of functional beverages include added sugars.

3. Percent juice: From looking at the label, many functional beverages give you the impression they contain a lot of fruit and/or vegetable juice. Check near the nutrition label for a statement about how much real juice is in the product, given as a percent—such as 10 percent juice.

If you are looking to consume phytochemicals and antioxidants, you are likely better off eating foods such as fruits and vegetables.

SUMMARY

1. Functional beverages include sports drinks, fitness drinks, energy drinks, and enhanced drinks.

2. Sports drinks contain small amounts of carbohydrate and electrolytes and are designed to be used during exercise lasting 60 minutes or more to enhance performance. Fitness waters fall somewhere between sports drinks and plain water—they have fewer kcalories, carbohydrate, and electrolytes than sports drinks and often include artificial sweeteners for taste and added nutrients.

3. Energy drinks contain varying amounts of caffeine.

4. Mixing alcohol with energy drinks may lead you to drink more than you normally would and make other poor decisions, because you aren't really aware how intoxicated you are. Energy drinks also have safety issues when consumed by children, adolescents, pregnant women, and adults with heart problems.

5. Some enhanced drinks contain quite a bit of sugar, and some contain ingredients that have not been studied enough to know if certain amounts will have healthy effects. Consider the kcalories, sweeteners, and percent juice of any product you choose.

CAFFEINE

Check out how much you know or don't know about caffeine. Are the following true or false?

1. Tea has more caffeine than coffee.

2. Brewed coffee has more caffeine than instant coffee.

3. Some nonprescription drugs contain caffeine.

4. Caffeine is a nervous system stimulant.

5. Withdrawing from regular caffeine use causes physical symptoms.

Check your answers as you read on.

Caffeine is a naturally occurring stimulant found in the leaves, seeds, or fruits of over 60 plants around the world. Caffeine appears in the coffee bean in Arabia, the tea leaf in China, the kola nut in West Africa, and the cocoa bean in Mexico. Extensive research has been conducted on how caffeine affects health. Most experts agree that moderate use of caffeine (up to 300 milligrams or about 3 cups of coffee) is not likely to cause health problems.

Caffeine is best known for its stimulant, or "wake-up," effect. Once you drink a cup of coffee, the caffeine is readily absorbed by the body and carried around in the bloodstream, where its level peaks in about one hour. Caffeine mildly stimulates the nervous and cardiovascular systems. Caffeine also affects the brain and results in an elevated mood, decreased fatigue, and increased attentiveness. You can now think more clearly and work harder. It increases your

Caffeine A naturally occurring stimulant found in the leaves, seeds, or fruits of over 60 plants around the world—such as the coffee bean, tea leaf, cocoa bean, and kola nut. Is mildly addicting and has side effects at high doses.

TABLE 7-1   Caffeine in Foods, Beverages, and Medications

FOOD/BEVERAGE/MEDICATION	CAFFEINE (MILLIGRAMS) PER SERVING
Coffee Beverages	
Espresso coffee, brewed, 1 fluid ounce (fl. oz.)	64 mg
Coffee, brewed, 16 fl. oz.	190
Coffee, instant, 16 fl. oz.	128
Coffee, brewed, decaffeinated, 16 fl. oz.	6
Coffee, instant, decaffeinated, 16 fl. oz.	6
Starbucks brewed coffee, 16 fl. oz.	330
Starbucks brewed decaffeinated coffee, 16 fl. oz.	25
Starbucks Caffè Americano, 16 fl. oz.	225
Starbucks flavored latte, 16 fl. oz.	150
Starbucks Caffè Latte, 16 fl. oz.	150
Starbucks Caffè Vanilla Frappuccino®, 16 fl. oz.	95
Tea Beverages: Hot	
Tea, black, brewed, 8 fl. oz.	47
Tea, brewed, decaffeinated, 8 fl oz.	2
Tea, herbal, brewed, 8 fl. oz.	0
Tea, instant, 8 fl. oz.	26
Starbucks Chai Tea Latte, 16 fl. oz.	95
Starbucks Tazo® Awake Brewed Tea, 16 fl. oz	90–135
Starbucks Tazo® Calm Brewed Tea, 16 fl. oz.	0
Teas: Cold	
Lipton Brisk Lemon Tea, 16 fl. oz.	17
Snapple Lemon Iced Tea, 16 fl. oz	42
Arizona Iced Tea, black, 16 fl. oz.	30
Starbucks Shaken Iced Passion Tea, 16 fl. oz.	0
Chocolate Beverages	
Hot chocolate, 16 fl. oz.	10
Chocolate milk, whole milk, 16 fl. oz.	10
Starbucks, hot chocolate, 16 fl. oz.	25
Soft Drinks	
Cola, 12 oz. can	29
Coca-Cola Classic, 12 oz. can	34
Diet Coke, 12 oz. can	46
Pepsi, 12 oz. can	38
Diet Pepsi, 12 oz. can	35
Dr Pepper, regular or diet, 12 oz. can	42
Root beer, regular or diet, 12 oz. can	0
Barq's Root Beer, 12 oz. can	22
Lemon-lime soda, regular or diet, 12 oz. can	0
7-Up, regular or diet, 12 oz. can	0
Lemon-lime soda, with caffeine, 12 oz. can	55
Mountain Dew, 12 oz. can	54

FOOD/BEVERAGE/MEDICATION	CAFFEINE (MILLIGRAMS) PER SERVING
Energy Drinks	
Venom Energy Drinks, 17 oz. can	170
Monster Energy Drink, 16 oz. can	160
Java Monster, 15 oz. can	160
Red Bull, 8.5 oz. can	80
5-Hour Energy Shot, 2-ounce bottle	138
Chocolate	
Hershey's Special Dark Chocolate Bar, 1.45 oz. bar	26
Milk chocolate bar, 1.55 oz. bar	10
Hershey's semi-sweet chocolate chips, 1 tablespoon	9
M & M milk chocolate candies, 1.69 oz.	7
Over-the-Counter Drugs	
NoDoz Maximum Strength, 1 tablet	200
Excedrin Extra Strength, 2 tablets	130
Anacin Maximum Strength, 2 tablets	64

Sources: U.S. Department of Agriculture, Agricultural Research Service, 2011, USDA Nutrient Database for Standard Reference, Release 24. Nutrient Data Laboratory Home Page: www.nal.usda.gov/fnic/foodcomp and manufacturers.

heart rate, blood flow, respiratory rate, and metabolic rate for several hours. When taken close to bedtime, caffeine interferes with getting to sleep.

Exactly how caffeine will affect you, and for how long, depends on many factors such as the amount of caffeine ingested, whether you are male or female, your height and weight, your age, and whether you are pregnant or smoke. For pregnant women, caffeine is metabolized slowly and stays in the body longer than normal. For smokers, caffeine is metabolized faster than normal.

With frequent use, tolerance to many of the effects of caffeine develops. At high doses of 600 milligrams (about 6 cups of coffee) or more daily, caffeine can cause nervousness, sweating, tenseness, an upset stomach, anxiety, and insomnia. It can also prevent clear thinking and increase the side effects of certain medications. This level of caffeine intake represents a health risk.

Caffeine can be mildly addicting. Even when moderate amounts of caffeine are withdrawn for 18 to 24 hours, you may feel symptoms such as headache, fatigue, irritability, depression, and poor concentration. The symptoms peak on days 1 or 2 and progressively decrease over the course of a week. To minimize withdrawal symptoms, experts recommend reducing caffeine intake gradually.

Table 7-1 displays the amount of caffeine in foods. An 8-ounce cup of drip-brewed coffee has about 100 milligrams of caffeine, whereas the same amount of brewed tea contains about 50 milligrams. Twelve-ounce cans of cola soft drinks provide about 35 to 45 milligrams of caffeine.

The caffeine content of coffee and tea depends on the variety of the coffee bean or tea leaf, the particle size, the brewing method, and the length of brewing or steeping time. Brewed coffee has more caffeine than instant coffee, and espresso has more caffeine than brewed coffee. Espresso is made by forcing hot pressurized water through finely ground dark-roast beans. Because it is brewed with less water, it contains more caffeine than regular coffee per fluid ounce.

Numerous prescription and nonprescription drugs also contain caffeine. It is often used in headache and pain-relief remedies, cold products, and alertness or stay-awake tablets. When caffeine is an ingredient, it must be listed on the product label. Caffeine increases the ability of aspirin and other painkillers to do their job. Some herbal products, such as guarana, yerba mate, kola nut and green tea extract, contain caffeine.

Moderate use of caffeine is not likely to cause any health problems. Moderate use of caffeine is considered to be 300 milligrams, which is equal to about three cups of coffee daily. Women who are pregnant or trying to become pregnant should consume no more than 200 milligrams (mg) of caffeine per day—or about two cups of coffee. Higher levels of caffeine may cause miscarriages or preterm births.

SUMMARY

1. Caffeine is a natural stimulant found in coffee beans, tea leaves, kola nuts, and cocoa beans. Caffeine stimulates the nervous and cardiovascular systems, causing you to feel less fatigue and increased attentiveness. It can interfere with sleep when taken close to bedtime.
2. Moderate use of caffeine (up to 300 mg, or about three cups of coffee) is not likely to cause health problems. Caffeine can be mildly addicting. At high doses of 600 milligrams, it can cause nervousness, tenseness, an upset stomach, and insomnia.
3. Women who are pregnant or trying to become pregnant should limit caffeine to 200 mg/day to avoid miscarriages or preterm births.

OVERVIEW OF MINERALS

Major minerals Minerals needed in relatively large amounts in the diet—over 100 milligrams daily.

Trace minerals Minerals needed in smaller amounts in the diet—less than 100 milligrams daily.

If you weighed all the minerals in your body, they would amount to only 4 or 5 pounds. You need only small amounts of minerals in your diet, but they perform enormously important jobs in your body: building bones and teeth, regulating your heartbeat, and transporting oxygen from the lungs to the tissues, to name a few.

Some minerals are needed in relatively large amounts in the diet—over 100 milligrams daily. These minerals, called **major minerals**, include calcium, chloride, magnesium, phosphorus, potassium, sodium, and sulfur. Other minerals, called **trace minerals** or trace elements, are needed in smaller amounts—less than 100 milligrams daily. Iron, fluoride, and zinc are examples of trace minerals. Table 7-2 lists the major and trace minerals.

Minerals have some distinctive properties that are not shared by other nutrients. For example, whereas over 90 percent of dietary carbohydrates, fats, and proteins are absorbed into the body, the percentage of minerals that is absorbed varies tremendously. Only about 20 percent of the iron in your diet is normally absorbed, about 30 percent of calcium is absorbed, and almost all the sodium you eat is absorbed.

Minerals in animal foods tend to be absorbed better than do those in plant foods, because plant foods contain fiber and other substances that bind minerals, preventing them from being

TABLE 7-2	Major and Trace Minerals
MAJOR MINERALS	**TRACE MINERALS**
Calcium	Chromium
Chloride	Copper
Magnesium	Fluoride
Phosphorus	Iodine
Potassium	Iron
Sodium	Manganese
Sulfur	Molybdenum
	Selenium
	Zinc

absorbed. The degree to which a nutrient is absorbed and is available to be used in the body is called **bioavailability**. Sometimes minerals compete with each other for absorption in the gastrointestinal tract.

Unlike vitamins, minerals are not destroyed in food storage or preparation. They are, however, water-soluble, and so there is some loss in cooking liquids.

Like vitamins, minerals can be toxic when consumed in excessive amounts. Unlike vitamins, many trace minerals are toxic at levels only several times higher than the recommended levels. Like vitamins, the presence or absence of one mineral can affect whether another mineral is absorbed or excreted. For example, a high sodium intake can increase calcium losses.

Among the seven major minerals, at least two are likely to be consumed in amounts low enough to be of concern. Few Americans, including all age groups, consume potassium in amounts equal to or greater than the AI. Adequate calcium is important for optimal bone health. Groups for which calcium intake is of particular concern include children ages 9 year and older, adolescent girls, adult women, as well as adults ages 51 and older. Magnesium may also be lacking in the American diet.

Bioavailability The degree to which a nutrient is absorbed and available to be used in the body.

SUMMARY

1. You need only small amounts of minerals but they perform very important jobs.
2. Table 7-2 lists the major and trace minerals.
3. The percentage of minerals that is absorbed varies tremendously. Minerals in animal foods tend to be absorbed better than do those in plant foods due to fiber and other substances that bind minerals. The degree to which a nutrient is absorbed and available to be used in the body is called bioavailability.
4. Minerals are not destroyed in food preparation—but they are water-soluble.
5. Minerals can be toxic when consumed in excess, which may be only slightly higher than the recommended level. Excessive intake of some minerals can affect whether another mineral is absorbed or excreted.
6. Americans are likely to be consuming too little of potassium, calcium, and magnesium.

CALCIUM AND PHOSPHORUS

Calcium is the most abundant mineral in the body. Most of the calcium and phosphorus in your body is found in your bones to give them strength.

WHAT CALCIUM AND PHOSPHORUS DO

Calcium and phosphorus are used for building bones and teeth. Approximately 99 percent of total body calcium is found in bones and teeth. Calcium complexes with phosphorus to form part of the crystals that give bone its strength. Bone is being rebuilt every day, with new bone being formed and old bone being taken apart. The calcium in bones is therefore in a state of constant change.

Calcium plays a similar role in teeth. The turnover of minerals in teeth is not as rapid as it is in bones. Fluoride, another mineral, hardens and stabilizes the crystals of teeth, and this decreases the turnover of minerals.

Calcium also circulates in the blood, where a constant level is maintained so that it is always available for use. Calcium helps to:

- Squeeze and relax muscles
- Send and receive nerve signals
- Clot blood
- Keep a normal heartbeat

Calcium also seems to help in lowering blood pressure. In cases of inadequate dietary intake, calcium is taken out of the bones to maintain adequate blood levels.

Bones are the framework for your body. Bone is living tissue that changes constantly, with bits of old bone being removed and replaced by new bone. You can think of bone as a bank account, where you make "deposits" and "withdrawals" of bone tissue.

During childhood and adolescence, much more bone is deposited than withdrawn, so the skeleton grows in both size and density. Up to 90 percent of peak bone mass is acquired by age 18 in girls and by age 20 in boys, which makes youth the best time to "invest" in bone health by taking in plenty of calcium.

The amount of bone tissue in your skeleton, known as your bone mass, can keep growing until around age 30. At that point, bones have reached their maximum strength and density, known as peak bone mass. Women tend to experience minimal change in total bone mass between age 30 and menopause. But in the first few years after menopause, most women go through rapid bone loss, which then slows but continues throughout the postmenopausal

TABLE 7-3	Sources of Calcium	
	CALCIUM: MILLIGRAMS (MG) PER SERVING	PERCENTAGE ADULT RDA (19–50 YEARS OLD)
Yogurt, plain, low-fat, 8 oz.	415 mg	42%
Orange juice, calcium-fortified, 6 fl. oz.	375	38%
Yogurt, fruit, low fat, 8 oz.	338–384	34–38%
Mozzarella, part-skim, 1.5 oz.	333	33%
Cheddar cheese, 1.5 oz.	307	31%
Milk, nonfat, 8 fl. oz.	299	30%
Milk, reduced-fat (2% milk fat), 8 fl. oz.	293	29%
Milk, whole, 8 fl. oz.	276	28%
Tofu, firm, made with calcium sulfate, ½ cup	253	25%
Salmon, pink, canned, solids with bones, 3 oz.	181	18%
Baked beans, 1 cup	154	15%
Cottage cheese, 1% milk fat, 1 cup	138	14%
Tofu, soft, made with calcium sulfate, ½ cup	138	14%
Frozen yogurt, vanilla, soft serve, ½ cup	103	10%
Ready-to-eat cereal, calcium-fortified, 1cup	100–1,000	10–100%
Turnip greens, fresh, boiled, ½ cup	99	10%
Kale, fresh, cooked, 1 cup	94	9%
Ice cream, vanilla, ½ cup	84	8%
Soy beverage, calcium-fortified, 8 oz.	80–500	8–50%
Chinese cabbage, bok choy, raw, shredded, 1 cup	74	7%
Bread, white, 1 slice	73	7%
Pudding, chocolate, ready to eat, 4 oz.	55	6%
Tortilla, corn, one 6" diameter	46	5%
Tortilla, flour, one 6" diameter	32	3%
Sour cream, reduced fat, 2 tablespoons	31	3%
Bread, whole wheat, 1 slice	30	3%
Broccoli, raw, ½ cup	21	2%
Cheese, cream, regular, 1 tablespoon	14	1%

Source: U.S. Department of Agriculture, Agricultural Research Service, 2011, USDA Nutrient Database for Standard Reference, Release 24. Nutrient Data Laboratory Home Page: www.nal.usda.gov/fnic/foodcomp

years. This loss of bone mass can lead to **osteoporosis**, a disease in which the bones become weak due to a loss of minerals and are more likely to break—especially in the hip, spine, and wrist.

Like calcium, phosphorus circulates in the blood and has many functions. Phosphorus:

- Is involved in energy metabolism
- Is a part of DNA (genetic material) and is therefore needed for growth
- Buffers both acids and bases in all the body's cells

Osteoporosis A disease characterized by loss of bone density and strength, it is associated with debilitating fractures, especially in people 45 and older, due to a tremendous loss of bone tissue in midlife. More common in women than men.

WHERE TO GET CALCIUM AND PHOSPHORUS

Milk and milk products provides more than 70 percent of the calcium consumed by Americans. Not all milk products are as rich in calcium as milk is (see Table 7-3). As a matter of fact, butter, cream, and cream cheese contain little calcium. Some plant foods contribute calcium, but the large quantities you would have to eat to provide the equivalent amount of calcium found in one cup of milk may be difficult. According to MyPlate, everyone 9 years of age and older need to drink 3 cups of milk or its equivalent daily. In general, 1 cup of milk, yogurt, or soymilk (soy beverage); 1½ ounces of natural cheese; or 2 ounces of processed cheese can be considered as 1 cup from the Dairy Group.

Without milk or milk products in your diet, it might be difficult to get enough calcium. Other good sources of calcium include calcium-fortified foods such as orange juice, soymilk, and cereal; tofu made with calcium carbonate; some mineral waters; and several greens, such as broccoli, kale, mustard greens, turnip greens, and bok choy. Other greens, such as spinach and Swiss chard, are calcium-rich but also contain binders that prevent most calcium from being absorbed. Dried beans and peas, breads, some nuts (such as almonds) contain moderate amounts of calcium that can contribute to your calcium intake when eaten in sufficient quantities.

About 30 percent of the calcium you eat is absorbed. The body absorbs more calcium (up to 60 percent) during growth and pregnancy, when additional calcium is needed. Both stomach acid and vitamin D help calcium get absorbed.

Phosphorus is widely distributed in foods. Milk and milk products are excellent sources of phosphorus, just as they are of calcium (Figure 7-3). Other good sources of phosphorus are protein-rich foods including meat, poultry, fish, eggs, and legumes.

FIGURE 7-3 Milk and milk products are excellent sources of calcium and phosphorus.

Photo by Peter Pioppo.

AM I GETTING ENOUGH?

You are probably getting enough calcium unless you belong to one of these groups: children, adolescent girls, adult women, and adults 51 years of age and older. Unfortunately, a low calcium intake during the growing years results in bones that are not as dense as they could be, leaving one more at risk for osteoporosis and bone fractures later in life. In contrast, postmenopausal women taking supplements may be getting too much calcium. Over half of women over age 60 are taking supplemental calcium. This increases their risk for developing kidney stones.

Phosphorus is rarely lacking in the American diet.

CAN CALCIUM AND PHOSPHORUS BE HARMFUL?

Calcium can be toxic when large doses of supplements are taken. The Tolerable Upper Intake Level for calcium is 2,500 milligrams per day. Amounts above that can contribute to the development of calcium deposits in the kidneys and other organs, kidney failure, and other problems. A Tolerable Upper Intake Level is also set for phosphorus, but adverse effects are extremely rare.

SUMMARY

CALCIUM		
FUNCTIONS	**SOURCES**	**RDA**
1. Builds and rebuild bones and teeth	Milk/milk products	Men: 1,000 mg (19–70 years old)
2. Squeezes and relaxes muscles	Calcium fortified soymilk, orange juice, cereal, and tofu	1,200 mg (71+ years)
3. Sends and receives nerve signals	Some mineral waters	Women: 1,000 mg (19–50 years old)
4. Clots blood	Greens such as broccoli, kale, and turnip greens (not spinach)	1,200 mg (51+ years)
5. Keeps a normal heartbeat	Dried beans and peas	
6. Lowers blood pressure	Breads	
	Some nuts such as almonds	

PHOSPHORUS		
FUNCTIONS	**SOURCES**	**RDA**
1. Builds and rebuilds bones and teeth	Milk/milk products	700 mg (19+ years old)
2. Energy metabolism	Meat, poultry, fish, eggs	
3. A part of DNA and needed for growth	Dried beans and peas	
4. Buffers acids and bases in the body		

 MAGNESIUM

Magnesium is found in all body tissues, with about 50 percent in the bones and the remainder in the soft tissues, such as muscles, and the blood.

WHAT MAGNESIUM DOES

The body works very hard to keep blood levels of magnesium constant because it has so many important functions in the body. Magnesium is essential to:

- Keeping bones and teeth strong
- Keeping muscles and nerves functioning
- Keeping the immune system working properly
- Maintaining enzyme systems responsible for energy metabolism

Evidence suggests that magnesium seems to play an important role in reducing blood pressure and strokes. Diets that provide plenty of fruits and vegetables, which are good sources of magnesium and potassium, are consistently associated with lower blood pressure. Magnesium appears to be a heart healthy nutrient, but more research needs to be done.

WHERE TO GET MAGNESIUM

Good sources of magnesium include the following (Figure 7-4):

- Beans (including soybeans)
- Nuts (especially almonds and cashews)
- Seeds
- Whole-grain breads and cereals
- Some fruits and vegetables such as bananas and potatoes
- Some seafood

Magnesium is a part of chlorophyll, the green pigment found in plants, so good sources also include green leafy vegetables.

The magnesium content of refined foods is usually low. Whole-wheat bread, for example, has twice as much magnesium as does white bread because it contains the magnesium-rich germ and bran, which are removed when white flour is processed.

Although magnesium is present in many foods (see Table 7-4), it usually occurs in small amounts. As with most nutrients, daily needs for magnesium cannot be met from a single food. Eating a wide variety of foods, including lots of fruits and vegetables, legumes, and whole grains, helps ensure an adequate intake of magnesium. If you live in an area with hard water, your water provides both magnesium and calcium.

AM I GETTING ENOUGH?

Even though dietary surveys suggest that many American do not consume magnesium in the recommended amounts, deficiency symptoms are rarely seen in adults in the United States. When magnesium deficiency does occur, it is usually due to disease or medications.

CAN MAGNESIUM BE HARMFUL?

Magnesium is only toxic from supplement sources, not food. It is rare but can be very serious.

FIGURE 7-4 Magnesium is found in legumes, nuts, seeds, whole-grain breads and cereals, some seafood, some fruits, and green leafy vegetables.

Courtesy of the US Department of Agriculture.

TABLE 7-4	Sources of Magnesium	
	MAGNESIUM: MILLIGRAMS (MG) PER SERVING	**PERCENTAGE OF ADULT RDA**
Wheat bran, crude, ¼ cup	89 mg	21%
Almonds, dry roasted, 1 oz.	80	19%
Spinach, frozen, cooked, ½ cup	78	19%
Raisin bran cereal, 1 cup	77	18%
Cashews, dry roasted, 1 oz.	74	18%
Soybeans, mature, cooked, ½ cup	74	18%
Wheat germ, ¼ cup	69	16%
Nuts, mixed, dry roasted, 1 oz.	64	15%
Bran flakes cereal, ¾ cup	64	15%
Shredded wheat cereal, 2 rectangular biscuits	61	15%
Oatmeal, instant, prepared with water, 1 cup	61	15%
Peanuts, dry roasted, 1 oz.	50	12%
Peanut butter, smooth, 2 tablespoons	49	12%
Potato, baked with skin, 1 medium	48	11%
Black-eyed peas, cooked, ½ cup	46	11%
Pinto beans, cooked, ½ cup	43	10%
Rice, brown, long-grained, cooked, ½ cup	42	10%
Lentils, cooked, ½ cup	36	9%
Vegetarian baked beans, ½ cup	35	8%
Kidney beans, canned, ½ cup	35	8%
Chocolate milk, low-fat, 1 cup	33	8%
Banana, raw, 1 medium	32	8%
Yogurt, fruit, low-fat, 8 fl. oz.	32	8%
Milk, low-fat or nonfat, 1 cup	27	6%
Raisins, seedless, ½ cup packed	26	6%
Halibut, cooked, 3 oz.	24	6%
Bread, whole wheat, 1 slice	23	5%
Avocado, cubes, ½ cup	22	5%
Chocolate pudding, ready-to-eat, 4 oz.	19	5%
Orange, fresh, large	18	5%

Source: US Department of Agriculture, Agricultural Research Service, 2011, USDA Nutrient Database for Standard Reference, Release 24.
Nutrient Data Laboratory Home Page: www.nal.usda.gov/fnic/foodcomp

SUMMARY

MAGNESIUM		
FUNCTIONS	**SOURCES**	**RDA**
1. Keeps bones and teeth strong	Beans	Men
2. Supports normal muscle and nerve function	Nuts (especially almonds and cashews)	400 mg (19–30 years old)
3. Supports immune system function	Seeds	420 mg (31+ years old)
4. Promotes enzyme systems involved in energy metabolism	Whole-grain breads and cereals	Women
5. May reduce blood pressure	Some fruits and veggies such as bananas and potatoes	310 mg (19–30 years old)
	Some seafood	320 mg (31+ years old)

 SODIUM

A key to healthy eating is choosing foods lower in salt and sodium. Most Americans consume more salt than they need. This section discusses the roles of sodium in the body as well as ways to consume less sodium.

WHAT SODIUM DOES

Sodium, potassium, and chloride are collectively referred to as **electrolytes** because when dissolved in body fluids, they separate into positively or negatively charged particles called **ions**. Potassium, which is positively charged, is found mainly within the cells. Sodium (positively charged) and chloride (negatively charged) are found mostly in the fluid outside the cells.

The electrolytes maintain two critical balancing acts in the body:

- **Fluid balance**
- **Acid-base balance**

Fluid balance involves maintaining the proper amount of water in each of the body's three "compartments": inside the cells, outside the cells, and in the blood vessels. Electrolytes maintain fluid balance by getting water to move into and out of the three compartments as needed. Electrolytes are also able to buffer, or neutralize, various acids and bases in the body that are produced by normal body processes, so that the blood stays neutral.

In addition to its role in fluid and acid-base balance, sodium is needed for muscle contraction and transmission of nerve impulses.

WHERE TO GET SODIUM

The major source of sodium in the diet is salt, a compound made of sodium and chloride. Salt by weight is 39 percent sodium, and 1 teaspoon of salt contains 2,300 milligrams (a little more than 2 grams) of sodium (Figure 7-5). Sodium is used to enhance the flavor of foods; it also enhances the texture of food and serves as a preservative. High amounts of sodium are found

Electrolytes Minerals in your body that carry an electric charge when dissolved in water. Examples include sodium, potassium, and chloride.

Ion An atom or group of atoms carrying a positive or negative electrical charge.

Fluid balance The process of maintaining the proper amount of water in each of the three body compartments: inside the cells, outside the cells, and in the blood vessels.

Acid-base balance The process by which the body buffers the acids and bases normally produced in the body so that the blood is neither too acidic nor too basic.

FIGURE 7-5 One teaspoon of salt contains 2,300 milligrams of sodium.

Photo by Peter Pioppo.

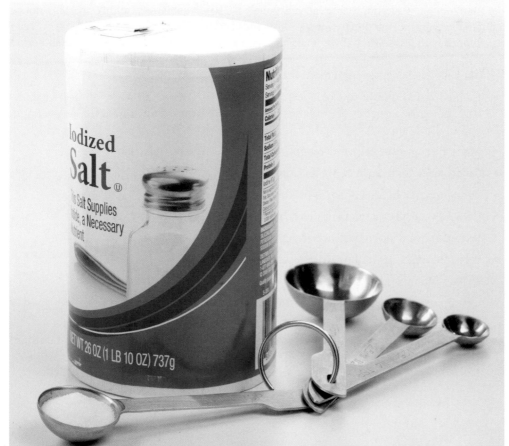

in processed foods and restaurant food, and it is estimated that these foods provide 80 percent of the sodium in most people's diets. Very little of the sodium in foods is naturally occurring—most of it is added as it is being processed or prepared by the food industry.

AM I GETTING ENOUGH?

Virtually all Americans consume more sodium than they need. The estimated average intake of sodium for all Americans ages 2 years and older is approximately 3,500 mg per day. The sodium AI for individuals ages 9 to 50 years is 1,500 mg per day. Lower sodium AIs are set for children and older adults because their kcalorie requirements are lower. For everyone 14 years and older, the Tolerable Upper Intake Level (UL) is set at 2,300 mg per day. As you can see, Americans are taking in more sodium than the UL.

CAN SODIUM BE HARMFUL?

On average, the higher your sodium intake, the higher your blood pressure will be. Research shows in adults that as sodium intake decreases, so does blood pressure. Keeping blood pressure in the normal range is very important because it reduces your risk of cardiovascular disease, stroke, and kidney disease.

Americans should reduce their sodium intake to less than 2,300 mg or 1,500 mg per day, depending on age and other individual characteristics. African Americans; individuals with high blood pressure, diabetes or chronic kidney disease; and individuals 51 and older need to be watchful of their sodium intake. Although nearly everyone benefits from reducing their sodium intake, the blood pressure of these individuals tends to be even more responsive to the blood pressure raising effects of sodium than others, so they are advised to reduce their intake to 1,500 mg per day. Other Americans should try to reduce their sodium intake to less than 2,300 mg, or about 1 teaspoon of salt per day. The preference for salty taste can be changed.

HOW TO LOWER SODIUM INTAKE

For many foods, if sodium was reduced as much as 25 percent, you wouldn't probably notice a difference in taste. The preference for salty foods can be changed. It takes time, but you can enjoy foods with less sodium if you make the changes gradually and consistently over a period of time.

The following are processed foods that are high in sodium.

- Canned, cured, and/or smoked meats and fish such as bacon, salt pork, sausage, scrapple, ham, bologna, corned beef, frankfurters, luncheon meats, canned tuna fish and salmon, and smoked salmon
- Many cheeses, especially processed cheeses such as American cheese
- Salted snack foods such as potato chips, pretzels, popcorn, nuts, and crackers
- Food prepared in brine, such as pickles, olives, and sauerkraut
- Canned vegetables, tomato products, soups, and vegetable juices
- Frozen convenience foods such as pizza and entrées
- Prepared mixes for stuffings, rice dishes, and breading
- Dried soup mixes and bouillon cubes
- Certain seasonings, such as salt, soy sauce, garlic salt, onion salt, celery salt, monosodium glutamate, and seasoned salt
- Condiments and sauces such as Worcestershire sauce, horseradish, ketchup, and mustard

Many of these foods are available in lower-sodium forms.

The following are tips to help you lower the sodium in your diet:

1. When food shopping, choose more unprocessed and minimally processed foods instead of processed and prepared foods. For example, choose more fresh or frozen vegetables instead of canned vegetables.

2. Sodium levels vary widely for the same or similar items—compare sodium levels carefully. For example, in the cold cereal aisle, you can buy a puffed wheat cereal with 0 milligrams of sodium/cup or a raisin bran cereal with more than 300 milligrams/cup.

3. Choose reduced sodium or low-sodium versions of your favorite canned foods such as canned soup and tomato sauce. Once your taste buds adjust to less sodium, the high-sodium versions will taste too salty.

4. Choose reduced sodium or low-sodium versions of your favorite lunch meats, cheeses, and hot dogs. There are many excellent choices available in the deli section.

5. Select unsalted versions of nuts, seeds, pretzels, and chips.

6. Eat more fruits and vegetables and fill half your plate with fruits and vegetables.

7. Choose frozen vegetables without sauces—sauces add a lot of sodium to each serving. Frozen (and fresh) vegetables are naturally low in sodium.

8. Instead of salt, learn to use spices and herbs to enhance the natural flavor of food. Also, many condiments are high in sodium. Look for reduced-sodium versions of condiments such as soy sauce and ketchup.

9. Make your own salad dressings using healthy oils, vinegar, herbs, and spices. Low-fat salad dressings are pumped up with sodium to make up for the lost flavor in the oil. Healthy oils, such as olive oil, are very low in sodium.

10. Use cooking methods that are flavorful and don't rely on sodium. Try searing mushrooms, sautéing chicken, and roasting vegetables.

Table 7-5 is a guide to the sodium content of foods.

The **DASH diet** stands for Dietary Approaches to Stop Hypertension (also called high blood pressure). The DASH diet emphasizes vegetables, fruits, and low-fat milk and milk products; includes whole grains, poultry, seafood, and nuts; and is lower in sodium, red and processed meats, sweets, and sugar-containing beverages than typical intakes in the United States. A DASH-style pattern (see Appendix C) lowers blood pressure and reduces risk for cardiovascular disease.

DASH diet (Dietary Approaches to Stop Hypertension) A diet that helps reduce blood pressure and emphasizes vegetables, fruits, and low-fat milk and milk products; includes whole grains, poultry, seafood, and nuts; and is lower in sodium, red and processed meats, sweets, and sugar-containing beverages than the typical American diet.

SUMMARY

SODIUM		
FUNCTIONS	**SOURCES**	**AL**
1. Helps maintain fluid balance in body	Salt (1 tsp. = 2,300 mg sodium)	1,500 mg (19–50 years old)
2. Buffers acids and bases in the body	Processed foods	1,300 mg (51–70 years old)
3. Needed for muscle contraction and transmission of nerve impulses	Restaurant foods	1,200 mg (70+ years old)
		Due to overconsumption of sodium and the fact that it raises blood pressure, Americans are advised to reduce sodium intake. Tips are given at the top of the page.

TABLE 7-5 | Sodium Content of Foods

	SODIUM MILLIGRAMS (MG) PER SERVING	ADEQUATE INTAKE (%), AGES 9–50
Table salt, 1 tsp.	2325 mg	155%
Tomato sauce, canned, 1 cup	1284	86%
McDonald's quarter pounder with cheese	1190	79%
McDonald's bacon, egg, and cheese biscuit	1160	77%
Baked beans, canned, plain or vegetarian, 1 cup	1106	74%
Chicken noodle soup, canned, 1 cup	841	56%
Salami, 2 slices	822	55%
McDonald's grilled chicken classic sandwich	820	55%
McDonald's cheeseburger	750	50%
Corn, sweet yellow, canned, cream style, 1 cup	730	49%
Ham, sliced, prepackaged, 2 slices	682	45%
Minestrone coup, canned, 1 cup	612	41%
Marinara sauce, ½ cup	524	35%
American cheese, 1 oz.	474	32%
Frankfurter, beef, 1	467	31%
Tomato soup, canned, 1 cup	464	31%
Peas, green, canned, drained, 1 cup	464	31%
Bagel, plain, 4 in.	460	31%
Cornbread stuffing, from dry mix, ½ cup	455	30%
Chocolate pudding, made from instant dry mix, prepared with 2% mix, ½ cup	417	28%
Tuna fish salad, ½ cup	412	27%
Fast food French fries, 1 medium	389	26%
Pretzels, hard, salted, 1 oz.	385	26%
Corn muffin, commercially prepared, 1	365	24%
Biscuit, plain or buttermilk, prepared from recipe, 2½-in. biscuit	348	23%
Cherry pie, prepared from recipe, 1 slice	344	23%
Soy sauce, 1 tsp.	335	22%
Shake, fast food, chocolate, 16 fl. oz.	323	22%
Feta cheese, 1 oz.	316	21%
Dill pickle, 1 spear	306	20%
Chocolate cake, prepared from recipe, 1 piece	299	20%
Wheaties cereal, 1 cup	211	14%
Cheddar cheese, 1 oz.	176	12%
Cheerios cereal, 1 cup	171	11%
Catsup, 1 Tbsp.	167	11%
Potato chips, salted, 1 oz.	136	9%
Whole wheat bread, 1 slice	132	9%
Salad dressing, French, 1 Tbsp.	130	9%
Bread, white, 1 slice	130	9%
Milk, all varieties, 1 cup	103–115	7–8%

Source: US Department of Agriculture, Agricultural Research Service, 2011, USDA Nutrient Database for Standard Reference, Release 24. Nutrient Data Laboratory Home Page: www.nal.usda.gov/fnic/foodcomp and restaurants.

POTASSIUM

Whereas too much sodium contributes to high blood pressure, the opposite is true for potassium. A diet high in potassium from high potassium foods, such as many fruits and vegetables as well as dairy and meat, is helpful to lower blood pressure.

WHAT POTASSIUM DOES

Potassium, an electrolyte found mainly in the fluid inside individual body cells, helps do the following:

- Maintain fluid balance and acid-base balance along with sodium
- Muscles contract, including maintaining a normal heartbeat
- Send nerve impulses

A diet rich in potassium blunts the effects of salt on blood pressure and lowers blood pressure.

WHERE TO GET POTASSIUM

Potassium is distributed widely in foods, both plant and animal (Figure 7-6). Unprocessed whole foods such as fruits and vegetables (especially winter squash, potatoes, oranges, grapefruits, and bananas), milk, yogurt, legumes, meats, and seafood are excellent sources of potassium (Table 7-6).

AM I GETTING ENOUGH?

Few Americans, including all age-gender groups, consume enough potassium. In view of the health benefits of adequate potassium intake, increased intake of good sources of potassium is necessary. Individuals with kidney disease and those who take certain medications should consult with their physician for specific guidance on potassium intake.

FIGURE 7-6 Excellent sources of potassium are unprocessed whole foods such as fruits and vegetables (especially potatoes, oranges, grapefruits, bananas, and raisins), milk, legumes, meats, and seafood.

Photo by Peter Pioppo.

TABLE 7-6	Sources of Potassium	
	POTASSIUM MILLIGRAMS (MG) PER SERVING	PERCENT ADEQUATE INTAKE FOR ADULTS
Fruits and Vegetables		
Potato, baked, 1 large potato	1,081 mg	23%
Potato, boiled, without skin, 1 potato	443	9%
Banana, raw, 1 banana	422	9%
Spinach, cooked, ½ cup	420	9%
Tomato juice, canned, ½ cup	278	6%
Raisins, seedless, ¼ cup	272	6%
Squash, winter, cooked, ½ cup	247	5%
Oranges, raw, all varieties, 1	237	5%
Orange juice, ½ cup	222	5%
Melons, cantaloupe, raw, ½ cup	214	5%
Tomatoes, red, ½ cup	214	5%
Grapefruit, raw, ½	170	4%
Dairy		
Yogurt, fruit, low-fat, 8 oz. container	443	9%
Milk, nonfat, 1%, 2%, whole, 1 cup	322–382	7–8%
Yogurt, plain, whole milk, 8 oz. container	352	7%
Grains		
Kellogg's All Bran, ½ cup	306	7%
Muffins, oat bran, 1 muffin	289	6%
Granola with raisins, ½ cup	216	5%
Kellogg's Frosted Mini-Wheats, 1 cup	195	4%
Cheerios, 1 cup	183	4%
Meat, Poultry, Fish, and Legumes		
Halibut, cooked, 3 oz.	449	10%
Refried beans, canned, ½ cup	424	9%
Beans, pinto, cooked, ½ cup	373	8%
Peas, split, cooked, ½ cup	355	8%
Pork, fresh, loin, braised 3 oz.	318	7%
Fats and Sweets		
Potato chips, plain, 1 oz.	466	10%
Pumpkin pie, 1 slice	222	5%

Source: US Department of Agriculture, Agricultural Research Service, 2011, USDA Nutrient Database for Standard Reference, Release 24. Nutrient Data Laboratory Home Page: www.nal.usda.gov/fnic/foodcomp

Deficiency symptoms include weakness, an increase in blood pressure, and irregular heartbeat.

CAN POTASSIUM BE HARMFUL?

There is no evidence of chronic excess intakes of potassium in healthy individuals, and therefore no Tolerable Upper Intake Level has been set. Toxic levels of potassium in the body can be caused by disease or overconsumption of supplements.

SUMMARY

POTASSIUM		
FUNCTIONS	**SOURCES**	**AL**
1. Maintains fluid balance and acid-base balance	Unprocessed whole foods such as:	4700 mg (19+ years old)
2. Contracts muscles, including maintaining a normal heartbeat	fruits and vegetables	Few Americans consume the AI.
	milk and yogurt	
3. Sends nerve impulses	legumes	
4. Lowers blood pressure	meats	
	seafood	

IRON

The iron in your red blood cells makes it possible for oxygen that you breathe in to be transported to all the cells of your body. We must eat some iron-rich foods daily to keep our red blood cells healthy.

WHAT IRON DOES

Iron, one of the most abundant metals in the universe and one of the most important in the body, is a key component of **hemoglobin**, a part of red blood cells that carries oxygen to the cells in the body. Cells require oxygen to break down glucose and produce energy.

Iron is also part of **myoglobin**, a muscle protein that stores and carries oxygen that the muscles use to contract. Iron works with many enzymes in energy metabolism and is therefore necessary for the body to produce energy.

You store some iron in the liver and other places for future needs. Iron stores are mobilized when dietary intake is inadequate.

Hemoglobin A protein in red blood cells that carries oxygen to the body's cells.

Myoglobin A muscle protein that stores and carries oxygen that the muscles will use to contract.

WHERE TO GET IRON

Meat, poultry, and fish are good sources of iron. Whole-grain, enriched, and fortified breads and cereals also supply iron. You can also get iron from legumes, green leafy vegetables, eggs, and dried fruit (Table 7-7 and Figure 7-7).

Only about 15 percent of dietary iron is absorbed in healthy individuals. Iron absorption significantly increases when body stores of iron are low. When iron stores are high, absorption decreases to help protect against iron overload. The body also absorbs iron more efficiently when there is a high need for red blood cells, such as during growth spurts or pregnancy or when there has been blood loss.

The ability of the body to absorb and use iron from different foods varies. Most iron in animal foods, called **heme iron**, is absorbed much better than is iron in plant foods, called **nonheme iron**. Animal foods also contain some nonheme iron. The presence of vitamin C in a meal increases nonheme iron absorption, as does consuming meat, poultry, and fish at the same meal.

The following factors decrease the absorption of nonheme iron:

- Calcium
- Polyphenols found in tea and coffee
- Phytic acid found in legumes, grains, and rice

Some proteins found in legumes and soybeans also inhibit nonheme iron absorption.

Heme iron The main form of iron in animal foods; it is absorbed and used more readily than iron in plant foods.

Nonheme iron A form of iron found in all plant sources of iron and also as some of the iron in animal foods.

TABLE 7-7	Sources of Iron		
	IRON MILLIGRAMS (MG) PER SERVING	PERCENT RDA FOR WOMEN 19–50 YEARS OLD	PERCENT RDA FOR MEN 19+ YEARS OLD
Grains			
Total cereal, General Mills, ¾ cup	18.0 mg	100%	225%
Cheerios, General Mills, 1 cup	9.5	53%	123%
Special K, 1 cup	8.7	48%	109%
Cream of wheat, ½ cup	4.7	26%	59%
Rice, white, long-grain, parboiled, enriched, ½ cup	1.6	9%	20%
Hamburger roll	1.5	8%	19%
White bread, 1 slice	0.9	5%	11%
Whole-wheat bread, 1 slice	0.7	4%	9%
Meats, Poultry, Fish, Legumes, & Nuts			
Beans, kidney, ½ cup	2.6	14%	33%
Beef, ground, 85% lean, cooked, 3 oz.	2.2	12%	28%
Refried beans, canned, ½ cup	2.1	12%	26%
Beef, rib, roasted, 3 oz.	2.0	11%	25%
Cashew nuts, 1 oz.	1.7	9%	21%
Soymilk, original and vanilla, 1 cup	1.6	9%	20%
Chickpeas, canned, ½ cup	1.5	8%	19%
Tofu, firm, 3 oz.	1.3	7%	16%
Tuna fish salad, ½ cup	1.0	6%	13%
Chicken breast, roasted, 3 oz.	0.9	5%	11%
Egg, boiled, 1 large	0.6	3%	8%
Flounder, cooked, 3 oz.	0.2	1%	3%
Fruits and Vegetables			
Spinach, boiled, ½ cup	3.2	18%	40%
Tomatoes, stewed, ½ cup	1.7	9%	21%
Dried peaches, 3 halves	1.6	9%	20%
Turnip greens, frozen, cooked, ½ cup	1.6	9%	20%
Peas, frozen, cooked, ½ cup	1.2	7%	15%
Raisins, ½ cup	0.7	4%	9%
Romaine, 1 cup shredded	0.5	3%	6%
Fats and Sweets			
Pumpkin pie, 1 piece	2.0	11%	25%
Granola bar, hard, plain, 1	0.9	5%	11%

Source: US Department of Agriculture, Agricultural Research Service, 2011, USDA Nutrient Database for Standard Reference, Release 24.
Nutrient Data Laboratory Home Page: www.nal.usda.gov/fnic/foodcomp

AM I GETTING ENOUGH?

Iron deficiency A condition in which iron stores are used up.

Substantial numbers of adolescent females and women of childbearing age, as well as toddlers, have **iron deficiency**, meaning that their stores of iron are used up. Iron deficiency is the most common nutritional deficiency. This condition commonly results from repeated blood loss from menstruation and/or inadequate intake of iron. Iron deficiency results in tiredness.

FIGURE 7-7 Good sources of iron include meat, fish, and poultry. You also get iron from whole-grain and enriched breads and cereals as well as legumes, green leafy vegetables, eggs, and dried fruit.

Photo by Peter Pioppo.

When iron stores become severely depleted, you can develop **iron-deficiency anemia**. In iron-deficiency anemia, red blood cells are smaller than usual and carry less hemoglobin. Women of childbearing age, pregnant women, infants and young children, and teenage girls are at the greatest risk of developing iron-deficiency anemia because they have the greatest needs. Signs of iron-deficiency anemia include feeling tired and weak, decreased work and school performance, slow cognitive and social development during childhood, difficulty maintaining body temperature, and decreased immune function.

Another group at risk of low iron intake is vegetarians. The bioavailability of iron from vegetarian diets is lower than from nonvegetarian diets because their diets are mostly plant based. Therefore, vegetarians are advised to take in 1.8 times more iron than the RDA. So for vegetarian men, the RDA increases from 8 mg/day to 14 mg/day. For vegetarian women (19–50 years old), the RDA increases from 18 mg to 32 mg/day.

People who are at risk for being iron deficient may need iron supplements in addition to a diet rich in iron and should consult a health care provider. Supplemental iron is available in two forms: ferrous and ferric. Ferrous iron salts (ferrous fumarate, ferrous sulfate, and ferrous gluconate) are the best-absorbed forms of iron supplements. It is recommended that most people take their prescribed daily iron supplement between meals and with liquids other than milk, tea, and coffee (which contain substances that decrease iron absorption).

Iron-deficiency anemia A condition in which red blood cells are smaller than usual and carry less hemoglobin; may result from severely depleted iron stores or blood loss. Symptoms include fatigue, decreased work and school performance, and decreased immune function.

CAN IRON BE HARMFUL?

Although the body generally avoids absorbing huge amounts of iron, some people do absorb large amounts. The problem with iron is that once it is in the body, it is hard to get rid of. In individuals who absorb too much iron, iron supplements can damage the liver and do other damage, a condition called **iron overload** or **hemochromatosis**. This condition is usually caused by a genetic disorder and is much more prevalent in men than in women. Symptoms include fatigue, joint pain, skin that turns gray or bronze, and symptoms of liver disease such as nausea and stomach pain.

It is especially important to keep iron supplements away from children, because they are so toxic that they can kill. Consuming 1 to 3 grams of iron can be fatal to children under 6, and lower doses can cause severe symptoms such as vomiting, diarrhea, and weak pulse. It is important to keep iron supplements in childproof containers and out of children's reach.

Iron overload (hemochromatosis) A common genetic disease in which individuals absorb too much iron from their food and supplements that can damage the liver. It is more common in men than women.

IRON		
FUNCTIONS	**SOURCES**	**RDA**
1. Key component of hemoglobin and myoglobin—carries oxygen to cells 2. Needed for the body to make energy	Meat, poultry, fish Whole-grain, enriched, and fortified breads and cereals Dried beans and peas Green leafy vegetables, eggs, and dried fruit	Men 8 mg (19+ years old) Women 18 mg (19–50 years old) 8 mg (51+ years old) Iron deficiency and iron-deficiency anemia are more common in women of child-bearing age and vegetarians.

ZINC

Zinc is everywhere in the body. It is in every cell and has well over 100 functions, including aiding growth of the body.

WHAT ZINC DOES

Cofactor A substance that is necessary for the activity of an enzyme.

Zinc is in every cell in the body. It is a **cofactor** (a substance that is necessary for the activity of an enzyme) for nearly 100 enzymes. Some of the functions of zinc are as follows:

- Protein, carbohydrate, and fat metabolism
- DNA synthesis
- Immune function and wound healing
- Growth and development
- Taste perception

The effect of zinc treatments on the severity or duration of cold symptoms is inconclusive. Additional research is needed to determine whether zinc compounds have any positive effect on the common cold.

WHERE TO GET ZINC

Protein-containing foods are all good sources of zinc, particularly shellfish (especially oysters), meat, and poultry (see Table 7-8). Legumes, dairy foods, whole grains, and fortified cereals are good sources as well. Zinc is much more readily available, or is absorbed better, from animal foods. Like iron, zinc is more likely to be absorbed when animal sources are eaten and when the body needs it. Only about 40 percent of the zinc we eat is absorbed into the body.

AM I GETTING ENOUGH?

Deficiencies are more likely to show up in pregnant women, the young, the elderly, and vegetarians. Adults deficient in zinc may have symptoms such as poor appetite, diarrhea, skin rash, and hair loss. Signs of severe deficiency in children include growth retardation, delayed sexual maturation, decreased sense of taste, poor appetite, delayed wound healing, and immune deficiencies. Marginal deficiencies occur in the United States.

TABLE 7-8   Sources of Zinc

	CALCIUM: ZINC (MG) PER SERVING	PERCENT ADULT RDA
Oysters, cooked, breaded and fried, 3 oz.	74.0 mg	673%
Beef chuck roast, braised, 3 oz.	7.0	64%
Crab, Alaska king, cooked, 3 oz.	6.5	59%
Beef patty, broiled, 3 oz.	5.3	48%
Breakfast cereal, fortified with 25% of DV for zinc, ¾ cup servings	3.8	35%
Lobster, cooked, 3 oz.	3.4	31%
Pork chop, loin, cooked, 3 oz.	2.9	26%
Baked beans, canned, plain or vegetarian, ½ cup	2.9	26%
Chicken, dark meat, cooked, 3 oz.	2.4	22%
Yogurt, fruit, low fat, 8 oz.	1.7	15%
Cashews, dry roasted, 1 oz.	1.6	15%
Chickpeas, cooked, ½ cup	1.3	12%
Cheese, Swiss, 1 oz.	1.2	11%
Oatmeal, instant, plain, prepared with water, 1 packet	1.1	10%
Milk, low-fat or non-fat, 1 cup	1.0	9%
Almonds, dry roasted, 1 oz.	0.9	8%
Kidney beans, cooked, ½ cup	0.9	8%
Chicken breast, roasted, skin removed, ½ breast	0.9	8%
Cheese, cheddar or mozzarella, 1 oz.	0.9	8%
Peas, green, frozen, cooked, ½ cup	0.5	5%
Bread, whole wheat, 1 slice	0.5	5%
Flounder or sole, cooked, 3 oz.	0.3	3%

Source: US Department of Agriculture, Agricultural Research Service, 2011, USDA Nutrient Database for Standard Reference, Release 24.
Nutrient Data Laboratory Home Page: www.nal.usda.gov/fnic/foodcomp

CAN ZINC BE HARMFUL?

Long-term ingestion of zinc at or above the Tolerable Upper Intake Level (40 milligrams) results in a copper deficiency, impaired immunity, and other concerns. Since zinc supplements can be fatal at lower levels than is the case with many of the other trace minerals, zinc supplements should only be used under a physician's guidance.

SUMMARY

ZINC		
FUNCTIONS	**SOURCES**	**RDA**
1. Acts as cofactor for nearly 100 enzymes	Protein foods	Men
2. Involved in protein, carbohydrate, and fat metabolism	Shellfish, meat, poultry	11 mg (19+ years old)
3. Involved in DNA synthesis	Legumes, dairy	Women
4. Helps immune function and wound healing	Whole grain and fortified cereals	8 mg (19+ years old)
5. Helps children grow and develop		
6. Taste foods		

IODINE

When you think of iodine, think of iodized salt. Iodized salt has iodine added to it and was developed to fight iodine deficiencies almost 100 years ago. Iodine is a part of two important hormones (made in the thyroid gland) that maintain a normal level of metabolism in your body and are essential for normal growth and development, body temperature, nerve and muscle function, and much more.

CULINARY FOCUS: NUTS AND SEEDS

Nuts and seeds are versatile, and nutritious, ingredients. From their use whole in breakfast granolas or snacks, or ground in entrees and side dishes, nuts and seeds play many roles in the kitchen.

PRODUCT

Nuts and seeds pack quite a few vitamins (such as folate and vitamin E) and minerals, along with fiber and protein, in their small sizes. Nuts in particular also contain quite a bit of fat. Luckily, most of the fat (except in walnuts) is monounsaturated. Walnuts and flaxseed are rich in the omega-3 fatty acid linolenic acid, an essential fatty acid. One ounce of many nuts contains from 13 to 18 grams of fat, making nuts a relatively high-kcalorie food. By comparison, seeds contain less fat, more fiber, and still quite a few kcalories, but they can be easily worked into any diet. The fat and fiber they contain will, in any case, make you feel full longer.

Large studies have confirmed the link between eating nuts and a reduction in heart disease. The monounsaturated fat in nuts helps lower low-density lipoprotein cholesterol, the bad kind. Nuts (and seeds) are also excellent sources of phytochemicals.

PREPARATION

Roasting or sautéing nuts is a technique that enhances flavor by extracting the oils through heat. This results in a crisp texture once cooled. Toasting nuts brings out the natural oils adding a rich fragrant flavor. To oven-roast nuts, such as almonds, pecans, walnuts, or brazil nuts, spread a single layer on a sheet pan coated with vegetable oil spray. Bake at 250°F for about 10–20 minutes, or until the nuts are lightly colored. Stir once or twice until lightly toasted. Remember that the nuts will continue to cook if left on the hot sheet pan. Remove from the pan to cool. Small seeds, such as sesame seeds, can be toasted the same way—only watch the timing carefully at they cook quicker and continue to cook from residual heat.

To fry nuts, heat a nonstick sauté pan over low heat. Add in 1 teaspoon of vegetable oil per pound of nuts. Swirl the oil around the pan, add in the nuts, and toss as needed to avoid burning. When the nuts have turned a golden color, pull them off the heat since they will continue to cook a little more. When nuts are warm, they are easier to slice because they are not as brittle as when they are cooled.

Nuts and seeds turn rancid easily due to their fat content. To keep them as fresh as possible, store in an airtight container in the refrigerator for up to six months, or about one year in the freezer. Not only will rancidity affect taste, but rancid oils are not healthy.

MENUING AND PRESENTATION

- Figures 7-8 and 7-9 show a variety of nuts and seeds that can be used in diverse preparations.

FIGURE 7-8 Nuts. Cashews are in the middle. Clockwise from center top: Brazil nuts, filberts (hazelnuts), pecans, walnuts, sliced almonds, pine nuts, pistachios, and macadamia nuts.

Photo by Peter Pioppo.

- Nuts, chopped or ground, are a great addition to baked items (Figure 7-10) as a part of the flour. Including nuts in items such as muffins, scones, quick breads, cookies, doughs, cakes, toppings, and crusts adds dimension and character. Menu items such as honey-almond or strawberry pecan muffins; macadamia oat crust, hazelnut, and dried cranberry scones; or banana, date, and walnut bread are great ways to feature nuts adding an enticing description.

- Nuts and seeds work well in granolas, give a unique crunch and flavor to vegetable stews, add interest to simple and complex salads such as fennel, orange, watercress with toasted walnuts or roasted apples; mâche, frisée, with candied pecans; or wheat berry, roasted vegetables, and spiced almonds.

- Nuts create interesting and flavorful breading for fish, poultry, and pork. Mix equal parts bread crumbs and finely chopped or ground raw nuts of your choice. Add herbs/spices to accent your flavor profile if desired, such as basil, thyme, cayenne pepper, granulated garlic, onion powder, chili, or cumin. Marinate your protein in egg whites then place into the seasoned crumb mixture, pressing to coat. Place on a sheet pan sprayed with vegetable oil and oven bake for best results, flipping your product as needed to create an even browning.

FIGURE 7-9 Seeds. Clockwise from top: Golden flax seeds, pumpkin seeds, sunflower seeds, alfafa seeds, and brown flax seeds.

Photo by Peter Pioppo.

FIGURE 7-10 Nuts added to baked goods add dimension and character.

Photo by Peter Pioppo.

- Pumpkin seeds are native to Mexico and used in many Latin American cuisines. The seeds are excellent sources of minerals, vitamins, and antioxidants. Roasted unsalted pumpkin seeds flavored with chili and lime can be enjoyed as a healthy snack. Pumpkin seeds are often an ingredient added to a variety of Mole sauces, adding a nutty flavor and thick texture. Just like other nuts and seeds, pumpkin seeds can also be used in granolas, biscuits, breads, cookies, stews, casseroles, or baked goods. Additionally the seeds can be sprinkled over fruits/vegetables or green salads. They create a nice component to desserts as a topping or filling.

- Pumpkin seed oil is used in salad dressing as well as in cooking. It is best drizzled over a finished dish to add a unique flavor profile with a controlled amount of fat. Pumpkin seeds can be pulverized into a thick powder or paste and used as a thickener or toasted and used as a crust, adding a delicious flavor to variety of preparations.

- Sunflower seeds can help make dishes healthier, their mild nutty flavor and firm texture makes them perfect for a quick snack, or they can be used in casseroles, baked goods, stews, vegetables, stuffings, and salads.

- Sesame seeds and caraway seeds are often used in baking, flavoring breads, cookies, rolls, and bagels. Caraway seeds are traditional in a variety of vegetable preparations mostly in German style cooking. Toasted sesame seeds can be sprinkled on soups, fish, stuffings, breading and cooked vegetables, adding flavor, presentation, and texture.

- In cooking, flaxseeds add a pleasant, nutty flavor. The attractive, oval reddish-brown seeds of flax (just a little bigger than sesame seeds) can be used whole or ground. The seed adds extra texture and provide nutritional benefits to breads, baked goods, stews, and many other preparations. This powerful food has been around for centuries possessing many health components such as soluble fiber, omega-3 fatty acids, and protein. Whole seeds provide fiber, but the body can't break down the whole seed so you don't get any of the benefits of the omega-3 fatty acids and protein inside. For ground flaxseed, grind whole seeds in a coffee grinder to your desired powder.

- Add whole seeds to bread dough, pancake, muffin, or cookie mixes. When sprinkled on top of any of these before baking, they add crunch, taste, and eye appeal.

- Use ground flax to enhance the flavor of oatmeal, breadings, coating for a variety of proteins, a crisp crumb topping, or incorporated into a bottom crust in baked goods. Offer flaxseeds as a toppings for smoothies and yogurt or as an option to enhance a juice bar, fruit bar, and salad bar selection.

- Ground flaxseed can substitute for fat or egg yolk in many recipes. Use 3 tablespoons ground flaxseed to replace 1 tablespoon of fat, or use 1 tablespoon ground flaxseed mixed with 3 tablespoons of water to replace 1 egg. Keep in mind that flax will cause your baked items to brown more quickly, and may result in a slightly chewier texture with less volume depending on the recipe or substitution used.

CHECK-OUT QUIZ

1. About 25 percent of your body's weight is water.
   a. True
   b. False

2. Two cups of yogurt contain about as much calcium as 1 cup of milk.
   a. True
   b. False

3. Sodium and potassium are involved in muscle contraction and transmission of nerve impulses.
   a. True
   b. False

4. One teaspoon of salt contains 1,500 mg of sodium.
   a. True
   b. False

5. Calcium and potassium are the minerals most likely to be deficient in the American diet.
   a. True
   b. False

6. Few minerals are toxic in excess.
   a. True
   b. False

7. Nearly all foods contain water.
   a. True
   b. False

8. A deficiency of iodine causes symptoms of diabetes.
   a. True
   b. False

9. Sodium, potassium, and calcium are referred to as electrolytes.
   a. True
   b. False

10. Name the mineral(s) referred to by the following descriptions:
    a. This mineral is in your bones.
    b. This mineral is found mostly in milk and milk products.
    c. These minerals are electrolytes that maintain fluid balance and acid-base balance.
    d. This mineral is found mostly in processed and restaurant foods.
    e. Bananas and potatoes are rich sources of this mineral.
    f. This mineral is found in salt.
    g. A deficiency of this mineral causes a form of anemia.
    h. This mineral may decrease blood pressure.
    i. This mineral increases blood pressure.
    j. This nutrient is better absorbed from animal foods than plant foods.
    k. This is a cofactor for nearly 100 enzymes.
    l. A deficiency of this mineral can cause low levels of thyroid hormones and lower your metabolism.

NUTRITION WEB EXPLORER

For a complete list of websites for the following activities, please visit the companion book page at www.wiley.com/college/drummond.

OREGON DAIRY COUNCIL

Click on "Calcium Check-Up," then click on "Use Quick Check" to see how much calcium is in your diet. Write up how your calcium intake compares to your calcium goal. List five ways you can get more calcium.

MEDLINE PLUS: DIETARY SUPPLEMENTS

Click on a topic under "Latest News" and write a summary of what you read.

PLAY SODIUM SENSE—HEALTHY FAMILIES BC

Select your food items for breakfast, lunch, and dinner onto the plates. How many milligrams of sodium did you consume total?

CONSUMER REPORTS—TEST YOUR SODIUM SMARTS

Take this quiz to see how much you know about sodium in food. Out of the 12 questions, how many did you get right?

NATIONAL CENTER FOR COMPLEMENTARY AND ALTERNATIVE MEDICINE

The National Center for Complementary and Alternative Medicine is dedicated to exploring complementary and alternative healing practices in the context of rigorous science. Read "What Is CAM?" to learn about this new area, and write up a summary on what you learned.

IN THE KITCHEN

Minerals on a Salad Bar

You will be given ingredients to prepare 2 salads. The first salad must include ingredients that are good sources of iron, potassium, and zinc. The second salad must include ingredients that are good sources of calcium, phosphorus, and magnesium. Once your salads are prepared, you will be given a worksheet to identify which ingredients in the salads are good sources of specific minerals. You will also evaluate each salad based on taste and appearance.

HOT TOPIC

DIETARY SUPPLEMENTS

Surveys show that more than half of the US adult population uses dietary supplements. Most adults take only one supplement—often a multivitamin-multimineral supplement. Annual sales of dietary supplements in the United States have continued to grow, and many new products are developed each year. Traditionally, the term "dietary supplement" referred to products made of one or more of the essential nutrients, such as vitamins, minerals, and protein. But the 1994 Dietary Supplement Health and Education Act (DSHEA) broadened the definition to include, with some exceptions, any product intended for ingestion as a supplement to the diet. In addition to vitamins, minerals, and herbs or other botanicals, dietary supplements may include amino acids and substances such as enzymes, organ tissues, glandulars, and metabolites.

It's easy to spot a supplement, because the DSHEA requires manufacturers to include the words "dietary supplement" on product labels. Manufacturers must also include a list of ingredients. In addition, a Supplement Facts panel is required on the labels of most dietary supplements. This label must identify each dietary ingredient contained in the product (Figure 7-11).

Dietary supplements come in many forms, including tablets, capsules, powders, softgels, gelcaps, and liquids. Though commonly associated with health-food stores, dietary supplements also are sold in grocery, drug, and national discount chain stores, as well

as through TV programs, the Internet, and direct sales.

Unlike prescription and over-the-counter drugs that must be proven safe and effective for their intended use before being approved by the FDA, the FDA does not "approve" dietary supplements for safety or effectiveness before they are sold.

Once a dietary supplement is sold, the FDA has to show that a dietary

supplement is "unsafe" before it can take action to restrict the product's use or remove the product from the marketplace. However, manufacturers and distributors of dietary supplements must record, investigate, and forward to FDA any reports they receive of serious adverse events associated with the use of their products.

Claims that taking a supplement will provide certain health benefits have

Supplement Facts
Serving Size 1 Tablet

	Amount Per Serving	%Daily Value
Vitamin A (as retinyl acetate and 50% as beta-carotene)	5000 IU	100%
Vitamin C (as ascorbic acid)	60 mg	100%
Vitamin D (as cholecalciferol)	400 IU	100%
Vitamin E (as dl-alpha tocopheryl acetate)	30 IU	100%
Thiamin (as thiamin mononitrate)	1.5 mg	100%
Riboflavin	1.7 mg	100%
Niacin (as niacinamide)	20 mg	100%
Vitamin B_6 (as pyridoxine hydrochloride)	2.0 mg	100%
Folate (as folic acid)	400 mcg	100%
Vitamin B_{12} (as cyanocobalamin)	6 mcg	100%
Biotin	30 mcg	10%
Pantothenic Acid (as calcium pantothenate)	10 mg	100%

Other ingredients: Gelatin, lactose, magnesium stearate, microcrystalline cellulose, FD&C Yellow No. 6, propylene glycol, propylparaben, and sodium benzoate.

FIGURE 7-11 Supplement facts label.

Courtesy of the US Food and Drug Administration.

always been a controversial feature of dietary supplements. Manufacturers often rely on these claims to sell their products. Under the DSHEA and previous food-labeling laws, supplement manufacturers are allowed to use, when appropriate, three types of claims: nutrient claims, health claims, and nutrition support claims, which include "structure-function claims."

Nutrient claims describe the level of a nutrient in a food or dietary supplement and are discussed in Chapter 2. For example, a supplement containing at least 200 milligrams of calcium per serving could carry the claim "high in calcium."

Health claims show a link between a food or substance and a disease or health-related condition. The FDA authorizes these claims based on a review of the scientific evidence. For example, a claim may show a link between folate in the product and a decreased risk of neural tube defects in pregnancy if the supplement contains enough folate. Table 2-9 lists allowable health claims.

Nutrition support claims describe a link between a nutrient and the deficiency disease that can result if the nutrient is lacking in the diet. For example, the label of a vitamin C supplement could state that vitamin C prevents scurvy. When these types of claims are used, the label must mention the prevalence of the nutrient-deficiency disease in the United States.

These claims also can refer to the supplement's effect on the body's structure or function, including its overall effect on a person's well-being. These are known as structure-function claims. Examples include:

- Calcium builds strong bones.
- Antioxidants maintain cell integrity.
- Fiber maintains bowel regularity.

Manufacturers can use structure-function claims without FDA approval. They base their claims on their review and interpretation of the scientific literature. Like all label claims, they must be true and not misleading. They must also be accompanied by the disclaimer "This statement has not been evaluated by the Food and Drug Administration. This product is not intended to diagnose, treat, cure, or prevent any disease."

Manufacturers that plan to use a structure-function claim must inform the FDA of the use of the claim no later than 30 days after the product is first marketed. While the manufacturer must be able to substantiate its claim, it does not have to share the substantiation with the FDA or make it publicly available. If the submitted claim promotes the product as a drug instead of a supplement, the FDA can advise the manufacturer to change or delete the claim.

At your local drug store, you not only find dietary supplements and over-the-counter drugs such as cough medicine, but you also find products labeled "homeopathic."

The term *homeopathy* is derived from the Greek words *homeo* (similar) and *pathos* (suffering or disease). The first basic principles of homeopathy were formulated by Samuel Hahnemann in the late 1700s. The practice of homeopathy is based on the belief that the body can heal itself, so if a substance causes a symptom in a healthy person, giving the person a very small amount of the same substance may cure the illness. Until recently, homeopathic drugs had been marketed on a limited scale, but today the homeopathic drug market has grown to become a multimillion-dollar industry in the United States. Some critics say that many homeopathic products sold in stores are not really homeopathic.

Like dietary supplements, homeopathic remedies are not approved for safety or effectiveness by FDA before they are sold. Whereas dietary supplements cannot claim to treat, prevent, or cure a disease or condition (only drugs can do that with FDA approval), homeopathic remedies can. For example, the zinc supplement Cold-Eeze is a homeopathic medicine that claims to be a cold remedy. If you are not sure whether a product is a dietary supplement or homeopathic remedy, just look on the label for the terms *dietary supplement* or *homeopathic*.

Dietary supplements also are not replacements for conventional diets.

Supplements do not provide all the known—and perhaps unknown—nutritional benefits of conventional food.

To help protect yourself, do the following:

- Look for ingredients in products with the USP notation, which indicates that the manufacturer followed standards established by the US Pharmacopoeia.
- Avoid substances that are not known nutrients.
- Limit your intake of vitamins and minerals to the doses recommended in the DRIs. Keep in mind that there is really no compelling evidence that taking dietary supplements above and beyond a normal dietary intake is helpful in any way. In fact, recent research shows that taking multivitamins is not sufficient to prevent cancer or heart disease.
- Taking vitamins and minerals in excess of the RDA can be harmful. Some vitamins, such as A and D, and certain minerals are harmful in higher doses.
- Consider the name of the manufacturer or distributor. Supplements made by a nationally known food and drug manufacturer have probably been made under tighter controls, because these companies already have in place manufacturing standards for their other products.
- Write to the supplement manufacturer for more information.
- Be sure to consult a doctor before purchasing or taking any supplement.
- Ask the pharmacist about possible interactions between supplements and prescription and over-the-counter (OTC) medicines. Taking a combination of supplements or using these products together with prescription or OTC drugs could, under certain circumstances, produce adverse effects, some of which could be life-threatening. For example, St. John's wort may reduce the effectiveness of prescription drugs for heart

disease, depression, seizures, or certain cancers; it may also diminish the effectiveness of oral contraceptives.

- Just because an herb is natural does not mean that it is safe. In addition, herbal products may be contaminated with hazardous substances such as pesticides and disease-causing microorganisms.

- Beware of fraudulent dietary supplements that are sold for weight loss, sexual enhancement, or bodybuilding. The FDA has warned in the past that some of these products are "masquerading as dietary supplements—they may look like dietary supplements but they are not legal dietary supplements." Some of these products contain prescription drug ingredients at high levels that are not listed on the label.

- The FDA has also questioned the safety of these herbal ingredients: bitter orange, chaparral, colloidal silver, comfrey, country mallow, germanium, kava, lobelia, and yohimbe.

- Consumers should also be sure to tell their health-care providers about the supplements they take.

Although most Americans can get needed vitamins and minerals through food, situations occur when supplements may be needed:

- Women in their childbearing years and pregnant or lactating women, who may need additional iron and other nutrients as determined in consultation with her physician

- People with known nutrient deficiencies, such as iron-deficient women

- Elderly people who are eating poorly, have problems chewing, or have other concerns

- Drug addicts or alcoholics

- People eating less than 1,200 kcalories a day—such as dieters—who may need supplements because it is hard to get enough nutrients in such low-calorie diets

- People on certain medications or with certain diseases

If you really feel you need additional nutrients, your best bet is to buy a multivitamin and multimineral supplement that supplies 100 percent of the RDA or AI. It can't hurt, and it may act as a safety net for individuals who eat haphazardly. But keep in mind that more is not always better and that no supplement can take the place of food and serve as a permanent method for improving a poor diet. In other words, use these products to supplement a good diet, not to substitute for a poor diet.

PART

2

BALANCED COOKING AND MENUS

BALANCED COOKING METHODS AND TECHNIQUES

CHAPTER 8

LEARNING OBJECTIVES

- Explain the difference between a seasoning and a flavoring ingredient and give examples of each.

- Identify appropriate times for adding seasoning and flavoring ingredients to the cooking process for best flavor.

- Identify common herbs, spices, and blends used in the kitchen and be familiar with each one's aroma, flavor, and effect on food.

- Discuss how to develop a flavor profile for a menu item, including five examples of flavor builders you could use.

- Explain how to use the following techniques to add flavor: reduction, searing, deglazing, sweating, pureeing, rubs, and marinades.

- Describe how to use the following cooking methods in balanced cooking: sauté and dry sauté, stir-fry, roast, broil, grill, steam, poach, and braise.

- Explain the functions of basic baking ingredients and techniques to make healthier baked goods.

INTRODUCTION

As the 78 million baby boomers (those born between 1946 and 1964) continue to age, balanced food choices coupled with their active lifestyles has been a focus of attention. Yet, baby boomers are not alone: A wide range of ages are paying serious attention to what they eat and drink. Although Americans have been accused of asking for balanced foods and then returning to their less healthy alternatives, the tide seems to be turning as consumers embrace healthier, fresher, cleaner, restaurant choices. More and more restaurants are using organic, locally farmed produce and dairy; organic grains and legumes; sustainable fish and seafood; and free-range, antibiotic/hormone free, heritage meats and poultry in creative ways with community responsibility as the driving factor. They have eliminated trans fats, using quality-first pressed oils for flavoring or resourceful fat substitutes in baking such as applesauce, fruit purées, and flaxseed. Many chefs have shifted to artesian, hand-crafted products from small producers such as high-quality vinegars, first-origin chocolates, pickles, preserves, breads, cheeses, sausages, cured meats, and smoked products while maintaining a value-driven experience embracing global and regional flavors creating interesting options.

As foodservice professionals, we have a responsibility to our clients to understand contemporary cooking techniques that are balanced, moderate in rich ingredients, and well prepared to possess nutrition stability. I prefer to call this, "A Way of Life" or "Lifestyle" rather than a diet, because a diet indicates a short-term fix, usually with the stigma of constant hunger and unappetizing foods. Your clients are eating out more today than ever, typically not cooking at home as much as they did years ago; furthermore their children aren't cooking, either. Our very hectic, demanding lifestyles force people to rely on take-out meals or fast casual dining at least several times a week. The typical American purchases a meal, whether breakfast, lunch, dinner, or snack from a foodservice operation about three to five times a week. You have a captive audience of people who depends on the chef, cooks and foodservice employee to prepare nutritious food for them with the limits and balance they require to maintain their current lifestyles.

With over 950,000 restaurants in large cities, small towns, rural areas across America, the foodservice industry presents consumers more than ever before with more menu choices that are part of health and wellness. The vast majority of operators promote balanced choices by adding more fruits, grains, beans and vegetables to substituting a sauce that is lower in sodium, fat, and cholesterol. From appetizer to dessert, quality ingredients, taste, and presentation are important for menu items targeting all ages (Figure 8-1).

FIGURE 8-1 Quality ingredients and taste are important considerations for balanced menu items.

Photo by Peter Pioppo.

FLAVOR

A solid foundation in foods and cooking is necessary to develop balanced menus and recipes. You are expected to know basic culinary terminology and techniques and have a working knowledge of ingredients, from spices, herbs, vegetables, fruits, proteins, fish, and dairy to grains (the edible seeds of a variety of the grass family) and legumes such as beans and lentils.

A basic culinary skill that needs further refinement and attention for balanced cooking is to understand and develop flavor. Because you can't rely on more than moderate amounts of fat, salt, and sugar for taste and flavor, you will need to develop excellent flavor-building skills.

Seasonings and **flavorings** are very important in balanced nutritional cooking, because they help replace ingredients such as fat, sugar and salt. Seasonings are used to enhance and build flavor profiles that are already present in a dish, and flavorings are used to add a new flavor dynamic or modify the original flavor.

Seasonings Substances used in cooking to bring out a flavor that is already present.

Flavorings Substances used in cooking to add a new flavor or modify the original flavor.

HERBS AND SPICES

Herbs and spices are key flavoring ingredients in nutritional menu planning and execution as well as the backbone of most menu items, lending themselves to cultural and regional food styles. Good, sound nutritional cooking can be virtually equal to classical cooking in terms of technique, creative seasoning, flavor blending, and presentation. When you are moderating fat, cholesterol, and sugars it is helpful to enhance recipes with an abundance of spices such as cinnamon, chilies, nutmeg, mace, coriander, ginger, fennel, star anise, juniper, cardamom, and cumin. These spices provide a sense of sweet satisfaction, heat, depth as well as bold character to a variety of recipes.

Herbs are the leafy parts of certain plants that grow in temperate climates. **Spices** are the roots, bark, seeds, flowers, buds, and fruits of certain tropical plants. Figures 8-2 and 8-3 show a number of herbs and spices. Herbs are generally available fresh and dried. Spices are mostly available in a dried form either ground, cracked, whole pods or seeds.

The use of herbs in recipes changes the flavor direction to reflect the herbs prominent essence. For instance, the use of basil, oregano, and thyme in a tomato vinaigrette points the dish toward an Italian, Mediterranean direction. Take the same dish and add cilantro and lime juice and you move south of the border; add fresh chopped tarragon, mustard, and shallots and you're in France. There is no end to your creative abilities once you understand the basics of global flavors, traditional and classic combinations in cooking using balanced ingredients.

Fresh herbs, as opposed to dry, are by far superior and more versatile for creating recipes. Herbs commonly available fresh include flat and curly parsley, cilantro, basil, dill, chives,

Herbs The leafy parts of certain plants that grow in temperate climates; they are used to season and flavor foods.

Spices The roots, bark, seeds, flowers, buds, and fruits of certain tropical plants; they are used to season and flavor foods.

FIGURE 8-2 Fresh herbs.
Reprinted with permission of John Wiley & Sons, Inc.

Basil

Chives

Cilantro

Dill

Fennel

Oregano

Italian parsley

Rosemary

Sage

Tarragon

(a)

FIGURE 8-3 **(a)** Spices. Top row, left to right: black peppercorns, green pepper-corns, pink peppercorns. Bottom row, left to right: white peppercorns, Sichuan pepper-corns. **(b)** Spices. Top row, left to right: cloves, nutmeg, allspice, cinnamon sticks. Bottom row, left to right: juniper berries, cardamom, saffron, star anise. **(c)** Spices. Top row, left to right: celery seed, dill seed, coriander seed, caraway seed. Bottom row, left to right: fennel seed, cumin seed, anise seed.

Source: Wayne Gisslen, *Professional Cooking,* 7th ed. Copyright John Wiley and Sons, Inc. 2011. Photo by J. Gerard Smith.

(b)

(c)

tarragon, mint, thyme, chervil, rosemary and oregano. Fresh herbs are necessary to improve flavors of many preparations, adding a clean, fresh taste. They can only withstand about 30 minutes of cooking, so they work best for finishing dishes, tossing in before serving or in other accompaniments such as fresh salsas, relishes, chutneys, compotes, dressings, simple and complex salads.

When the use of fresh herbs is not possible, dried herbs can be substituted with better than average results. Just remember that you need less dried herbs because they are much more potent. Dried herbs work well in longer cooking applications, such as in stocks, stews, braising, marinating, dry rubs and sauces. You can use dried herbs in the beginning of a cooking process, then finish with fresh herbs toward the end of cooking to get a rich and clean flavor.

The primary purpose of herbs and spices is to build a depth of flavor in your dishes, not to rescue, remedy or mask the existing combination of ingredients. Herbs and spices should be cooked in the dish during preparation so their essence blends in smoothly, giving character and intensity. Herbs also may be added in as a finishing step with the right combination of fresh herbs depending on your desired end result.

Learning to identify the innumerable different herbs and spices requires a keen sense of taste and smell, visual identity is simply not enough. Taste, aroma, texture and experimentation in preparation is the most efficient way you gain experience. Their aromatic quality not only adds intense flavor to the food, but heightens the anticipation of the diner when the item is served possessing a powerful fragrance and savor.

There are many herbs and spices. Let's identify those most likely to be found in a typical kitchen. To help you understand them, we'll sort them into groups, starting with pepper.

Pepper comes in three forms: black, white, and green. White and black pepper both come from the oriental pepper plant. Black peppercorns are picked when slightly under ripe and then air-dried; this results in their dark color. White peppercorns are fully ripe berries that have been soaked in water and hulled, which produces a slightly fermented taste. Green peppercorns are picked before ripeness and preserved.

Black pepper comes in four forms: whole black peppercorns, crushed, butcher's grind, and table-ground pepper. Whole pepper is used as a flavor builder during cooking, as in making stocks. As with all spices, whole spices take longer to release flavors than ground spices. Crushed black pepper can function as a flavor builder during cooking as well, or it can be added as flavoring to a finished dish. Many Americans enjoy the flavor contrast of fresh crushed peppercorns straight from the pepper mill on a crisp green salad. The flavor of ground black pepper is characteristic of certain cuisines and certain parts of the country. Cooks who cater to these clienteles are likely to add this flavor as they season food.

As a seasoning, black pepper is used only in dark-colored foods; it spoils the appearance of light-colored foods. *White pepper* is used in light-colored foods because its presence is concealed. White pepper comes in two forms: whole peppercorns and ground white pepper. White peppercorns are used in the same ways as black. Ground white pepper is good for all-around seasoning. It lends itself to white dishes both in appearance and in flavor, and it has the strength necessary to season dark dishes. It also stands up to high heat better than black pepper does. Ground white pepper is chosen by most good cooks as the true seasoning pepper. It is seldom used as a table pepper.

Green peppercorns are preserved either by packing them in liquid (such as vinegar or brine) or by drying them. They have a fresh taste that's less pungent than that of the other types. They pair well with vegetables and legumes, and found in a variety of sauces, adding a pleasant bold bite, slightly spicy and pungent.

Pink peppercorns are not true peppercorns, but they look like peppercorns and have a sweet, slightly peppery taste. They are native to South America, but they are sometimes mistakenly called Japanese peppers because they are one of the few spices used in Japanese cooking. They are aromatic but not peppery and often are included in mixes mostly for color. Possible adverse reactions to pink peppercorns have been reported when they are added generously to dishes, and so they should be used in small amounts.

Red pepper, also called *cayenne*, is completely unrelated to white or black pepper. It comes from dried red pepper pods. It possesses a spicy hot profile and can be easily overused. When added with moderation to a variety of preparations, it can lend a lasting spiced undertone that adds character and boldness to dishes.

Let's examine a variety of herbs and spices that are used often for their distinctive flavor, availability and enrichment of taste. As a flavor builder, each goes beyond the subtlety of the stocks, herbs, and spices—even though you can't single out the essence in the dish as a whole. Used in quantities large enough to taste, they become major flavors rather than builders. Basil, oregano, and tarragon are available fresh and also come in the form of crushed dried leaves. They look somewhat alike, but their tastes are very different.

Basil has a warm, sweet flavor that is welcome in many soups, sauces, entrées, relishes, salsas, and dressings as well as with vegetables such as tomatoes, peppers, eggplant, and squash. It blends especially well with citrus, papaya, mango, and pineapple. It can also be used as a fresh addition to greens in a salad. Familiar varieties include Opal, Thai, Lemon and Green, each with a unique character and flavor. Like many other herbs, it has symbolism: In India basil expresses reverence for the dead; in Italy it is a symbol of love.

Oregano belongs to the same herb family as basil, but it makes a very different contribution to a dish—a strong bittersweet taste and perfume aroma you may have experienced in many Italian, Mediterranean, Spanish, South American, and Mexican dishes. Common varieties are Cuban and Puerto Rican oregano.

Tarragon has a pleasant perfume like flavor that is somehow mild and strong at the same time. It possesses licorice qualities, grown as French tarragon, the traditional robust high quality flavor and Russian Tarragon that has hardy leaves but lacks the usual flavor profile. Tarragon is most often used in poultry and fish dishes as well as in salads, sauces, and cold dressings.

Rosemary, like bay leaf, is often used in dishes in which a liquid is involved: soups, stocks, sauces, stews, and braised foods. The leaf of an evergreen shrub of the mint family, it has a pungent, hardy flavor and fragrance. Some of the fresh include soft perfume needles such as Tuscan Blue, Salem, Spice Island, Arp and Common variety. All types may be used with poultry, beef, lamb, pork, tomato preparations, sachets, and compound butters. Dried rosemary is used in cooked preparations and should be strained because it is not as palatable due to its dry hard needles. It is used mostly with meats, game, poultry, mushrooms, and ragouts.

Dill has a classic flavor that is very familiar in dill-flavored pickles. Fresh or dried dill leaves, often called dill weed, are additionally used in soups, fish dishes, stews, complex salads, and butters. Fresh dill lends a delicate flavor, with flowery undertones to the many preparations discussed.

Mustard seed has a familiar flavor mostly associated with the hot dog condiment, prepared mustard. Dry mustard, a powdered spice made from the seed of the mustard plant, comes in three varieties: white, yellow, and brown. The brown variety has the sharpest most pungent flavor, then yellow and white the mildest. All are used to flavor sauces, dips, dressings, and entrées. Prepared mustards are made from all the varieties and serve as excellent condiments, enhancing flavor, texture, and taste.

Paprika is another powdered spice that comes in two flavors: mild and hot. Both kinds are made from dried pods of the same pepper family as red pepper and cayenne and look something like the seasoning peppers, but they do not lend the same dominance as their relatives. Hungarian paprika is the hot spicy one; Spanish paprika is mild in flavor, and its vibrant red color has incredible eye appeal. Hungarian paprika is used to make goulash and other braised meats and poultry. Spanish paprika is used for coloring, blending with rubs, and mild seasonings. Paprika is sensitive to heat and will turn brown if exposed to excessive direct heat.

Another branch of the pepper pod family gives us *chili peppers*, a relative of sweet peppers, but containing a compound called capsaicin, which gives them their heat profile. Colors range from red, burgundy to green, and flavors from mild to hot. Chili peppers are used in mostly Mexican, Asian, Thai, Peruvian, Indian, Cuban, and South American cuisines. Many chili powder blends are available, from smoked, dark, mild, and hot blends to single varietal made from toasted ground or smoked dried chili peppers, such as Arbol, Pasilla, Guajillo, Ancho, Casabel, Chipotle, Morita, Aji Panca, and Fresno.

Curry powder is a blend of spices, sometime containing as many as 15 different kinds. In India and South Asia, where it originated, cooks blend their own curry powders, which may vary considerably. In the United States, curry powder comes premixed in various blends from mild to hot. Curry powders usually include cloves, black and red peppers, cumin, garlic, ginger, clove, cinnamon, coriander, fennel, cardamom, fenugreek, mustard seed, mace, nutmeg, turmeric (which provides the characteristic yellow color), and sometimes other spices, depending on the history or region.

Sweet aromatic spices are from the tropics and frequently used in baking, dessert cookery, sauces, stews, marinades, rubs, vegetables, and entrées. Among these spices are cinnamon, nutmeg, mace, ginger, and allspice. *Cinnamon* comes from the dried bark of the cinnamon or cassia tree, *nutmeg* and *mace* from the seed of the nutmeg tree, and *ginger* (fresh or dried) from the root of the ginger plant. Ground cinnamon and ginger are used in a variety of cuisines, both sweet and savory. Cinnamon is also available in sticks. Nutmeg flavor goes well with desserts, potatoes, dumplings, spinach, quiche, and some soups and entrées. Mace, a somewhat paler alternative to nutmeg, has a similar flavor and is used in bratwurst, savory dishes, baked goods, and pâtés. *Allspice,* the dried fruit berry of a Central American tree, has a pungent aromatic bouquet (with flavor notes of cinnamon, clove and nutmeg) and is used in jerk spice, mole sauces, and pickling. The finest allspice comes from Jamaica.

Mint is a sweet herb with the familiar flavor you have mostly experienced in toothpaste and chewing gum. As a plant, mint is available in a number of varieties, such as apple, banana, black, English and spearmint. Others include chocolate, licorice, orange, and pineapple mint. The flavor of a mint sauce is a traditional and refreshing complement to lamb. Fresh mint makes a good flavoring and garnish for fruits, vegetables, salsas, relishes, salads, dressings, mixed greens, iced tea, desserts, and sorbets.

TOASTING SPICES AND USING SPICE BLENDS

To get maximum flavor from whole spices, you can give them a quick dry toast that releases their natural oils. The following are some whole spices that can be toasted:

- Mustard seed
- Peppercorns
- Fennel
- Coriander
- Cloves
- Star anise
- Cardamom
- Caraway
- Cumin
- Juniper
- Allspice

To toast, place a sauté pan on medium to high heat for about 1 minute, add the spices in the dry pan, and toss quickly being careful not to burn, until a nutty aroma is extracted from the spices. Once cool, finely grind in a coffee grinder or professional spice grinder and store in an airtight container. Use as a seasoning agent in applications such as marinades, rubs, salad dressings, soups, stews, ragouts, salsas, and relishes.

Many herbs and spices can be combined to produce blends with distinctive flavors. For example, a cattleman's blend with paprika, peppers, chilies, and other dried herbs can be used as a steak seasoning. A seed blend, such as ground cardamom, fennel, anise, star anise, cumin, and coriander, is excellent in soups, marinades, rubs, compotes, vegetables, and chutneys. Always check the salt and sugar content of premade blends. Ethnic blends with additional ingredients such as the following also present many flavoring possibilities, reducing, moderating and/or eliminating the need for salts, cream, butter, oils, and sugars:

- Italian: garlic, onion, basil, oregano, crushed red pepper, thyme
- Asian: ginger, five-spice powder, garlic, scallion, thai basil, cilantro
- French: tarragon, mustard, chive, chervil, shallot
- South American: chili peppers, lime juice, garlic, cilantro, onion
- Indian: ground nutmeg, fennel, coriander, cinnamon, fenugreek, curry
- Mediterranean: oregano, marjoram, thyme, pepper, coriander, onion, garlic

Flavor combinations such as these add depth to foundations for starting or finishing a dish.

Table 8-1 is a reference chart for herbs and Table 8-2 is a reference chart for spices.

TABLE 8-1

PRODUCT	MARKET FORMS	DESCRIPTION	USES
Basil	Fresh, dried: crushed leaves	Bright green tender leaves of an herb in the mint family. Sweet, slightly peppery flavor.	Tomatoes, eggplant, squash, carrots, peas, soups, stews, poultry, red sauces.
Bay leaves	Whole, ground	Long, dark green brittle leaves from the bay tree, a small tree from Asia. Pungent, warm flavor when leaves are broken. Always remove bay leaves before serving (due to toughness).	Stocks, sauces, soups, braised dishes, stews, marinades.
Chervil	Fresh, dried: crushed leaves	Fernlike leaves of a plant in the parsley family. Like parsley with slight pepper taste; smells like anise.	Stocks, soups, sauces, salads, egg and cheese dishes, French cooking.
Chives	Fresh, dried	Thin grass like leaves of a plant in the onion family. Mild onion flavor.	Poultry, seafood, potatoes, salads, soups, egg and cheese dishes.
Dill weed	Fresh, dried: crushed	Delicate, green leaves of dill plant. Dill pickle flavor.	Salads, dressings, sauces, vegetables, fish, dips.
Epazote	Fresh	Coarse leaves of a wild plant that grows in the Americas. Strong, exotic flavor.	Mexican and Southwestern cooking.
Lemongrass	Fresh stalks, dried: chopped	Tropical and subtropical scented grass. White leaf stalks and lower part are used. Bright lemon flavor.	Soups, marinades, stir-fries, curries, salads, Southeast Asian cooking.
Marjoram	Fresh, dried: crushed leaves	Leaves from a plant in the mint family. Mild flavor similar to oregano with a hint of mint.	Lamb, poultry, stuffing, sauces, vegetables, soups, stews.
Mint	Fresh, dried: crushed leaves	A family of plants that include many species and flavors, such as spearmint, peppermint, and chocolate. Cool, minty flavor.	Lamb, fruits, some vegetables, tea.
Oregano	Fresh leaves, dried: crushed	Dark green leaves of oregano plant. Pungent, pepper-like flavor.	Italian foods such as tomato sauce, meats, sauces, Mexican cooking.
Parsley	Fresh, dried flakes	Green leaves and stalks of several varieties of parsley plant. Mild, sweet flavor.	Bouquet garni, fines herbs, almost any food.
Rosemary	Fresh, dried leaves	Stiff green leaves that look like pine needles of a shrub. Strong flavor like pine.	Roasted and grilled meats, such as lam, sauces such as tomato, soup, stuffing.
Saffron	Whole (threads), ground	Dried flower stigmas of a member of the crocus family. Used in very small amounts, has a bitter yet sweet taste, and colors foods yellow. Most expensive spice. Mix threads with hot liquid before using.	Paella, risotto, Indian cooking, seafood, poultry, baked products.
Sage	Whole (fresh or dried), rubbed (chopped), ground	Gray-green leaves and blue flowers of a member of the mint family. Strong, musty flavor.	Pork, sausage, stuffing, salads, beans.
Savory	Crushed leaves	Small, narrow leaves of plant in mint family. Summer savory is preferable to winter savory. Bitter flavor.	Meat, poultry, sausage, fish, vegetables, beans.
Tarragon	Fresh, dried: crushed	Small plant with long narrow leaves and gray flowers. Delicate sweet flavor with hint of licorice.	Béarnaise sauce, vinegars, dressings, poultry, fish, salads.
Thyme	Fresh, dried: crushed or ground	Tiny leaves and purple flowers of a short, bushy plant. Spicy, slightly pungent flavor.	Bouquet garni, meat, poultry, fish, soups, sauces, tomatoes.

TABLE 8-2 | Spice Reference Chart

PRODUCT	MARKET FORMS	DESCRIPTION	USES
Allspice	Whole, ground	Dried, dark brown berries of an evergreen tree indigenous to the West Indies and Central and South American. Smells of cloves, nutmeg, and cinnamon.	Braised meats, curries, baked goods, puddings, cooked fruit.
Anise seed	Whole, ground	Tiny dried seed from a plant native to eastern Mediterranean. Licorice (sweet) flavor.	Baked goods, fish, shellfish, soups, sauces.
Caraway seed	Whole, ground	Crescent-shaped brown seed of a European plant. Slightly peppery flavor.	German and East European cooking (such as sauerkraut and coleslaw), rye bread, pork.
Cardamom	Whole pod, ground seeds	Tiny seeds inside green or white pods that grow on a bush of the ginger family. Sweet and spicy flavor. Very expensive.	Poultry, curries, and other Indian cooking, Scandinavian breads and pastries, puddings, fruits.
Cayenne, red pepper	Ground	Finely ground powder from several hot types of dried red chili peppers. Very hot. Use in small amounts.	In small amounts in meats, poultry, seafood, sauces, and egg and cheese dishes.
Celery seed	Whole, ground, ground mixed with salt or pepper	Small gray-brown seeds produced by celery plant. Distinctive celery flavor.	Dressings, sauces, soups, salads, tomatoes, fish.
Cinnamon	Stick, ground	Aromatic bark of the cinnamon tree, a small evergreen tree of the laurel family. Sweet, warm flavor.	Baked goods, desserts, fruits, lamb, ham, rice, carrots, sweet potatoes, beverages.
Cloves	Whole, ground	Dried flower buds of a tropical evergreen tree. Sweetly pungent and very aromatic flavor.	Stocks, marinades, sauces, braised meats, ham, baked goods, fruits.
Coriander seeds	Whole, ground	Small seeds from the cilantro plant. Mild, slightly sweet and musty flavor.	Pork, pickling, soups, sauces, chutney, casseroles, Indian cooking.
Cumin seed	Whole, ground	Seed of a small plant in the parsley family. Looks like caraway seed but lighter in color. Pungent, strong, earthy flavor.	Used to make curry and chili powders. Cooking of India, Middle East, North Africa, and Mexico; sausage, Muenster cheese, sauerkraut.
Curry powder (spice blend)	Ground blend	A blend of up to 20 spices. Often includes black pepper, cloves, coriander, cumin, ginger, mace, and turmeric. Depending on brand, flavor and hotness can vary tremendously.	Indian cooking, eggs, beans, soups, rice.
Dill seed	Whole, ground	Small, brown seed of dill plant. Bitter flavor—much stronger than dill weed.	Pickling, sour dishes, sauerkraut, fish.
Fennel seed	Whole, ground	Oval, green-brown seeds of a plant in the parsley family. Mild licorice flavor.	Fish, pork, Italian sausage, tomato sauce, pickles, pastries.
Fenugreek seed	Whole, ground	Small beige seeds of a plant in the pea family. Bittersweet flavor. Smells like curry.	Indian cooking such as curries and chutneys.
Ginger	Fresh whole, dried whole, dry ground (also candied or crystallized)	Tan root of the tropical ginger plant. Hot but sweet flavor.	Asian dishes such as curries, baked goods, fruits, beverages.

TABLE 8-2   **Spice Reference Chart** (*Continued*)

PRODUCT	MARKET FORMS	DESCRIPTION	USES
Juniper berries	Whole	Purple berries of an evergreen bush. Pine-like flavor.	Venison and other game dishes, pork, lamb, marinades.
Mace	Whole blades or ground	Lacy orange covering on nutmeg. Like nutmeg in flavor but less sweet.	Baked products, fruits, pork, poultry, some vegetables.
Mustard seed	Whole, ground (prepared mustard)	Tiny seeds of various mustard plants; seed may be white or yellow, brown or black. The darker the seed, the sharper and more pungent the flavor.	Meats, sauces, dressings, pickling spices, prepared mustard.
Nutmeg	Whole, ground	Large brown seed of the fruit from the nutmeg tree. Sweet, warm flavor.	Baked products, puddings, drinks, soups, sauces, many vegetables.
Paprika	Ground	Fine powder from mild varieties of red peppers. Two varieties: Spanish and Hungarian. Hungarian is darker in color and much stronger in flavor.	Spanish used mostly as garnish; Hungarian used in braised meats, sauces, gravies, some vegetables.
Peppercorns, black and white	Whole, crushed, ground	Dried, black or white hard berries from same tropical vine that are picked and handled in different ways. Black: pungent earthy flavor. White: similar but milder than black.	Almost any food.
Pink peppercorns	Whole	Dried or pickled red berries of an evergreen. Not related to black pepper. Bitter flavor, not as spicy as black pepper.	Use in small quantities in meat, poultry, and fish dishes and in whole pepper mixtures.
Poppy seed	Whole	Tiny cream-colored or deep blue seeds from the poppy plant. Nutty flavor.	On breads and rolls, in salads and noodles, ground poppy seed in pastries.
Sesame seeds	Whole	Creamy oval seeds of tall, tropical sesame plant. Nutty flavor.	On breads and rolls, ground seeds used to make tahini.
Star anise	Whole, ground	Dried, star-shaped fruit of an evergreen native to China. Dark red. Licorice-like flavor.	Chinese cooking.
Turmeric	Ground	Orange-yellow root of member of ginger family. Musky, peppery flavor. Colors foods yellow.	Curry powder, curries, chutney.

PUTTING IT ALL TOGETHER: FLAVOR PROFILES

Cooking is cooking is cooking! You must master cooking fundamentals before you can produce consistent flavorful and well-executed products. The many choices for building flavor in balanced dishes may seem extremely overwhelming. To become comfortable and proficient with using these ingredients, they need to be incorporated into your everyday style of cooking. You don't need to separate your cooking process from healthy balanced to unhealthy fattening dishes. You should basically be using the same methods of preparation as you do for any style of cooking. The exceptions are deep-frying, pan-frying, sautéing with oil, and confit (when preserved in salt and cooked in fat). When you are preparing dishes that limit the amounts of fat, it is smarter to use those fats at the end of your preparation rather than during the cooking process.

The best way to go about the dish you are preparing is to start with a basic series of steps. As your skills develop, these steps will become more natural and automatic, just like the basic cooking fundamentals you learn in the beginning of your career.

To plan your menu item, go through the following steps:

1. Identify the meal period.

2. Identify what flavor direction you want to go in, such as a sweet/sour profile for a dessert or a spiced Asian direction for an entrée.

3. Choose the main feature item with sides or garnishes.

4. Consider how the dish will be presented—we first eat with our eyes.

5. What combination of flavor ingredients? Here are some examples:

   - Hints of ginger, garlic with soy

   - Cilantro with lime, onion, and cumin

   - Garlic with lemon and basil

Now it's time to plan your dish strategically and identify how you are going to achieve the maximum flavor from the list of ingredients you have chosen. In the previous information this was referred to as flavor building. When a dish is created with limited amounts of butter, fat, salt, and sugar, other ingredients are needed to excite the palate, accomplishing an experience of taste and satisfaction. Use the tools provided in this chapter and experiment to get your own style and result, keeping in mind the guidelines of sound cooking with proper techniques. Some key points in this area include the following:

- Toasting and grinding spices

- Reduction of liquids to maximize flavor

- Marinating before cooking

- Fresh herbs

- Good fortified stocks

- Oils and vinegars infused with herbs and spices (Figure 8-4)

- Limited portion of protein/ more vegetables, greens and fruits

  - Picking the best cooking method is another key way to maximize the potential of each dish.

Using a **marinade** with your meats before cooking them is critical to impart flavor to the meat. Caramelizing (browning) with a slight spray of fat in a pan helps to seal in the flavor of the meat and add additional character to your dish. The addition of a well-prepared stock to braise an item is perfectly acceptable so long as the meat is well-trimmed and the stock has been defatted. In the case of braising a protein, the addition of a caramelized **mirepoix** with garlic and tomato product, thyme, rosemary, and bay leaf will add yet another level of flavor to your dish. When caramelizing your mirepoix, add some spices to give your dish the character you are trying to accomplish or flavor direction. For spices, think of using flavors such as curry blends, cumin, chili powders, fennel, coriander, fenugreek, while additionally adding in flavors such as garlic, shallots, soy sauce, mustards, vinegars, and lemon or lime juice, which are great additions when braising meats or preparing a sauce.

The creativity is up to you when you are designing your own dish or following a classic style recipe. To modify these recipes you need to limit the amount of oil, butter, and salt, and substitute with additional spices, herbs, acid, and vegetables to accomplish the same flavor profile. When preparing these dishes, the addition of fat is better at the end before serving so the first taste has the profile of the fat richness. A simple example would be cooking carrots. If you poach them in a flavorful vegetable or chicken stock, they will absorb the flavor. When you are ready to serve, heat them

Marinade A seasoned liquid used before cooking to flavor and moisten foods; usually with the foundation of an acidic ingredient.

Mirepoix A mixture of rough-cut or diced vegetables, herbs, and spices, used for flavoring stocks, sauces, soups, and other food preparations.

FIGURE 8-4 Infused vinegars and oils can lend flavor to a wide variety of dishes.

Photo by Peter Pioppo.

in that same stock, then sauté in a reduced stock with the small addition of butter to finish, folding in fresh herbs like dill or chives and few twists of fresh black pepper.

When cooking beans, legumes, and grains, use either chicken stock or vegetable stock to impart flavor, as well as shallots and fresh herbs such bay leaf, oregano, rosemary, or thyme. This will help the starches absorb flavor during preparation.

Your preparation will dictate how successfully your dish comes together. Planning is the key to success. You shouldn't start a project without a detailed plan of execution. This, together with your "mise en place," will allow you to create a dish or menu with the finesse of an experienced culinarian. You can never be overprepared. You want to have your ingredients at your fingertips; your products prepped and laid out so you can time them in the proper order. If you are preparing something that requires reduced orange juice, make sure that you have already gone through that process and have that set and ready to go. If you need to toast spices, chop herbs, or cut vegetables, all of these should be prepped just as you would for any other dish. Flavor building needs practice, as any cooking method, so read through this chapter time and time again to feel more comfortable with the many methods, styles, and ingredients you can use as fat, salt, and sugar substitutes.

The creations you discover can very easily become future menu items. You should include balanced choices on your menus. Good nutrition is all about balance, portion control, and quality ingredients. Preparing foods that reduce fat, salt, and sugar should never equate to reduced flavors (see Table 8-3 for powerhouses of flavor). This is the reputation that you as culinary professionals need to eliminate among your clients. We need to be educated and prepare our foods to accommodate special needs, diets, and allergies during peak service times without creating chaos in our operations. The tools provided in this chapter clearly describe how you can accomplish this mission in your kitchen. Enjoy, and cook with your heart.

TABLE 8-3	Powerhouses of Flavor

- Fresh herbs
- Toasted spices
- Herbs and spice blends
- Freshly ground pepper
- Citrus juices, citrus juice reductions
- Strong-flavored vinegars and vinaigrettes
- Wines
- Strong-flavored oils such as walnut oil and extra-virgin olive oil
- Infused vinegars and oils
- Reduced stock (glazes)
- Rubs and marinades
- Raw, roasted, or sautéed garlic
- Caramelized onions and shallots
- Roasted bell peppers
- Chili peppers
- Grilled or oven-roasted vegetables
- Coulis, salsas, relishes, chutneys, mojos
- Dried foods: tomatoes, cherries, cranberries, raisins
- Fruit and vegetable purees
- Condiments such as sambal, Worcestershire sauce, hot chili sauce, horseradish and Dijon mustard
- Natural, pure extracts

1. Seasonings are used to enhance and build flavor profiles that are already present in a dish, and flavorings are used to add a new flavor dynamic or modify the original flavor.
2. Herbs are the leafy parts of certain plants that grow in temperate climates. Spices are the roots, bark, seeds, flowers, buds, and fruits of certain tropical plants. Herbs are generally available fresh and dried. Spices are mostly available dried.
3. Fresh herbs can only withstand about 30 minutes of cooking, so they work best for finishing dishes, tossing in before serving, or in other accompaniments such as fresh salsas. Dried herbs work well in longer cooking applications.
4. Whole spices take longer to release flavors than ground spices.
5. Figures 8-2 and 8-3, and Tables 8-1, and 8-2, show common herbs and spices used in the kitchen as well as information on each one's aroma, flavor, and effect on food.
6. To toast a spice, place a sauté pan on medium to high heat for about 1 minute, add the spices in the dry pan, and toss quickly, being careful not to burn, until a nutty aroma is extracted from the spices.
7. Many herbs and spices can be combined to produce blends with global seasonings adding distinctive flavors.
8. To develop a flavor profile, you need to consider the meal period, the flavor direction you want to go in, the accompanying dishes, presentation, and your specific combination of ingredients, such as cilantro with lime, onion, and cumin. Consider the powerhouses of flavor listed in Table 8-3 when developing flavor profiles.

BALANCED PREPARATION

Reduction Boiling or simmering a liquid down to a smaller volume.

Searing Exposing the surfaces of a piece of protein to high heat in a hot pan with little or no oil, or in a hot oven, to give the meat color and a distinctive flavor.

Deglazing Adding liquid to the hot pan used in making sauces and meat dishes; any browned bits of food sticking to the pan are scraped up and added to the liquid, adding flavor and color.

Sweating Cooking slowly in a small amount of fat over low or moderate heat without browning. Vegetables and other foods can be sweated in stock, juice, or wine as well.

Puréeing Mashing or straining a food to a smooth pulp.

Rub A dry marinade made of herbs and spices (and other seasonings), sometimes moistened with a little oil, vinegar, mustard, or other flavoring liquid, is rubbed or patted on the surface of meat, poultry, or fish (which is then refrigerated and cooked at a later time).

Before looking at cooking methods, let's review some techniques that are used often to develop flavor in nutritional recipes:

- **Reduction** means boiling or simmering a liquid down to a smaller volume. In reducing, the simmering or boiling action causes some of the liquid to evaporate. The purpose may be to thicken the product, to concentrate the flavor, or both. A soup or sauce is often simmered for one or both reasons. The use of reduction eliminates thickeners and intensifies and increases the flavor so that you can serve a smaller portion.

- **Searing** means exposing the surfaces of a piece of protein to high heat in a hot pan with little or no oil or in a hot oven. Searing is done to give color and to produce a distinctive flavor. Dry searing can be done over high heat in a nonstick pan, using vegetable oil spray, or brushing with olive or canola oil.

- **Deglazing** means adding liquid to a hot pan with a fond—browned bits and caramelized drippings of meat and vegetables that are stuck to the bottom of a pan after sautéing or roasting (Figure 8-5). Any browned bits of food sticking to the pan are scraped up, adding flavor and color to the sauce.

- **Sweating** means cooking slowly in a small amount of fat over low or moderate heat without browning. You can sweat vegetables and other foods without fat. Instead, sweat in stock, juice, or wine.

- **Puréeing** of vegetables or starchy foods is commonly done as a way to thicken soups, stews, sauces, and other foods without using any fat. Although the food processor is useful for this process because you can slowly pulse the mixture, emulsion blenders and high-speed table blenders can make a very smooth purée with a silky texture (Figure 8-6).

Rubs combine dry ground spices such as coriander, paprika, and chili powder and finely cut herbs such as thyme, cilantro, and rosemary. Rubs may be dry or wet. Wet rubs, also called pastes, are mixed with liquid ingredients such as mustard and vinegar. Pastes produce

FIGURE 8-5 Deglazing a pan.
Reprinted with permission of John Wiley & Sons, Inc.

a crust on the food. Wet or dry rubs work particularly well with beef, chicken, and pork and can range from a smoked paprika barbecue seasoning to a Jamaican jerk rub. To make a rub, mix various seasonings together and spread or pat evenly on the meat, poultry, pork, lamb or fish—just before cooking for delicate items and at least 24 hours before for large cuts of meat. The larger the piece of meat or poultry, the longer the rub can stay on. The rub flavors the exterior of the meat as it cooks.

Marinades, seasoned liquids in which foods are soaked before cooking, are useful for adding flavor as well as for tenderizing meat and poultry. Marinades bring out the biggest flavors naturally so that you don't need to drown the food in fat, cream, or sauces. Marinades allow a food to stand on its own with a light dressing, chutney, sauce, or relish. Fish can also be marinated. Although fish is already tender, a short marinating time (about 30 minutes) can develop a unique flavor. A marinade usually contains an acidic ingredient such as wine, vinegar, citrus or tomato juice or plain nonfat yogurt to break down the tough meat or poultry. The other ingredients add flavor. Without the acidic ingredient, you can marinate fish for a few hours to instill flavor. A simple fish marinade consists of fish stock, lemon rind (optional), white wine, tarragon, thyme, dill, black pepper, shallots, Dijon mustard, and a few drops of oil. Oil is often used in marinades to carry flavor, but it isn't essential. There are many products available that have balanced ingredients and serve as a good base for marinating.

To give marinated foods flavor, try citrus zest, diced vegetables, fresh herbs, shallots, garlic, low-sodium soy sauce, mustard, and toasted spices. For example, citrus and pineapple marinades can be flavored with Asian seasonings such as ginger, garlic, star anise, and lemongrass. Tomato juice with allspice, Worcestershire sauce, garlic, cracked black pepper, mustard, fresh herbs, and coriander are great for flank steak. Besides tomato, adding chopped kiwi or pineapple to a marinade helps soften tough cuts because these ingredients contain enzymes that break down tough muscle.

FIGURE 8-6 Puréeing of vegetables to thicken soups, stews, and sauces can be done easily using an emulsion blender.

Reprinted with permission of John Wiley & Sons, Inc.

SUMMARY

1. Balanced preparation techniques include reduction, searing, deglazing, sweating, and pureeing of vegetables or starchy foods to thicken soups, etc.
2. Rubs combine dry ground spices and finely cut herbs. They may be dry or wet. Wet rubs (also called pastes) are mixed with liquids such as mustard and vinegar. Rubs work particularly well to add flavor to beef, chicken, and pork. Pat evenly on meat, poultry, or fish—just before cooking for delicate items and at least 24 hours before for large cuts of meat.
3. Marinades are useful for adding flavor as well as for tenderizing meat and poultry. A marinade usually contains an acidic ingredient such as vinegar to break down any toughness. Oil is often used in marinades to carry flavor but it isn't essential. To give marinated foods flavor, try citrus zest, diced vegetables, fresh herbs, shallots, garlic, low-sodium soy sauce, mustard, and toasted spices.

BALANCED COOKING METHODS

The basic cooking methods for balanced cooking are discussed here. They include both dry-heat and moist-heat methods.

DRY-HEAT COOKING METHODS

Dry-heat cooking methods are acceptable for balanced cooking when heat is transferred with little or no fat. Excesses fat is allowed to drip away from food being cooked. Both pan frying and deep frying add varying amounts of fat and kcalories, so they are not an acceptable cooking method.

Sauté and Dry Sauté

Sautéing Cooking foods in single portions or small pieces very quickly on high heat with a small amount of fat.

Sautéing can be made acceptable by using nonstick pans, healthy oils, and at times finishing with a small amount of butter. Sautéing, cooking food quickly in a small amount of fat over high heat, is used to cook tender foods that are in either single portions or small pieces. Sautéing can also be used as a step in a recipe to add flavor to foods such as vegetables by either cooking or reheating. Sautéing adds flavor in large part from the caramelization (browning) that occurs during cooking at relatively high temperatures.

When sautéing, use a shallow pan to let moisture escape and allow space between the food items in the pan. Use a well-seasoned or nonstick pan and add about half a teaspoon or two sprays of oil per serving after preheating the pan. Vegetable-oil cooking sprays come in a variety of flavors, such as butter, olive, Asian, Italian, and mesquite. A quick two-second spray adds about 1 gram of fat (9 kcalories) to the product. To use these, spray the preheated pan away from any open flame (the spray is flammable) and then add the food. Pump spray bottles are also available that allow you to fill them with oils of your choice.

Dry Sautéing Sautéing with a minimal amount of vegetable oil cooking spray or a small amount of liquid such as wine, flavored vinegar, or defatted stock.

For an even lower-in-fat cooking method, use the **dry sauté** technique. To use this technique, heat a nonstick pan, spray with vegetable-oil cooking spray, then wipe out the excess with a paper towel. Heat the pan again, then add the food.

If browning is not important, you can simmer the ingredient in a small amount of liquid such as wine, vermouth, flavored vinegar, juices or defatted stock to bring out the flavor. Vegetables naturally high in water content, such as tomatoes and mushrooms, can be cooked with little or no added fluid at a very high heat.

When classic sautéing or dry sautéing, add shallots, garlic, or other seasonings, then deglaze the pan with stock, wine, or another liquid and reduce to a sauce consistency or add a previously reduced sauce to accompany your dish.

- To add flavor, use marinades, herbs, and spices.
- Use high heat or moderately high heat and a well-seasoned or nonstick pan used only for dry sautéing.
- Pound pieces of meat or poultry flat to increase the surface area for cooking and thereby reduce the cooking time.
- To sauté vegetables, simmer them with a liquid such as stock, juice, or wine in a nonstick pan and add seasoning blends such as cardamom, coriander, and fennel or granulated garlic, onion powder and fresh herbs. Cook over moderately high heat. As the vegetables start to brown and stick to the pan, add more liquid and deglaze the pan. Once the vegetables are ready, add a little bit of butter or flavorful oil, perhaps 1 teaspoon for four servings, to give the product a rich satisfying flavor and a shiny texture.
- Another sauté method is to blanch vegetables in boiling water to the desired doneness and then shock them in ice water or lay out on towels and place immediately in the refrigerator. Dry sauté in a hot nonstick pan with stock, wine, fresh herbs, garlic, and shallots (chopped) and finish with fresh black pepper and extra virgin olive oil, butter, or a first pressed nut oil (1 teaspoon for four servings).
- Once prepared, serve sautéed foods immediately. They do not hold well.

CHEF'S TIPS FOR CLASSIC SAUTÉING AND DRY SAUTÉING

Stir-Fry

Stir-frying, cooking bite size foods over high heat in a small amount of oil, preserves the crisp texture and bright color of vegetables and cooks strips of poultry, meat, or fish quickly (Figure 8-7). Typically, stir-frying is done in a wok, which today come in nonstick versions. Steam-jacketed kettles and tilt frying pans can also be used to make large-quantity stir-fry menu items. Cut up the ingredients as appropriate into small pieces, thin strips, or diced portions. Many time vegetables need to be blanched to ensure even cooking. Proteins are cooked separately, then combined.

Stir-frying An Asian cooking method involving a quick sauté over high heat, occasionally followed by a brief steam in a flavored sauce.

FIGURE 8-7 Stir-frying preserves the crisp texture and bright color of vegetables and cooks strips of poultry, meat, or fish quickly.

Courtesy of 3445128471/Shutterstock.

CHEF'S TIPS FOR STIR-FRYING

- Coat the cooking surface with a thin layer of oil. Peanut oil works well because it has a strong flavor (so that you can use a small amount) and a high smoking point. You can also use vegetable oil cooking spray and wipe away any excess.

- Have all your ingredients ready and next to you because this process is fast.

- Partially blanch the thick vegetables first (such as carrots or broccoli) so that they will cook completely without excessive browning.

- Preheat the equipment to a high temperature.

- Foods that require the longest cooking times—usually meat and poultry—should be the first ingredients you start to cook.

- Stir the food rapidly during cooking and don't overfill the pan.

- Use garlic, scallion, ginger, rice wine vinegar, low-sodium soy sauce, and chicken/vegetable stock for flavor.

- Add a little sesame oil at the end for taste (about 1 teaspoon per four servings).

- Thin sauces are usually thickened with a cornstarch slurry, creating a silky smooth consistency.

Roast

Roasting Cooking with heated dry air such as in an oven.

Roasting, cooking with heated dry air, is an excellent method for cooking larger, tender cuts of meat, poultry, and fish that will provide multiple servings. When roasting, always place meats and poultry on a rack so that the drippings fall to the bottom of the pan and the meat therefore doesn't cook in its own juices. Also, cooking on a rack allows for air circulation and more even cooking. Roasting adds rich flavors to meat, poultry, and seafood.

For an accompaniment to meat or poultry, you may want to simply use the natural fond in the bottom of the pan, with the addition of stock to make a jus. To give the jus additional flavor, add a mirepoix to the roasting pan during the last 30 to 40 minutes. To remove most of the fat from the jus, you can use a fat-separator or skim off the fat with a ladle. If time permits, you can refrigerate the jus and the fat will congeal at the top.

Jus lié Juices from a roast thickened slightly with a starch slurry.

If you prefer to thicken the natural jus to make **jus lié**, first remove the fat from the jus. Because you will be using the roasting pan to make this product, also pat out with paper towels the fat from the bottom of the pan. Add some vegetables and cook at a moderately high heat so that they brown or caramelize. Add stock to avoid burning and a little wine to deglaze the pan. Stir the ingredients to release the food from the pan because you want that flavor. Continue to add stock to cover, then reduce the jus until the color is appropriate. If there is not enough time to reduce the jus to the proper consistency, you can thicken it with a cornstarch slurry (starch and cold water mixed to a syrupy consistency). Another way to thicken the natural jus is to add starchy vegetables such as carrots, parsnips, or potatoes to the jus during the last 45 minutes of cooking. These vegetables can then be pureed to naturally bind the jus.

Roasting is a great method for cooking halibut, swordfish, sea bass, monkfish, and other meaty fish. Use vegetable spray on the bottom of a roasting pan, then place some onions and mushrooms in the pan. Season the fish with a rub or marinade combinations discussed earlier and place in the pan. Top with tomatoes and a few olives or another combination of vegetables. Then finish with a little olive oil, white wine, citrus zest, and stock, and you have a great buffet or a la carte item.

To develop flavor, rubs and marinades for meat, poultry, and fish are two possibilities that have already have been discussed. Smoking can also be used to complement the taste of meat, poultry, or fish. Hardwoods or fruitwoods such as the following are best for producing quality results:

- Fruit (apple, cherry, peach) woods work well with light entrées such as poultry and fish.
- Hickory, oak, and sugar maple are more flavorful and work better with beef, pork, sausage, and salmon.
- Mesquite produces an aromatic smoke that works well with beef and pork.

To use hardwoods, you need to soak them in water for 30 to 40 minutes and then drain. This way the wood does not burn but smolders.

Smoke-roasting, also called pan-smoking, is done on top of the stove. It works best with smaller, tender items such as chicken breast, fish fillet, vegetables, and seafood but it also can be used to add flavor to larger pieces. Place about a half-inch of soaked wood chips in the bottom of a roasting pan or hotel pan lined with aluminum foil. Next, place the seasoned food on a rack sprayed with vegetable spray and reserve. Heat the pan over moderate-high heat until the wood starts to smoke, then lower the heat. Place the rack over the wood and cover the pan. Keep covered until the food has the desired smoke flavor, then complete the cooking in the oven if the food is not yet done. Be very careful when opening the lid, due to the sudden release of trapped smoke. Smoking for too long can cause undesirable results. Practice, experience, and training are needed to become proficient and knowledgeable at smoke-roasting.

Smoke-roasting Cooking with dry heat in the presence of wood smoke.

CHEF'S TIPS FOR ROASTING

- Trim excess fat before cooking.
- Roast on a rack and uncovered (otherwise you are steaming the food). Cook at an appropriate temperature to limit drying out. Basting during cooking will also reduce drying. Finally, don't slice the meat until it has rested sufficiently so that you don't lose valuable juices, and be sure to slice across the grain to maximize tenderness.
- To develop flavor in meats, poultry, and fish, use rubs and marinades. You can also stuff the food (with vegetables and grains, for example), sear and/or season the food before cooking, or smoke the food over hardwood chips.
- When smoking, add dried pineapple skins, dried grapevines, or dried rosemary sticks to the wood chips for added flavor.
- Accompaniments that add flavor with moderate or no fat are vegetable coulis, chutneys, vinaigrettes, salsas, compotes, and mojos.

Broil and Grill

Broiling, cooking with radiant heat from above, is wonderful for single servings of steak, chicken breast, and fish such as salmon, tuna, and swordfish that can be served immediately. The more well done you want the product or the thicker it is, the longer the cooking time and the farther from the heat source it should be. Otherwise, the outside of the food will be cooked but the inside will not be done.

Broiling Cooking with radiant heat from above.

Grilling, cooking over a heat source, is also an excellent method for cooking meat, poultry, seafood, and vegetables (Figure 8-8). Like broiling, grilling browns foods, and the resulting caramelization adds flavor.

Once considered a tasty way to prepare hamburgers, steaks, and fish, grilling is now used to prepare a wide variety of dishes from around the world. For example, chicken is grilled

Grilling Cooking on an open grid over a heat source.

FIGURE 8-8 Grilling is an excellent method for cooking meat, poultry, seafood, and vegetables.

Reprinted with permission of John Wiley and Sons, Inc.

to make fajitas (Tex-Mex style of cooking) or to make jerk barbecue (Jamaican style of cooking). Grilling is also an excellent method to bring out the flavors of many vegetables. Grill them dry to avoid burning, then marinate with a light dressing consisting of a little oil, vinegar, or lemon juice and selected spices and fresh herbs.

Grilling foods properly requires much cooking experience to get the grilling temperature and timing just right. Here are some general rules to follow:

1. Cook meat and poultry at higher temperatures than seafood and vegetables.

2. For speedy cooking, use boneless chicken that has been pounded flat and meats that are no more than a half-inch thick. Fat should be trimmed off meats.

3. Don't try to grill thin fish fillets, such as striped bass or flounder, because they will fall apart. Firm-fleshed thicker pieces of fish, such as swordfish, salmon, mahi mahi and tuna, do much better on the grill. Don't turn foods too quickly or they'll stick and tear. Mark foods by turning the food around 90 degrees without turning it over.

To flavor foods that will be broiled or grilled, consider marinades, rubs, herbs, and spices. Lean fish can be marinated, sprinkled with Japanese bread crumbs, and glazed in a broiler. If grilling, consider placing soaked hardwood directly on the coals to add a smoked essence to the food.

CHEF'S TIPS FOR BROILING AND GRILLING

- Keep the grill clean and properly seasoned to prevent sticking. The broiler must be kept clean and free of fat buildup; otherwise it will smoke.

- Marinating tender or delicately textured foods for broiling or grilling will firm up their texture so that they are less likely to fall apart during the cooking process.

- Use marinades, rubs, herbs, spices, and smoking to add flavor. When grilling, consider using hardwood chips.

- During broiling, butter has traditionally been used to prevent food from drying out from the intense direct heat. In its place, you may spray the food with olive oil or baste it with marinade, stock, wine, or reduced-fat vinaigrette during cooking. To finish, sprinkle a thin layer of Japanese bread crumbs and glaze under the broiler.

- To retain flavor during cooking, turn foods on the broiler or grill with tongs—not a fork, which causes loss of juices.

- If you are making kabobs, soak the wooden skewers in water ahead of time so that they can endure the cooking heat without excessive drying or burning.

- Prepare these foods to order and serve immediately.

- Serve with flavorful sauces such as chutneys, relishes, purees, mojos, grain and vegetable salads or salsas.

Moist-heat cooking methods involve water or a water-based liquid as the vehicle of heat transfer and are often used with secondary cuts of meat and fowl or legs and thighs of poultry. In moist-heat cooking of meat or poultry, the danger is that the fat in the meat or poultry will stay in the cooking liquid. This problem can be resolved to a large extent by chilling the cooking liquid so that the fat separates and is removed before the liquid is used. To chill the liquid properly, place in an ice bath and stir frequently to release steam.

Compared to most dry-heat cooking methods, these methods do not add the flavor that dry-heat cooked foods get from browning, deglazing, or reduction. For foods that use moist-heat cooking to be successful, you will need the following:

- Good fortified stocks, jus, or broth
- Well-trimmed seasoned meats and skinless poultry
- Seasonings such as wine, fresh herbs, spices, and aromatic vegetables

Steam

Steaming has been the traditional method of cooking vegetables in many high-quantity kitchens because it is quick and retains flavor, moisture, and nutrients. It provides a balanced healthy alternative, because it requires no fat. The best candidates for steaming include foods with a delicate texture such as fish, shellfish, chicken breasts and vegetables. Be sure to use absolutely fresh ingredients to come out with a quality product.

Fish is great when steamed **en papillote** (in parchment) or in grape, spinach, or cabbage leaves. The covering helps retain moisture, flavor, and nutrients. Consider marinating fish beforehand to add moisture and flavor.

It is possible to introduce flavor into steamed foods (as well as poached foods) by adding herbs, spices, citrus juices, and other flavorful ingredients to the water (e.g., steam halibut on a bed of baby bok choy with snap peas, julienne carrots, daikon, and red peppers). Steam over a broth seasoned with lemon, bay leaf, onion, celery, ginger, allspice, and thyme to impart flavor and aroma to the finished dish. Steamed foods also continue to cook after they come out of the steamer, so allow for this in your cooking time.

Steaming Cooking by direct contact with steam.

En papillote Allowing food that is sealed in a folded packet to be steamed so that it cooks in its own flavorful liquids.

Poach

Poaching, cooking a food submerged in liquid at a temperature of 160° to 180°F (71° to 82°C), is used to cook fish, tender pieces of poultry, eggs, and some fruits and vegetables. To add flavor, you can poach the foods in liquids such as chicken stock, fish stock, wine, citrus juice, or vinegar flavored with fresh herbs, spices, ginger, garlic, shallots, and a mirepoix of vegetables to complement your item. For sweet poaching liquids, use fruit juices, wine, honey, whole spices, such as allspice, star anise, and cinnamon. Serve poached foods with flavorful sauces or accompaniments.

Fish is often poached in a flavored liquid known as **court bouillon**. A court bouillon is a flavorful liquid used to poach a variety of foods. Court bouillon adds flavor to the item that is cooking rather than taking away the items flavor profile. The liquid is simmered with vegetables (such as onions, leeks, carrots, and celery), seasonings (such as herbs), and an acidic product, such as wine or lemon juice (Figure 8-9). Cook the court bouillon for about 30 minutes, strain out the vegetables and herbs, and then add the fish to infuse these flavors. Place the fish to be poached on a rack or wrap it up in cheesecloth. Court bouillon may be used to cook fish to be served either hot or cold, but it is not generally used in making sauces.

Poaching Cooking a food submerged in liquid at a temperature of 160° to 180°F (71° to 82°C).

Court bouillon Water containing herbs, seasonings, and an acidic product used to cook fish.

Braise

Braising, or stewing (stewing usually refers to smaller pieces of meat or poultry), involves two steps: searing or browning the food (usually meat) in a small amount of oil or its own fat and then adding liquid and simmering until done. Foods to be braised are also often marinated before searing to develop flavor and tenderize the meat. When browning meat for braising, sear it in as little fat as possible without scorching and then place it in a covered braising pan to simmer in a small amount of liquid. To add flavor, place roasted vegetables, herbs, spices, and other flavoring ingredients in the bottom of the braising pan before adding the liquid.

Braising Searing or browning food (usually meat) in a small amount of oil or its own fat and then adding liquid and simmering until done.

FIGURE 8-9 Court bouillon ingredients.

Source: Wayne Gisslen, *Professional Cooking,* 7th ed. Copyright John Wiley and Sons, Inc. 2011. Photo by J. Gerard Smith.

Following are the steps for braising meat or poultry:

1. Trim the fat. Season the meat or poultry.
2. In a small amount of oil in a brazier, sear the meat or poultry to brown it to develop flavor.
3. Remove the meat and any excess oil from the pan.
4. Put the pan back on the heat. Add vegetables and caramelize (brown) them.
5. Deglaze the pan with wine and stock, being sure to scrape the fond (the flavorful drippings at the bottom of the pan).
6. Add tomato paste, wine, stock, and aromatic vegetables. Reduce.
7. Return the meat to the brazier, cover, and put into the oven, covered, to simmer. Make sure the meat is three-quarters covered with liquid. This allows the meat to cook more evenly and prevents the bottom from getting scorched.
8. When the meat is done (fork tender), strain the juices. Skim off fat and reduce juices. Puree the vegetables from the juices and use them (or you can use cornstarch slurry) to thicken the jus, making a light gravy.

Microwave

Microwaving is an additional method for cooking vegetables, because no fat is necessary and the vegetables' color, flavor, texture, and nutrients are retained, due to the rapid cooking time. Boiling or simmering vegetables is not nearly as desirable as steaming or microwaving, because nutrients are lost in the cooking water and more time is required. Microwaving is not as applicable in a high-volume cooking situation, because it works best for single or small servings.

CHEF'S TIPS FOR MOIST-HEAT COOKING

- Marinate the protein several days in advance to add some flavor. A dry rub or wet rub is best for this, without any salt.
- Braising is a great method to use with larger cuts of meat that you cook a day ahead of time. Take the cooked protein out of the liquid and cool. Meanwhile, reduce and skim the liquid to the proper consistency or thicken with vegetables. Cool the liquid, remove the fat, and pour over the meat to keep moist.
- Remove skin from poultry before using moist-heat cooking methods. The skin will add unnecessary fat to the cooking liquid.
- Use the stock appropriate for the protein you are cooking. Chicken stock has the most neutral flavor for all applications.
- Save the court bouillon stock for several uses within a week's time as long as you cool it properly and boil it each time; it gets more intense with each use. Or freeze for several weeks and use when needed, then discard. You can reduce this liquid as well and use it to season other fish preparations.

SUMMARY

1. Dry-heat cooking methods are acceptable for balanced cooking when heat is transferred with little or no fat. Excess fat is allowed to drip away from food being cooked. Both pan frying and deep frying add varying amounts of fat and kcalories, so they are not an acceptable cooking method.

2. Sautéing can be made acceptable by using nonstick pans, healthy oils, and at times finishing with a small amount of butter. You can also use two sprays of vegetable oil cooking spray.

3. To dry sauté, heat a nonstick pan, spray with vegetable-oil cooking spray, then wipe out the excess with a paper towel. Heat the pan again, then add the food. If browning is not important, you can simmer the ingredient in a small amount of liquid such as wine, vermouth, flavored vinegar, juices, or defatted stock to bring out the flavor. Add shallots, garlic, or other seasonings, then deglaze the pan with stock, wine, or another liquid and reduce to a sauce.

4. To stir-fry, coat the cooking surface with a thin layer of oil such as peanut oil, which has a strong flavor so you can use a small amount. Have all your ingredients ready. Preheat the pan to a high temperature. Stir the food rapidly during cooking and don't overfill the pan.

5. When roasting, always place meats and poultry on a rack so they don't cook in their own drippings. You can add flavor when roasting, broiling, or grilling by using rubs, marinades, or smoking, and also by deglazing the roasting pan to make either jus or jus lié.

6. For roasted, broiled, or grilled foods, consider flavorful sauces such as chutneys, relishes, purees, mojos, coulis, salsas, compotes, or vinaigrettes.

7. Moist-heat cooking methods involve water or a water-based liquid as the vehicle of heat transfer. In moist-heat cooking of meat or poultry, the danger is that the fat in the meat or poultry will stay in the cooking liquid. This problem can be resolved to a large extent by chilling the cooking liquid so that the fat separates and can be removed.

8. Moist-heat cooking methods do not add the flavor that dry-heat cooked foods get from browning, deglazing, or reduction. To be successful, you will need good fortified stocks or jus, well-trimmed seasoned meats and skinless poultry, and seasonings such as wine and fresh herbs.

9. Introduce flavor into steamed and poached foods by adding herbs, spices, citrus juices, and other flavorful ingredients to the water. For example, fish is often poached in a court bouillon, a flavorful liquid.

10. Foods to be braised are also often marinated before searing to develop flavor and tenderize the meat. When browning meat for braising, sear it in as little fat as possible without scorching and then place it in a covered braising pan to simmer in a small amount of liquid. To add flavor, place roasted vegetables, herbs, spices, and other flavoring ingredients in the bottom of the braising pan before adding the liquid.

11. Microwaving is an excellent way to cook small amounts of vegetables.

12. Many Chef's Tips are given in this section.

HOT TOPIC

HEALTHY BAKING

Baked goods range from those high in fat and sugar, such as cakes and pastries, to those with little or no fat or sugar, such as most breads. Healthier, yet still satisfying, versions of many baked goods are possible, and more recipes and tips on healthy baking are available now than ever before.

Before looking at healthy baking, let's review the functions of basic baking ingredients.

- *Flour* provides gluten, a protein that gives baked products structure. Different flours contain different amounts of gluten. Bread flour has the most protein that is useful for breads and pizza dough when you want structure. Cake flour is used at times in healthy baking because its fine-textured flour contains little protein to help produce a lighter, fluffier product with a tender crumb. This is useful when you decrease the type of fat in a recipe—because then there is less fat to tenderize.

- *Eggs* are high in protein, so they are also very important to give baked goods structure. Eggs also contribute flavor and color, as well as tenderness because the egg yolk contains fat.

- *Fats*, like butter and oil, provide moisture and help give baked goods their tender crumb. Reducing the amount of fat in a recipe results in a tougher product because gluten develops more freely.

- Besides providing sweetness, *sugars* absorb liquid, which keeps the baked good moist, and also makes baked goods tender. Sugars also help with browning and color.

Healthy baking means finding a balance between structure and moisture. The main ingredients responsible for structure (and toughness) are flour and eggs, while the main ingredients responsible for tenderness and moisture are fats and sugars (Figure 8-10). Of course, wet ingredients, such as water and milk, also act as moisteners, and dry ingredients, such as flour and milk solids, act as dryers. If you want to decrease or eliminate an ingredient from a recipe, you have to consider what the ingredient does and do one of the following:

1. Use another ingredient that performs the same job (such as using more sugar to tenderize when decreasing fat).

2. Reduce the amount of ingredients that have the opposite effect (such as using fewer eggs when decreasing fat).

Following are tips on how to reduce fat and/or saturated fat in a recipe:

- Since oils have less saturated fat than butter or shortening, and they provide tenderness and moistness to baked goods, they can be substituted for fats in baked goods such as muffins and quickbreads. However, oils can't be substituted

for solid fats when creaming is essential to leavening, unless you are substituting oil for just part of the fat. For example, when making drop cookies like chocolate chip, you can replace half of the butter with oil, such as canola. If you also increase the salt a smidge (1/8 teaspoon), the extra salt draws out more buttery flavor for a satisfying cookie. If you need to make up for lost leavening power, you can increase the baking powder or fold in whipped egg whites. It is more difficult to use oil in rolled cookies, and oil can't be used at all in laminated doughs such as croissant and puff pastry.

- If you reduce the amount of fat in a recipe you need to increase other tenderizing ingredients (sometimes in place of the fat), decrease ingredients that toughen the product, or do both.

  - To increase tenderizers, consider using sugars—especially liquid sweeteners such as honey—or fruit or vegetable purees such as applesauce or banana puree. Also dairy products such as fat-free buttermilk, sour cream, or yogurt can replace some of the fat well because they increase moisture and their acidity tenderizes gluten.

  - To reduce ingredients that toughen the texture, you can use flours with less gluten, such

FIGURE 8-10 In baking, the main ingredients responsible for structure (and toughness) are flour and eggs, while the main ingredients responsible for tenderness and moisture are fats and sugars.

Reprinted with permission of John Wiley and Sons, Inc.

as cake flour or pastry flour, or use fewer eggs, in which case you will need to increase other liquids. Also during preparation, be careful not to overmix or overbake—both will toughen the product.

- Another way to reduce fat, especially saturated fat, in a recipe is by substituting lower-in-fat dairy products for full-fat products. In many recipes, low-fat or nonfat milk can be used successfully to replace whole milk. Fat-free yogurt can often work in place of regular sour cream, as well as light or fat-free sour cream.

Egg yolks are also a source of saturated fat—remember that egg whites are fat-free. When eggs are used as binders, you can use egg whites instead of the whole egg as long as you use equal weights (or 1 whole egg = 2 egg whites in small quantity cooking). Egg whites also work when used as leavening, but can't be relied on to provide a lot of structure to a baked good. To completely replace eggs, ground flaxseed works well because it is rich in gums and has binding power. Whisk together 1 teaspoon of flaxseed powder with ¼ cup of water for each egg to be replaced. In quantity recipes, mix 1 tablespoon ground flaxseed with each

4 ounces of flour. Puréed fruit or pureed silken tofu can also be experimented with to replace eggs.

To reduce sugar in a recipe, first experiment with using less sugar than called for. You can often reduce the amount of sugar by 10 percent or more. To make up for the sugar, you can use flours with less gluten, or dairy products such as fat-free buttermilk, sour cream, or yogurt, as just discussed. Also use spices such as cinnamon, cloves, allspice, and nutmeg, or flavorings such as vanilla extract or almond flavoring to boost sweetness.

Following are some tips for preparing and baking a variety of baked goods.

1. **Cakes:** Light cakes rely less on fat and more on sugar and liquids for a soft, tender texture. The first step of the mixing process—creaming butter and sugar—whips air into the batter. Once flour and milk are added, beat only until combined. If you overbeat at this point, the cake will turn out tough. You want a fluffy batter—as that will give you a moist, tender cake. More sugar helps prevent gluten from toughening the texture, but it also causes excess browning, especially if the cake also contains dairy. Be very careful that you don't overbake—pull the cake out from the oven when a wooden pick inserted in the center comes out with a few moist crumbs on it.

2. **Pie crusts:** Making perfect pastry is harder when you are using less fat. When making pie crusts, it will take some time—sometimes five minutes—to cut the fat into the flour. You want to be sure you coat the flour proteins with fat so they will melt during cooking, giving off steam and creating moist layers. Don't overwork the dough—it will just make the crust tough.

3. **Cookies:** In cookie recipes you will at times use both oil and butter. Oil disperses better than butter and helps keep the cookies moist and tender, whereas butter keeps the cookies from spreading too much during baking. Don't use soft margarine, whipped butter, or diet margarines because they have too

little fat and too much water and air to make a good cookie. Light cookies require precise measuring of the flour (and other ingredients, too). Just a little too much flour will create hard cookies. Pull from the oven a bit sooner than regular cookies—they will go from just cooked to overbrowned quickly.

4. **Quick breads:** Quick breads are much more versatile than other baked goods in terms of making a variety of tasty, healthful versions.

Quick breads can easily be made into unlimited varieties using healthy ingredients such as whole wheat flour, bran, oatmeal, vegetable oils, nonfat dairy, and many kinds of fruits, nuts, and spices.

No matter what you are baking, flavoring is very important. There are many ways to add flavor into baked goods—consider Vietnamese cinnamon in apple muffins, orange juice glaze on orange-pecan tea bread, fresh dill and cheddar cheese in biscuits, peppermint extract in brownies, espresso powder in fudge cake, crystallized ginger in blueberry shortcakes, Cointreau (orange-flavored liqueur) in Boston cream pie, and whole nutmeg in flan.

Of course, there are a number of lower-in-fat desserts that bakers already make. Don't forget to use favorites such as angel food or sponge cakes, baked meringues, low-fat icings such as fondant or boiled icings, fruit cobblers and crisps, and custards and puddings using low-fat dairy products.

RECIPE MAKEOVERS

CHAPTER 9

- Explain at least three general ways to modify recipes to change the nutrient content.

- Discuss three considerations to keep in mind when you modify a recipe.

- Given a recipe that was modified to make it more balanced, identify and explain modifications that were made.

- Given a recipe, modify it on paper to meet a stated nutrition goal, and test the recipe.

INTRODUCTION

Classics, tradition, value, and comfort sell food today, especially when your items are nutritionally balanced. It is imperative as a chef/cook to be versed in all these styles of cooking. Most of the recipes in this book use classic methods that you can tweak as you add your own identity and personality. Many of the seasonings, preparations, and ingredients used in the pages ahead can be utilized in a variety of menu items.

It is ridiculous to believe you have to "cook two different ways: healthy or unhealthy cooking." The cooking is the same, but the ingredients in richer, unrestricted cooking contain more sodium, fat, cholesterol, and sugar. Even when cooking with no dietary restrictions, you should make a brandy cream for a rib eye steak or filet mignon by beginning with a great natural veal sauce, reduced to the proper consistency, with the addition of reduced cream, not a sauce of 80 percent cream and 20 percent veal demi-glace. Many chefs tend to add too much cream, butter, eggs, salt, sugar, and fats, believing they are the secret weapons that make a dish great. On the contrary, people leave your restaurant overstuffed, tired, and uncomfortable.

Some of the best restaurants today use vegetable reduction with a hint of butter, or herb- and spiced-flavored oils that are added just at the end of a preparation giving a clean, distinct flavor to the food you are consuming. There is a lot of accountability associated with cooking for your clients. After all, it is one of the few professions where people ingest what you are selling. For culinarians, this is honorable with tremendous moral and ethical responsibility.

There are different reasons to modify a recipe for nutritional and health purposes. For example, you might need to reduce kcalories, fat, saturated fat, cholesterol, carbohydrates, protein, sodium, sugar, or a combination, depending on your intended result or audience. Additionally, you might also wish to modify a recipe to provide more of a nutrient, such as fiber, vitamins, and/or minerals. Whether modifying a recipe to get more or less, there are four basic ways to go about it:

1. Change/add healthy techniques of preparation.

2. Change/add healthy cooking methods.

3. Change an ingredient by reducing it, eliminating it, or replacing it (see Table 9-1 for substitution possibilities).

4. Add a new ingredient(s), particularly to build flavor, such as dry rubs, toasted spices, fresh herbs, acids (vinegars/citrus juice), and condiments.

To modify a recipe, consider the following:

1. *Nutritional analysis.* Examine the product and decide what you want to accomplish and how you will achieve changing its nutrient profile. For example, in a meatloaf recipe, you may decide to decrease the fat content and increase the fiber content. Overall kcalories may be a focus for other modifications, or you might emphasize low sodium and cholesterol. Understanding the end result can help the success of the process.

TABLE 9-1 | **Recipe Substitutions**

IN PLACE OF	USE
Butter	Margarine (Light/low-fat margarines contain more water and may cause a baked product to be tough, so try decreasing regular margarine 1–2 Tbsp. first) In baking, use canola oil with a ratio of apple sauce
2% or whole milk	Nonfat or low-fat (1%) milk
Buttermilk	2% buttermilk, or 15 Tbsp. skim milk plus 1 Tbsp. lemon juice to make 1 cup
1 cup shortening	¾ cup vegetable or canola oil
1 cup heavy cream	1 cup evaporated skim milk—in soups and casseroles
	Baking—light cream or Half & Half
1 cup light cream	1 cup evaporated skim milk
1 cup sour cream	1 cup reduced-fat sour cream
Cream cheese	Light cream cheese or Neufchatel (Nonfat cream cheese produces dips and cake frosting that are very runny)
	Tofutti (creamy tofu base)
1 ounce baking chocolate	3 Tbsp. cocoa and 1 Tbsp. vegetable oil
1 egg	¼ cup egg substitute or 2 large egg whites
Whole-milk mozzarella	Part-skim mozzarella, low moisture
Whole-milk ricotta	Part-skim ricotta
Creamed cottage cheese	Low-fat cottage cheese or pot cheese
Cheddar cheese	Low-fat cheddar cheese (nonfat cheese does not melt) or ¾ cup very sharp cheddar cheese for 1 cup cheddar
Swiss cheese	Low-fat Swiss cheese
Ice cream	Ice milk/Pureed fruits with egg whites or frozen yogurt
Mayonnaise	Light or nonfat mayonnaise or nonfat plain yogurt (don't use nonfat versions if heating)
1 cup white flour	½ cup whole wheat flour and ½ cup white flour
1 cup oil in quick breads	½ cup pureed fruit plus ½ cup oil
1 cup chopped nuts	½ cup nuts toasted to bring out flavor
1 cup shredded coconut	½ cup toasted coconut or 1½ tsp. coconut extract

2. *Flavor.* Flavor needs to be a driving factor when substituting ingredients. There are some recipes that *cannot* be modified to a level of satisfaction that will represent favorable results. When evaluating, consider what you can add to the recipe to ensure maximum flavor. Should you try to mimic the taste of the original version, or will you have to introduce new flavors? A fresh herb, spice, or an acid may be needed to replicate a similar flavor that is taken away. Will you be able to produce this change without sacrificing taste, aroma, and satisfaction?

3. *Ingredient functions.* When modifying ingredients, think about what functions each ingredient performs in the recipe. Is it there for appearance, flavor, or texture? What will happen to the product if less of an ingredient is used or a new ingredient is substituted? You also have to consider adding additional flavoring ingredients to enhance the recipes, such as roasted vegetables, chilies, reductions, mustards, or citrus juices.

4. *Cooking techniques.* Can you use a healthier cooking technique, such as dry sauté, and still get a quality product? Consider the use of various preparation and cooking methods discussed in Chapter 8.

5. *Evaluate the recipe* to see whether it is acceptable. This step often leads to further modification and testing. Be prepared to test the recipe a number of times, and also be prepared for the fact that some modified recipes will never be acceptable, but each recipe modification becomes a learning experience that adds to your skill level and experience.

Of course, if you don't want to go through the trouble of modifying current recipes, you can select and test recipes from healthy cookbooks, magazines, or websites, or create your own version from scratch. When developing new recipes, be sure to choose quality, fresh ingredients, as well as appropriate cooking methods and techniques. Also, pay attention to developing a flavor profile, and finish cooking to order as much as possible using proper techniques of good, wholesome cooking. Recipe modification is an important process in healthy, balanced nutritional cooking; keep in mind that practice and experience are needed to become knowledgeable and proficient.

SUMMARY

1. There are four ways to modify a recipe: change/add healthy preparation techniques; change/add healthy cooking methods; change an ingredient by reducing it, eliminating it, or replacing it; and add a new ingredient(s)—especially to build flavor.
2. If you decide to modify a recipe, consider the following: exactly how you want to change the nutrient profile, how you will develop flavor, what function each ingredient plays, and how satisfying are the results.

APPETIZERS

CRAB CAKES

CHEF'S NOTES

There are numerous crab cake recipes, depending on the flavor profile of a specific cuisine, geographic region, or personal preference. Basic crab cakes consist of the following:

- A variety of crab meat
- Mayonnaise
- Eggs
- Bread crumbs
- Seasoning such as Old Bay, parsley, chives, and dry mustard

The mixture is then molded, standard breaded, and pan-fried golden brown. Crab cakes are served with sauces such as tartar, remoulade or aioli, with garnishes such as chopped tomato, capers, chives, and red onions. Modern presentations have shifted this delicacy into menu selections such as a popular appetizer, a topping on your favorite salad, a lunch sandwich choice, or as an accompaniment to an entrée. Versatile in appearance and easily paired with other foods, guests are attracted to this item on your menu even with the necessary higher price point.

There is so much creativity and variety with this preparation. Not only is it an American classic, it is also one of the most widely bought appetizer/entrée selections in restaurants of different styles across the board. This version is delicate and satisfying. You don't miss the mayonnaise, eggs, or the pan-fried cooking method at all. This dish is all about the flavor of the crabmeat with the traditional seasonings. The garnishes are endless, from salad greens to shaved fennel to an heirloom tomato salad with basil or cilantro to a fine julienne celery root slaw. This again, is where you as a culinarian can use your creativity to delight and surprise your guest with ingredients and presentations. With the dressing recipes provided, vegetable salads can be marinated to create a garnish that accompanies your dish, additionally using the many types of dried vegetables and potatoes to create a crisp, attractive garnish. This one is a definite on your menu (Figure 9-1).

Serves 5

1¼ pounds jumbo lump crabmeat	1½ egg whites
1½ teaspoons chives, chopped fine	¾ teaspoon lemon juice
1½ teaspoons parsley, chopped fine	1½ teaspoons white wine
1 tablespoon Old Bay Seasoning	½ egg white
1 teaspoon Dijon mustard	Japanese bread crumbs (Panko) as needed
¾ teaspoon fresh thyme, chopped	¾ teaspoon chives, chopped fine
5 ounces Yukon gold potatoes, cooked and riced	Vegetable oil cooking spray

STEPS

1. Pick crabmeat to remove bits of shell. Place crabmeat in a bowl.

2. In another bowl, mix together the chives, parsley, Old Bay Seasoning, mustard, thyme, and riced potatoes. Add crabmeat and fold together gently to avoid breaking up the crabmeat.

FIGURE 9-1 Crab Cakes.

Photo by Peter Pioppo.

3. In a stainless-steel bowl, place 1½ egg whites, lemon juice, and white wine. Whip to form stiff peaks.

4. Fold whipped egg whites into crab mixture and mold into 3-ounce crab cakes.

5. In a deep dish pan, place ½ egg white (whipped slightly to break up). Dip the crab cakes in the egg white, then in the crumbs mixed with chives.

6. In a nonstick pan sprayed with vegetable spray, sauté the crab cakes to a crisp golden brown. Serve with a fresh salsa or mojo of your choice with some spicy greens, tomato confit, fennel salad, or a crisp baked potato garnish.

● ● ● **NUTRITION COMPARISON**

	Kcalories	Protein (g)	Fat (g)	Carbo (g)	Sodium (mg)	Chol (mg)
Traditional Recipe	395	25	25	16	684	176
Balanced Recipe	150	24	1	10	607	48

WONTONS

CHEF'S NOTES

Dumplings, wontons, spring rolls, dim sum, and meat buns have gained increased popularity on a variety of menus, cuisines, and culinary styles. They can be filled with unlimited ingredients, adding different flavor profiles as an appetizer, hors d'oeuvre, in addition to an entrée, or even as a dessert choice. Typically wontons and dumplings are Asian in flavor profile, and are usually made with the following ingredients:

- Ground meat
- Julienne chopped vegetables
- Scallions, soy sauce, oyster sauce, ground ginger, and garlic

Wonton are often deep fried or steamed, and served with a variety of cornstarch thickened and/or soy-based dipping sauces. They are well-received menu items that can be prepared in advance, with delicious ingredients, having the ability to be frozen for ease of service.

Wontons, tortellonis, egg rolls, and dumplings are all produced today using a variety of egg, spring, soy, rice paper wrappers and wonton skins. They lend themselves easily to light selections. There are limitless numbers of flavored and textured fillings you can create. Use shredded vegetables with beef, pork, shrimp, and crab for an Asian-style wonton. Use mushroom, chicken, fennel, and caramelized onions with a variety of meats for an Indian, Italian, or Spanish-style cuisine. Use roasted fruits with skim milk ricotta, vanilla, and lavender honey as a great dessert option. These ideas are just the start of menu items adding versatility, style, and global appeal to your menu selections. In the following recipe, shown in Figure 9-2, the tortellonis are stuffed with a bountiful filling, poached in lightly boiling water, then served with fresh tomato or roasted pepper sauce.

Roast Chicken and Shredded Mozzarella Tortellonis

Serves 6 (3 dumplings each)

4 ounces chicken breast

½ teaspoon canola oil

Pinch fresh oregano, chopped

Pinch fresh basil, chopped

Pinch paprika

½ cup reduced-fat shredded mozzarella

½ cup skim-milk ricotta

¼ teaspoon fresh garlic, chopped

2 pinches freshly ground black pepper

1 teaspoon fresh basil, chopped

½ teaspoon fresh parsley, chopped

18 wonton skins

1 cup water

1 tomato, chopped

Freshly ground black pepper

18 arugula leaves

1 recipe Red Pepper Coulis

FIGURE 9-2 Roast Chicken and Shredded Mozzarella Tortellonis.
Photo by Peter Pioppo.

STEPS

1. **Preheat oven to 350°F.**

2. **Coat chicken breast with oil, herbs, and paprika. Roast about 5–10 minutes, or grill over hardwood for 5 minutes, until chicken reaches an internal temperature of 165°F. Let cool and fine-julienne. Reserve.**

3. **In bowl, mix cheeses, garlic, pepper, and herbs. Add chicken and mix well.**

4. **Lay out wonton skins and paint with water.**

5. Place 1 full teaspoon of chicken mixture in each skin (or more to use up all the filling).

6. Fold skin to make a triangle, then connect ends in opposite direction to make tortelloni shape.

7. Prepare chopped tomato garnish and toss with black pepper. Let juice accumulate. Paint leaves of arugula lightly with juice of tomato for flavor.

8. Poach tortellonis for 3 minutes in lightly boiling water.

9. Arrange each plate with a ladle of fresh tomato or roasted pepper coulis, arugula leaves and 3 poached tortelloni. Garnish with chopped tomato if desired.

• • • NUTRITION COMPARISON

	Kcalories	Protein (g)	Fat (g)	Carbo (g)	Sodium (mg)	Chol (mg)
Traditional Recipe	329	9	23	18	428	26
Balanced Recipe	151	11	4	17	233	25

ENTRÉES

MEATLOAF

CHEF'S NOTES

A favorite classic preparation is a great way to start the recipe modification process. Typically, traditional meatloaf ingredients include ground meat, veal, pork, or a combination; sautéed mirepoix; eggs; bread crumbs; tomato product; and seasonings. This crowd pleaser is fairly straightforward to begin the modification process. To start, think through the basics of what is needed in the replication of a meatloaf:

1. You need a ground meat of some kind: beef, pork, veal, turkey, or chicken, or a combination.

2. You need seasonings: garlic, onions, celery, carrots, pepper, jalapenos, salt, Worcestershire sauce, tomato product, Tabasco, chilis, or coriander, to name a few. Also consider fresh herbs, Italian parsley, oregano, cilantro, thyme, or chives.

3. You need a binder to hold the meatloaf together: eggs, egg whites, bread crumbs, rice, or other grains.

4. You need a liquid for moisture: such as stock, milk, or water.

The major ingredient in meatloaf is obviously the meat. Traditional meatloaf uses ground beef and/or ground pork or veal. Beef is available in different ratios of lean to fat, from 75 percent meat / 25 percent fat to 80/20, 90/10, or 95/5 blend.

Using a leaner ground meat cuts down fat, but this also creates a drier product. However, you can add back some of the lost moisture by adding grains or whole-wheat bread crumbs with a liquid to make panada. Another ingredient that adds in moisture is cooked, pureed eggplant. About ½ cup of pureed eggplant per pound of meat works well. Additionally, you can substitute cooked grains such as brown rice, quinoa, and amaranth, instead of the bread for a gluten-free alternative. Also, you can remove the whole egg (omitting

Panada A paste made by mixing bread crumbs and flour, and possibly other ingredients with water, milk, or stock. It is used as a binder or for making soups or thickening sauces.

additional saturated fat and cholesterol) by substituting whipped egg whites folded into the meat mixture.

Here is another opportunity to indulge in your creativity. Change your meatloaf to a slightly Asian profile with ginger, garlic, scallions, and cilantro with a hint of hoisin or soy sauce, or add freshly grated ginger, curry, cinnamon, cardamon, chili powder, and cumin with sautéed onions, fennel, lemon juice, and golden raisins. An additional variation is a barbecue meatloaf with Southwestern spices, sautéed peppers, smoked onions, and corn, served with barbeque sauce. You can use ground turkey, beef, chicken, pork, or a combination to create your own desired result.

Serves 25

12 ounces onions, fine dice	1 tablespoon salt
8 ounces celery, fine dice	½ teaspoon black pepper
2 ounces oil	2 tablespoon Jalapenos, brunoise
3 tablespoons chopped garlic	2 tablespoon chili powder
12 ounces diced whole-wheat bread	1 tablespoon ground cumin
4 ounces skim milk or vegetable stock	10 egg whites, beaten until they begin to foam
6 ounces tomato ketchup	
5 pounds ground beef 90/10	½ cup chopped cilantro
2½ pounds ground turkey	

STEPS

1. Sauté the onions and celery in oil for 2 minutes. Add in garlic and finish sautéing until translucent. Remove from pan and cool.

2. In a large bowl, soak the bread crumbs in milk or stock, let sit to absorb for about 1 hour.

3. Add into the bread mixture the sautéed vegetables, ketchup, meat, salt and pepper, jalapenos, chili powder, and cumin. Mix gently until evenly combined.

4. Next fold in the egg whites and the cilantro, mixing gently to completely incorporate.

5. Form the mixture into two or three loaves on a baking pan, or fill loaf pans.

6. Bake at 325°F about 1½ hours until done (internal temperature of at least 165°F).

• • • NUTRITION COMPARISON

	Kcalories	Protein (g)	Fat (g)	Carbo (g)	Sodium (mg)	Chol (mg)
Traditional Recipe	463	26	33	13	487	133
Balanced Recipe	398	28	23	11	556	123

CHEF'S NOTES

A hearty stew is another comfort food that will be a favorite menu item during the colder months. Position this great choice as a classic, rather then as a healthy item, which may be perceived as bland and unsatisfying. The basic ingredients of stew are as follows:

1. A protein (beef, pork, or veal), usually a tougher cut based on the slow cooking method. Poultry can be used as long as the method is adjusted to avoid overcooking.

2. The product is then dredged in seasoned flour and seared in oil, with the addition of onions and garlic, adding extra flour if needed to create the roux for thickening.

3. A tomato product is then added and caramelized for flavor and color.

4. Stock is used to cover the meat, and then braised with a sachet until tender.

5. Vegetables for garnish are usually cooked separately to keep their color and crisp, appetizing appearance.

Note: In this recipe, a more tender cut of meat is used, so a quicker cooking method will be used in this preparation.

This is where I would argue that classical cooking fundamentals are key to the success of this dish and many others like it. The process starts with a flavorful, marinated cut of meat, a well-crafted stock, and an appropriate evenly diced or classically cut vegetable garnish. The procedure for a well-balanced stew and a heavier stew is based on preparation and ingredients. Instead of traditional thickeners such as roux, reductions and light alternatives such as cornstarch and agar-agar are used. Stocks are properly degreased and the seared meats are rendered and strained of the fat before being incorporated into the stock. There is little or no addition of salt—just the natural flavors achieved by good-quality ingredients and proper cooking methods. Served with whipped cauliflower puree or baked sweet potato fries, this dish will easily become a menu favorite with or without promoting the balanced nutrition feature.

There are many ingredients you can use to create endless signature dishes. Try pork, lamb, chicken, or beef with a variety of different vegetables, marinades, seasonings, or garnishes to achieve the flavor or cuisine you are capturing. With the same methodology and different ingredients, you can create a variety of new and exciting menu items. Everyone loves the word *stew*—it's one of the original home-slow-cooked comfort foods.

Agar-agar A seaweed derivative that is flavorless and becomes gelatinous when dissolved in water, heated, and then cooled. Used in cooking as a thickener.

Serves 20

MARINADE

4 cups chicken broth	Nonstick cooking spray
1½ cups white wine	10 ounces onions, medium dice
1 ounce olive oil	2½ cups red wine
1 bay leaf	20 ounces carrots, medium dice, steamed
1 ounce fresh herbs	20 ounces zucchini, medium dice, steamed
1 lemon	20 ounces tomatoes, medium dice, seeded, steamed
1 lime	
5 pounds filet tip or lean cut of beef, easily sautéed	20 ounces fennel, half-moon sliced, steamed

VEGETABLE SEASONING

1 ounce granulated garlic

½ ounce dried thyme

½ ounce dried oregano

½ ounce onion powder

1 tablespoon paprika

2 teaspoons red pepper flakes

(This seasoning will make extra that can be reserved and stored in an airtight container for future use.)

20 ounces green beans, cut 1½-inch length, steamed

3 tablespoons fresh thyme, cleaned and chopped

10 cups brown veal stock, reduce by half

½ cup fresh basil, chiffonade

STEPS

1. Make marinade by heating the chicken broth and reducing by one-third. Add in wine, olive oil, bay leaf, and fresh herbs. Stir. Squeeze in lemon and lime juices. Simmer 5 minutes. Chill overnight.

2. Trim beef of any fat and cut into 1½-inch pieces. Marinate overnight.

3. Season beef with 5 tablespoons vegetable seasoning, and let sit 1 hour.

4. Spray nonstick skillet with nonstick cooking spray. Sear beef with onions in the hot skillet, then remove beef.

5. Add red wine and reduce by half. Add steamed vegetables to the same skillet and season with thyme. Add in reduced veal stock, then add beef back to pan and cook for about 5 minutes or until cooked. Add in basil. Remove and serve. This dish can be prepared without the vegetables, but keep the onions, and using them as a decorative seasoned garnish during plating.

NUTRITION COMPARISON

	Kcalories	Protein (g)	Fat (g)	Carbo (g)	Sodium (mg)	Chol (mg)
Traditional Recipe	320	25	14	22	592	71
Balanced Recipe	277	28	9	17	197	67

HAMBURGER

Can you say American! Food on a bun is a huge draw for a menu item, attracting children as well as adults. Traditionally, burgers were made from ground beef using tougher cuts of scrap such as chuck or brisket, with seasonings added, and finished with numerous toppings, such as cheddar and bacon or mushroom and blue cheese, then served on a great roll, often with a lettuce, red onion, and tomato garnish. Today we see burgers on menus made with everything from ground fish, lobster, shrimp, sausage, and turkey to mushroom, grains, chicken, and duck, with the addition of unique spices, garnishes, and global flavors. The direction over the past decade for the classic beef burger has become more serious, such as custom artisanal meat blends with cuts from sustainable heirloom cattle.

CHEF'S NOTES

Let's face it: Everyone loves a burger, and removing it from most menus is just not an option. But recreating this menu staple with different ingredients is an alternative to offer options to your customers. Once you come up with your own signature burgers, you could make them into sliders, adding them as an appetizer or bar menu selection. The options are endless, adding great alternatives to this American favorite.

In the following balanced recipes, ground chicken or turkey breast (no skin) is used to decrease the overall fat and also saturated fat (see Figure 9-3). Stock, egg whites (which are fat free), and riced potatoes add moisture so the chicken burger is still juicy.

Grilled Chicken Burger

Serves 32

10 pounds ground chicken breast, no skin

1½ pounds cooked riced potato

1 cup chopped parsley

2 tablespoons prepared seasoning blend with herbs and spices, no salt

½ teaspoon black pepper

1 cup sautéed diced white onions

½ ounce chicken stock

2 egg whites

TOPPING OPTIONS

32 whole-wheat hamburger buns

32 slices red onion

32 thick slices tomato

1/2 pound micro cilantro

2 quarts Asian slaw

32 each pepper, pickle, and olive skewered garnish

FIGURE 9-3 Grilled Chicken Burger.

Photo by Peter Pioppo.

STEPS

1. **In a mixing bowl, add all ingredients. Mix together to thoroughly incorporate, or place in mixer on slow speed with a dough hook.**

2. **Shape in ring mold, 4-inch diameter weighed out at about a 5-ounce chicken burger.**

3. Grill chicken burger until thoroughly cooked.

4. Place burger on whole-wheat bun. Arrange onion, tomato, and micro greens on plate or compose on burger and skewer with garnish. Serve with kale or eggplant chips.

Grilled Turkey Burger

Makes 32

1 cup whole wheat crumbs

¼ cup vegetable stock

8 pounds ground turkey breast

7 egg whites

1 cup cooked diced onions

6 stalks thin sliced scallions

4 garlic cloves, chopped (sautéed translucent with onions)

3 tablespoons Worcestershire sauce

2 tablespoons reduced-sodium soy sauce

1 tablespoon Tabasco

TOPPINGS

Roasted pepper dip

32 whole-wheat rolls

Sliced plum tomatoes

Arugula greens

STEPS

1. In a mixing bowl, moisten the whole-wheat crumbs with vegetable stock and let sit 20 minutes. Next, add all remaining ingredients. Mix together to incorporate all ingredients thoroughly.

2. Shape in ring mold (4-inch diameter) to make 5-ounce turkey patties.

3. Grill patties slowly until thoroughly cooked.

4. Spread roasted pepper dip on each side of the whole-wheat toasted bun. Place burger on bottom bun with tomato slice and arugula. Serve with kale or eggplant chips.

● ● ● **NUTRITION COMPARISON**

	Kcalories	Protein (g)	Fat (g)	Carbo (g)	Sodium (mg)	Chol (mg)
Traditional Recipe	620	30	41	31	441	133
Grilled Chicken Burger	303	38	5	27	280	82
Grilled Turkey Burger	326	30	12	26	335	98

CHICKEN QUESADILLAS

Served as an appetizer or lunch entrée, quesadillas are an easy item to produce in a commercial kitchen. Typically, quesadillas are made with the following:

- Flour tortillas
- Grilled chicken breast
- Sautéed onions and peppers, chili powder, ground cumin, cilantro
- Monterey jack and cheddar cheese

CHEF'S NOTES

The ingredients are then layered or rolled in the tortilla, browned on a flat top or plancha, and served with pica de gallo, sour cream, and guacamole. By using whole-wheat tortillas, part-skim cheeses, and lots of sautéed vegetables and seasonings, you can have a tasty menu item without excessive fat or kcalories. The addition of black bean salsa, pureed beans, or a little avocado cilantro salad nicely rounds out the composition of the dish. My vote is to keep quesadillas (Figure 9-4) as a staple on the menu as a light signature items that's better than its original version.

FIGURE 9-4 Chicken Quesadillas.

Photo by Peter Pioppo.

Serves 6

1 tablespoon olive oil

1 onion, medium sliced

1 large red, orange or yellow pepper, cut into ¼-inch strips

1 teaspoon garlic, minced

½ jalapeno pepper, minced

3 large mushrooms, cleaned and sliced (optional)

⅛ teaspoon cumin

⅛ teaspoon chili powder

2 tablespoons lime juice

1 tablespoon cilantro, chopped

1 pound skinned, trimmed chicken breast

6 whole-wheat 14-inch flour tortillas

3 ounces shredded part-skim mozzarella cheese

3 ounces shredded part-skim sharp cheddar cheese

Cooking spray

1 cup black bean relish—optional

2 cups greens such as mizuna and frisee

3 ounces avocado puree

STEPS

1. Heat olive oil in saucepan over medium heat. Add onion and pepper and stir frequently until onion starts to brown, about 5 minutes. Add garlic, jalapeno, and mushrooms, cook for several minutes, then add in cumin and chili powder.

2. Cook, stirring frequently, about 2 minutes. Add lime juice, and cook until most of the liquid evaporates, about 2 minutes.

3. Once finished cooking and slightly warm, add in chopped cilantro, cool, and reserve for next steps.

4. Grill chicken breast slowly until thoroughly cooked; place in oven to finish if breasts are a thick cut. Let rest, then cut into thin strips.

5. Over an open flame, warm the tortilla shells on both sides without browning.

6. Place vegetable mixture on the bottom three-fourths of the tortilla. Add chicken strips and two kinds of cheese. Start folding the quesadilla like a spring roll or sushi, but do not fold in ends. Make sure filling goes to each end. Cut some of the tortilla at the ends if there is too much wrapper for the size.

7. Coat large skillet, plancha, or griddle with cooking spray. Lightly brown and crisp tortillas on each side over medium heat.

8. Bake quesadillas in oven for 5 minutes if needed at 325°F.

9. Let quesadillas cool slightly and cut into wedges.

10. Serve with greens, black bean relish, and a swirl of avocado puree.

NUTRITION COMPARISON

	Kcalories	Protein (g)	Fat (g)	Carbo (g)	Sodium (mg)	Chol (mg)
Traditional Recipe	815	41	42	67	2,441	116
Balanced Recipe	410	31	12	47	726	55

VEGETABLE LASAGNA

Who doesn't love lasagna, but who can eat this heavy classic more then several times a year? The word *lasagna* indicates a rich, cheese-based pasta dish high in kcalories and fat. It's a well-recognized menu item that also speaks comfort, familiarity, and value. Basic ingredients for a vegetable lasagna include the following:

- Cooked lasagna noodles
- Whole milk drained ricotta
- Shredded or slice fresh whole-milk mozzarella
- Grated pecorino Romano cheese
- Cooked drained spinach, blanched or roasted broccoli florets, or other sautéed vegetables finished with tomato sauce

This recipe makeover takes all the great ingredients of lasagna, cheese, and vegetables and uses them to create a light alternative.

CHEF'S NOTES

You can easily use this recipe as an entrée accompaniment or as a lunch entrée with an arugula, roasted onion, and fennel salad. The vegetables in this recipe replace the pasta, so this is perfect for a gluten-free appetizer or a smaller version as an accompaniment to an entrée. If the cheese is replaced with a puree of white bean, roasted garlic, artichokes, and fresh herbs, this lasagna becomes a great vegan selection. You get the feeling of cheese and flavoring without the kcalories of whole-milk mozzarella and ricotta, not to mention rich pasta sheets. For this recipe we are using skim milk ricotta, feta cheese, and seasonings. The balanced recipe has great versatility on your menu design and can satisfy your guest with this variety of options.

Serves 8

MARINADE

4 cups chicken/vegetable stock

1½ cups white wine

1 ounce olive oil

1 bay leaf

1 ounce fresh herbs

1 lemon

1 lime

8 slices eggplant

8 slices zucchini

8 slices yellow pepper

8 slices red pepper

2 pounds skim-milk ricotta cheese

8 ounces feta cheese

4 tablespoons fresh oregano, chopped

1 tablespoon Mrs. Dash garlic herb seasoning or create your own

Cracked black pepper, to taste

8 ounces egg whites

32 ounces tomato sauce

Wilted spinach, for garnish

STEPS

1. Combine marinade ingredients. Dry grill vegetables, and while still hot, paint on marinade generously, but do not saturate.

2. Make cheese mixture by mixing together ricotta, feta, oregano, Mrs. Dash, black pepper, and egg whites until smooth.

3. Make one portion at a time to order. Layer vegetables and cheese mixture in an ovenproof mold sprayed with vegetable spray in this order: This can also be made in a larger vessel, depending on your application.

   1 slice eggplant

   2 tablespoons cheese mixture

   1 tablespoon tomato sauce

   1 slice zucchini

   2 tablespoons cheese mixture

   1 tablespoon tomato sauce

   1 slice yellow pepper

   2 tablespoons cheese mixture

   1 slice red pepper

4. When ready to serve, place in a 350°F oven for about 30 minutes. Let rest for 5 minutes. Unmold on a plate garnished with wilted spinach and top with hot tomato sauce.

NUTRITION COMPARISON

	Kcalories	Protein (g)	Fat (g)	Carbo (g)	Sodium (mg)	Chol (mg)
Traditional Recipe	417	24	19	39	44	64
Balanced Recipe	323	24	16	22	143	61

CHICKEN POT PIE

CHEF'S
NOTES

Chicken Pot Pie, another American classic, is usually made with:

1. A rich roux base, chicken velouté

2. Chicken meat

3. A variety of vegetables

Chicken pot pie is covered by a pie dough made with shortening and/or butter, then baked in an oven until golden brown and bubbling.

Vegetables are usually carrots, onion, potatoes, celery, and peas, but root vegetables, peppers, and squashes have become more common as this traditional favorite gets continually reinvented through the years. The variety of vegetables you choose for this dish is up to you. Parsnips, celery root, sweet potatoes, fennel, butternut squash, peppers, or fresh corn are all great choices that can work in this application.

This is another one of those winners with great flavor, classic look, and immense value. This can easily be a staple on your menu as a lunch, dinner or small side to fit the needs of your clients. Food doesn't need to be masked with butter, cream, salt, or fat to taste great. Note that this recipe uses a light velouté sauce (recipe on next page). The choice depends on what your clients prefer, and what you can execute in your kitchen.

Serves 6

2 teaspoons vegetable oil

1 small onion, small dice

½ cup carrots, small dice

½ cup celery, small dice

¼ cup parsnips, small dice

1 cup chicken stock

2 sprigs of thyme

½ sprig rosemary

1 bay leaf

½ cup potatoes, small diced and blanched

1 cup frozen peas

2 cups cooked chicken, medium dice

1 cup light velouté base

Salt and pepper, to taste

Pastry: ricotta chive crust made with skim milk ricotta, egg whites, herbs, flour, and butter

STEPS

1. **Heat oil in a saucepan. Add onion, carrots, celery, and parsnips. Add in ½ cup stock a little at a time. Make a bouquet garni with thyme, rosemary, and bay leaf and add to vegetables. Cook until soft, about 5–7 minutes. Add in blanched potatoes; cook additional 2 minutes**

2. **Turn oven to 375°F.**

3. Remove herbs from stock. Stir in peas and chicken and 1 cup of balanced velouté base. Cook an additional 5 minutes to desired consistency. Add salt and pepper to taste, if needed.

4. Spoon filling into a 6-ounce ovenproof dish. Place ricotta dough with vents on top and seal the edges. Brush lightly with milk.

5. Bake for about 30 minutes or until filling is bubbling and top is golden brown.

● ● ● **NUTRITION COMPARISON**

	Kcalories	Protein (g)	Fat (g)	Carbo (g)	Sodium (mg)	Chol (mg)
Traditional Recipe	414	20	24	29	541	55
Balanced Recipe	278	16	15	20	140	43

SAUCES AND DRESSINGS

VELOUTÉ SAUCE

CHEF'S NOTES

Velouté sauce is a classic mother sauce, used in a variety of preparations as a base for sauces, soups, and stews. It is made with a flavorful stock, usually chicken and fish, and thickened with a traditional butter and flour, cooked blond roux. The flavor and mouthfeel of this sauce is critical to a variety of popular dishes, including fricassee, a la king sauces, fish and oyster stew, as well as soups such as Billy Bi soup (Cream of Mussels), cream of chicken, mushroom, asparagus, and broccoli soup.

This alternative sauce can be the foundation of a lighter cooking style. A good, defatted stock with a lighter thickening option can be an accepted alternative to this classic sauce with extremely favorable results using high-quality fresh ingredients. This base sauce can be used in a variety of cooking applications, in sauce making (such as seven onion sauce, marsala, horseradish, and curry) or in soup preparation (such as corn and fish chowders, shrimp and lobster bisque, or cream of broccoli and spinach). Selections using this balanced sauce, such as pot pies; winter vegetable stews; veal and wild mushroom; pork with chilies, new potatoes, onions, and peppers; snapper or shellfish stews create interesting choices while still maintaining balance.

Makes 2 Cups

1 tablespoon arrowroot
2 cups chicken stock

sachet
Salt and white pepper

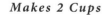

STEPS

1. Combine the arrowroot with just enough stock to form a paste.

2. Bring the remaining stock to a boil with sachet.

3. Whisk the arrowroot paste into the boiling stock and stir for a few minutes, just until thickened.

4. Salt and pepper if needed to taste. Strain and use.

NUTRITION COMPARISON

	Kcalories	Protein (g)	Fat (g)	Carbo (g)	Sodium (mg)	Chol (mg)
Traditional Recipe	76	2	6	3	136	16
Balanced Recipe	10	1	0.5	1	385	0

CREAMY DRESSING

CHEF'S NOTES

Popularity of thick, rich creamy salad dressings has expanded on menus from fast-food to fine-dining restaurants. Creamy dressings are traditionally made with these ingredients:

1. Mayonnaise, sour cream, and/or heavy cream

2. A variety of ingredients such as fresh herbs, spices, condiments, vinegar, and citrus juices that add flavor

3. Cheeses such as blue and grated Italian

From a classic Caesar dressing to a creamy blue cheese, to a seasoned ranch, Russian, or French dressing, all high-kcalorie, high-fat dressings add kcalories onto your often low-kcalorie balanced salads. To transform these creamy favorites into a more balanced choice, use tofu, pureed white beans, vegetables, nonfat yogurt, nonfat sour cream, or nonfat milk with the addition of bold seasonings to add flavor depth. These balanced preparations can be used for dips and sandwich spreads, adding a lighter option to a typically fattening product. With the right combination of fresh herbs, spices, good-quality cheeses, and vegetables, these dressings and sauces are great substitutes for the traditional versions.

Green Goddess Dressing

6 Servings

4 ounces soft tofu, well drained

½ cup cider vinegar

2 stalks celery

½ cup spinach leaves, washed and dried

⅓ cup fresh parsley (no stems)

1 tablespoon lemon juice

2 scallions

Fresh tarragon leaves, to taste

Fresh ground pepper, to taste

STEPS

1. Blend all ingredients in food processor until smooth. This dressing will last two to three days in the refrigerator.

Yogurt Dressing/Sauce

6 Servings

8 ounces plain nonfat yogurt

1 ounce skim milk

1 ounce white wine

½ ounce lime juice

1 ounce Dijon mustard

½ ounce honey

½ tablespoon curry

1 teaspoon turmeric

½ teaspoon cayenne

½ tablespoon vegetable seasoning

1 tablespoon cilantro, minced

1 teaspoon fresh mint, chopped

STEPS

1. Combine all ingredients in blender. Chill.

Cucumber Dill Dressing

Makes 2 Quarts

12 ounces soft light cream cheese

3 ounces farmer's cheese

1 pint skim milk

1 pound cucumbers, peeled and seeded

2 ounces fresh dill, finely chopped

3 tablespoons Dijon mustard

2 teaspoons garlic

½ teaspoon black pepper

1 teaspoon salt

2 ounces olive oil

3 ounces lemon juice

½ teaspoon Tabasco

STEPS

1. Blend cheeses together with skim milk.

2. Add other ingredients and blend until smooth.

Carbonara Sauce

Makes 1½ Quarts

1½ ounces corn-oil margarine

2½ ounces flour

1 quart skim milk

4 ounces white wine

6 ounces skim-milk ricotta cheese

Garlic herb seasoning and cracked black pepper, to taste

3 ounces Parmesan cheese

STEPS

1. Make roux with margarine and flour. Add skim milk, wine, ricotta, and seasonings.

2. Bring to a boil, simmering slowly until thickened. Add Parmesan and lightly simmer.

NUTRITIONAL COMPARISON (per 2 tablespoon serving)

	Kcalories	Protein (g)	Fat (g)	Carbo (g)	Sodium (mg)	Chol (mg)
Traditional Recipe	218	2	22	2	580	20
Green Goddess	25	2	0	2	147	0
Yogurt Dressing	45	2	0	6	194	0
Cucumber Dill Dressing	53	2	4	2	150	7
Carbonara Sauce	57	3	2	5	77	4

OIL AND VINEGAR DRESSINGS

Basic or classic French vinaigrette, oil and vinegar dressings typically have a ratio of three parts oil to one part vinegar/acid. Oils such as extra virgin olive or a nut oil, or a combination, are mixed with acids such as red or white wine; sherry, balsamic, or cider vinegars; lemon or lime juice; with the addition of salt and pepper. The base dressing can be flavored with a variety of ingredients, such as fresh herbs, spices, vegetables, fruits, or chilies, and can be emulsified with mustard depending on your application. To create a more balanced option, the oil ratio will have to be reduced and replaced with a slightly thickened stock or reduced juice to replicate the viscosity of oil. Since the amount of oil is reduced, using a strong flavorful oil such as extra virgin olive, pecan, pistachio, or hazelnut or a combination adds more depth to this lighter alternative.

You can substitute different flavored vinegars or virgin nut oils to create alternative dressing options easily adapted to your existing menus. Garnish these dressings with fresh fruits, such as raspberries, mango, or strawberries. Additionally, you can add finely diced vegetables and fresh herbs, such as fennel, sautéed zucchini, peppers, celery, chilies, basil, cilantro, chives, tarragon, or basil to create a variety of dressing options.

CHEF'S NOTES

Basic Herb Vinaigrette

Makes 1 Gallon

2 quarts plus 1 cup chicken stock or fresh vegetable stock

½ ounce cornstarch slurry (cornstarch with cold stock)

2 tablespoons fresh chopped garlic

2 teaspoons black pepper, coarsely ground

2 tablespoons fresh thyme, chopped

2 ounces fresh parsley, chopped

2 tablespoons, fresh oregano, chopped

1½ ounces fresh basil, chopped

2 ounces chives, chopped

1 quart balsamic vinegar

4 cups extra-virgin olive oil

STEPS

1. Heat stock to a rolling boil.

2. Thicken with cornstarch slurry to a nappe (to coat) consistency. Add garlic and incorporate.

3. Cool and add pepper, herbs, and vinegar.

4. Whisk olive oil into the mixture to emulsify.

5. Cool and store.

Orange Vinaigrette

Makes 3 Cups

½ cup orange juice concentrate

1 tablespoon Dijon mustard

1 tablespoon honey

1 tablespoon shallots, finely diced

½ teaspoon coriander, ground

½ teaspoon black pepper, ground

2 ounces white-wine vinegar

2 ounces extra-virgin olive oil

1 teaspoon thyme, fresh, chopped

1 teaspoon chives, fresh, chopped

1 teaspoon basil, fresh, chopped

STEPS

1. Place orange concentrate in food processor. Add mustard, honey, shallots, coriander, black pepper, and vinegar.

2. Start food processor and slowly add oil to emulsify.

3. Add fresh herbs last, but do not over puree.

4. Reserve in refrigerator until needed.

NUTRITION COMPARISON (per 2 tablespoon serving)

	Kcalories	Protein (g)	Fat (g)	Carbo (g)	Sodium (mg)	Chol (mg)
Traditional Recipe	342	0	34	0	1946	0
Herb Vinaigrette	141	1	14	4	15	0
Orange Vinaigrette	70	0	5	6	31	0

CARROT CAKE

Desserts are another opportunity to adapt traditional favorites with moderation of ingredients. Carrot cake typically has a batter consisting of:

- Shredded carrots
- Vegetable oil
- Whole eggs
- Brown sugar and honey
- Spices such as cinnamon, nutmeg, and ginger
- Nuts, such as walnuts or pecans, and raisins, dark or golden

Then the carrot cake is frosted with a cream cheese frosting, consisting of butter, sugar, and cream cheese.

This lighter version reduces the oil, the amount of eggs, and substitutes the cream cheese frosting with a nonfat yogurt sauce sweetened with maple syrup and thickened with gelatin. The use of egg whites helps to reduce the amount of saturated fat. Canola oil is a great source of monounsaturated fats. You may have thought that making a cake with whole-wheat flour would be tough and heavy, but it works quite well in this recipe. In most baking recipes, you can successfully replace half of the white flour with whole wheat, but with the combination of carrots, spices, honey, and egg whites, the whole-wheat flour blends very well. The addition of traditional spices, the sweetness of the honey, and the yogurt cream topping prove to create a satisfying option to the traditional version. Creatively, the presentation of this dessert will equally impress your guest while pleasantly surprising them with a good alternative to this craved classic (Figure 9-5).

CHEF'S NOTES

FIGURE 9-5 Carrot Cake.

Photo by Peter Pioppo.

Serves 12

NATURELLE BAVARIAN CREAM

¼ cup pure maple syrup

⅓ cup skimmed ricotta cheese

⅓ cup nonfat plain yogurt

¼ teaspoon vanilla extract

½ teaspoon gelatin

1 teaspoon water

2 cups whole wheat flour

1½ teaspoon baking powder

⅔ teaspoon cinnamon

¼ teaspoon baking soda

⅛ teaspoon nutmeg

2 eggs

2 tablespoons applesauce

1 cup canola oil

1¾ cups grated carrots

1½ cups egg whites

1½ cups honey

¼ cup apple juice

STEPS

1. Ahead of time, make Naturelle Bavarian Cream. To make Bavarian cream, mix maple syrup, ricotta cheese, yogurt, and vanilla in blender. Whip till smooth. Dissolve gelatin in warm water. Add a little dessert base to dissolved gelatin. Slowly mix gelatin into the rest of the base. Chill overnight.

2. Stir together dry ingredients. Combine eggs, applesauce, oil, and 1½ cups carrots in a separate bowl, reserving ¼ cup for garnish. Add dry ingredients to egg mixture; whip 3 minutes. Scraping sides and bottom of bowl.

3. Whip egg whites and honey until medium firm. Gently fold into carrot mixture. Pour into round cake pan. Bake at 350°F.

4. Slice cake into 12 wedges. Place a wedge on each plate.

5. Mix Naturelle Bavarian Cream with reserved ¼ cup grated carrots and apple juice to thin. Drizzle on cake. Serve with dried orange slices.

● ● ● **NUTRITION COMPARISON**

	Kcalories	Protein (g)	Fat (g)	Carbo (g)	Sodium (mg)	Chol (mg)
Traditional Recipe	650	5	33	89	287	107
Balanced Recipe	395	8	19	52	295	39

OATMEAL RAISIN COOKIES

CHEF'S NOTES

Although oatmeal cookies sound like they are healthy, with the fat to sugar ratio and addition of eggs, the traditional oatmeal cookie recipe is high in kcalories relative to the size of the cookie. Traditionally these cookies have some fundamental ingredients:

- Sweet butter
- Brown and white sugar
- Whole eggs
- Oatmeal
- Spices, such as cinnamon and nutmeg
- Raisins

With some basic modifications these cookies can be an answer to an occasional sweet tooth with little guilt and remorse.

This recipe uses mostly applesauce (with a tiny amount of canola oil) to replace the butter. This is an excellent cookie, with a naturally sweet taste due to the golden raisins and spices. Served warm, this modified recipe is as satisfying as the traditional favorite.

Makes 3½ Dozen or 42 Cookies

1 pound brown sugar
14 ounces applesauce
½ ounce canola oil
1 ounce egg whites
12 ounces all-purpose flour

7 ounces oatmeal
1 tablespoon baking soda
2 cups golden raisins
1 teaspoon cinnamon

STEPS

1. Mix together the sugar, applesauce, oil, and egg whites until smooth.
2. Combine the flour, oatmeal, baking soda, raisins, and cinnamon and add to the applesauce mixture. Mix until well blended.
3. Spoon onto a parchment-lined sheet pan and bake at 325°F until golden brown.

NUTRITION COMPARISON

	Kcalories	Protein (g)	Fat (g)	Carbo (g)	Sodium (mg)	Chol (mg)
Traditional Recipe	135	2	5	21	115	0
Balanced Recipe	128	2	0	29	96	0

SUMMARY

1. The recipe makeovers demonstrated a variety of ingredients, techniques, and cooking methods that can be used to prepare tasty balanced menu items.
2. In entrées/appetizers, some of the changes included the following:
   - Preparing lean proteins with marinades/rubs
   - Incorporating egg whites instead of whole eggs
   - Using alternative thickeners such as cornstarch and agar-agar
   - Using alternative cooking methods such as grilling or dry sautéing, or adding smaller amounts of oils such as olive oil
   - Incorporating whole grains and vegetables
   - Finishing dishes with butter or flavored oil at the end of cooking along with fresh herbs and spices
   - Adding a wide variety of herbs and other flavorings to develop unique flavor profiles served with tasty accompaniments and appropriate garnishes
3. By modifying a classic sauce with arrowroot, a quality chicken/fish stock, and seasonings, you can make an excellent low-kcalorie velouté sauce.
4. With a variety of high-quality flavorings, purees, and nonfat dairy products, you can create a variety of creamy salad dressings.
5. Samples of recipe modification such as in the classic herb and orange vinaigrette demonstrate the use of strong flavoring ingredients to decrease the amount of oil in the dressing.
6. The dessert recipes show how to incorporate ingredients such as applesauce, raisins, grated carrots, oatmeal, eggs whites, and spices to make delicious and satisfying dessert alternatives.

CHECK-OUT QUIZ

1. Explain at least three general ways to modify recipes to change the nutrient content.

2. Discuss three considerations to keep in mind when you modify a recipe.

3. Consider a pizza pie with sausage, pepperoni, and peppers. How do the ingredients for this balanced pizza differ?

   Pizza: pizza dough, cooking spray, 2 teaspoons olive oil, 4 ounces turkey Italian sausage, 1 cup sliced mushrooms, 1 cup sliced red bell pepper, 1 cup sliced orange bell pepper, 1 cup sliced onion, ¼ teaspoon crushed red pepper, 3 sliced garlic cloves, ¾ cup lower-sodium marinara sauce, 5 ounces shredded mozzarella cheese

4. Following is a traditional apple pie recipe. Modify the recipe to decrease the amount of fat and sugar.

   DEEP DISH APPLE PIE

   6 tart apples

   ½ cup sugar

   ½ cup brown sugar

   ½ teaspoon nutmeg

   Grated rind of 1 lemon

   Grated rind of 1 orange

   3 tablespoons butter or margarine

   ½ pastry recipe (below)

   Peel and core apples; cut into eights. Place in deep, greased baking dish. Combine sugar, brown sugar, nutmeg, lemon rind, and orange rind. Sprinkle over apples. Dot with butter or margarine. Top with thin sheet of pastry, pricked in a design. Bake at 425°F for 15 minutes. Reduce heat to 350°F and bake 40 to 45 minutes more or until crust is golden and filling is bubbly.

   PASTRY

   3 cups flour

   2 cups fat (butter or shortening)

   1 cup ice cold water

   Sift flour; measure. Mix and sift flour and salt. Cut in fat with two knives or pastry blender until flour shortening particles are about the size of small peas. Sprinkle 1 tablespoon cold water over mixture and mix in lightly with a fork. Continue adding in all the water in this fashion until pastry gathers around fork in a soft ball. Chill, then divide pastry in half and roll each half separately on lightly floured board to ⅛-inch thickness. Handle rolling pin very lightly. Makes enough for a two-crust 9-inch pie.

NUTRITION WEB EXPLORER

For a complete list of websites for the following activities, please visit the companion book page at www.wiley.com/college/drummond.

TASTE OF HOME

Click on "Healthy," then click on "Recipe Makeovers." Pick one of the many recipe makeovers on this website. Look at the "before" and "after" recipes and learn how the recipe was changed. Write a summary, including how flavor was developed in the new dish.

EATING WELL—HEALTHY SALAD DRESSING RECIPES

Look at the ingredients used in healthy salad dressing recipes. Pick a recipe you like that is a vinaigrette style dressing, and one that is a creamy style dressing. Make these dressings and evaluate them.

COOKING LIGHT—DESSERT MAKEOVERS

Pick one of the desserts that were made into a healthier version. Write up a summary describing what ingredients and techniques were changed to make the new dessert.

Developing and Testing Balanced Recipes

Your instructor will give you one of the following assignments.

Create your own creamy dressings with 3 different flavor profiles:

1. Vegetable flavored

2. Cheese flavored

3. Fresh herb and spice flavored

Using the basic vinaigrette recipe, create 3 different global flavorings, such as French, Asian, Italian, South American, or Mediterranean.

Using a traditional brownie, muffin, or cookie recipe, create your own modification and bake it, along with the traditional version.

Your instructor will give you a Worksheet to fill out with your recipe(s), including a nutrient analysis if possible. You will evaluate each product based on taste and appearance.

HOT TOPIC

GLUTEN-FREE BAKING

Celiac disease is an inherited autoimmune disease that usually affects several organs in the body before it is diagnosed and treated. When a person with celiac disease consumes any food, beverage, or medication containing gluten (found in wheat, barley, rye, or oats contaminated with these grains), his or her immune system is "triggered" and responds by damaging the lining of the intestinal tract.

Other people without celiac disease also have gastrointestinal symptoms after eating gluten. They have a condition called gluten sensitivity. Gluten sensitivity is not a wheat allergy or an autoimmune disease like celiac, and it does not seem to damage the intestines. Individuals with gluten sensitivity can often tolerate small to moderate amounts of gluten without symptoms, whereas individuals with celiac disease must eat a gluten-free diet.

FIGURE 9-6 Gluten-free baking ingredients. From left to right: brown rice flour, millet flour, xanthan gum, coconut flour, potato flour, almond meal flour, oat flour. At bottom: garbanzo bean flour.

Photo by Peter Pioppo.

Gluten is found in wheat and some other grains. To avoid gluten, you need to check food labels for these ingredients.

- Wheat—including durum wheat and flour, enriched flour, farina, graham flour, plain flour, white flour, wheat bran, wheat germ, semolina, kamut, and spelt wheat

- Barley

- Rye

- Triticale (a cross between wheat and rye)

- Oats (unless labeled gluten-free)

- Malt as in barley malt extract or flavoring or malt vinegar

- Brewer's yeast

- Dextrin

- Modified food starch

So which flours and starches can someone who is avoiding gluten eat? There actually are quite a few:

- Flours such as brown rice, sweet rice, coconut, sweet potato, tapioca, white rice, sorghum, teff, amaranth, millet, oat, buckwheat, almond, and bean flours such as black bean or garbanzo bean flour

- Starches/thickeners such as potato starch, tapioca starch, arrowroot, and cornstarch

Baking without gluten-free flours and starches can be challenging because gluten contributes important properties to various types of baked products like cookies, cakes, pastries, and breads. Gluten development is not as important for cookies as it is for cakes, so gluten-free flours can be substituted with similar results. Cakes and other types of batter-based products, like pancakes, need gluten for its gas-retaining ability that produces a light and airy interior structure and a tender crumb. Gums like xanthan gum and guar gum may be used to retain gas and give gluten-free baked goods the structure, volume, and texture that gluten would normally provide. Bread is perhaps the most challenging gluten-free baked product to make because gluten provides structure, creates a tender crumb, and retains gas. With experimentation and practice, a combination of gluten-free flours and gums can be used to create a loaf with good volume, softness, and texture.

Baking books and online resources frequently offer gluten-free flour blend formulations for use in making cookies, cakes, quick breads, and yeast breads. The formula might include three or four different types of flours and starches and make up to 12 cups of blended flour. Flours with stronger flavors (Table 9-2) typically make up no more than 25 percent to 30 percent of the total blend and are balanced by neutral flours and starches. Stronger-tasting flours (such as bean flours) generally are used in small quantities in recipes that feature delicate flavors. A higher percentage of these flours can be used in baked goods that include nuts, chocolate, or a high level of spice. Many chefs like to make their own gluten-free flour blend that can then be used in a variety of baked goods. One chef uses a ratio of 70 percent grain and nut flours to 30 percent starches to yield a good substitute for wheat flour. You can purchase ready-to-use gluten-free flour blends as well as gluten-free baking mixes.

Gluten-free flours also have less protein than wheat flour, and many are low in fiber. Choose mostly whole-grain flours to improve the fiber and nutrient content of your baked goods. Whole-grain flours, and especially bean flours, also contain more protein that will help give structure to your baked goods.

Gluten-free bread preparations have an entirely different look and feel than conventional breads made with wheat flour—more like batter than a chewy, elastic dough—and so the mixture must be baked in a walled container such as a loaf pan or muffin tin. If you've ever made a flourless chocolate cake, in which whipped eggs provide the structure, you'll be familiar with the way lightness and shape are brought into many gluten-free baked goods.

Table 9-3 lists additional gluten-free baking tips.

To avoid contamination when making gluten-free foods, clean all surfaces thoroughly. Keep gluten-free ingredients and foods separate from gluten-containing foods. Wheat flour can linger in the air in the kitchen for over 24 hours, so do your gluten-free baking first, and make sure you have scrubbed your kitchen and tools before starting baking. Use a different rolling pin and work surface than you do for conventional baking. Small amounts of gluten really do make a difference to individuals with celiac disease.

GLUTEN-FREE BLACK BOTTOM CUPCAKES

Yield: 12 Cupcakes

Filling

1 cup cream cheese, softened

½ cup sugar

⅛ teaspoon salt

1 egg

1 cup chocolate chunks

Batter

4 tablespoons rice flour

7 tablespoons tapioca starch

½ cup plus 3 tablespoons soy flour

1 cup plus 1 teaspoon sugar

7 tablespoons cocoa powder

1 teaspoon baking soda

½ teaspoon salt

1 cup water

⅓ cup oil

1 tablespoon vinegar

1 teaspoon vanilla

STEPS

1. **Preheat oven to 350°F. Grease a full-size 12-muffin tin.**

2. **Make the cream cheese filling: In a standing mixer with a paddle attachment or by hand with a wooden spoon, cream together the cream cheese, ½ cup sugar, and ⅛ teaspoon salt until smooth. Scrape the bowl; add the egg and continue to cream until light in color and as thick as sour cream. Fold in chopped chocolate chunks; set aside.**

TABLE 9-2	Characteristics of Gluten-Free Flours and Starches
FLOUR/STARCH	**CHARACTERISTICS**
Amaranth	Higher in protein than most grains Provides structure and binding capability Pleasant, peppery flavor Best used in combination with other gluten-free flours
Arrowroot	Used as thickener and in baking similarly to cornstarch
Bean/Legume	Legume flours include fava beans, garbanzo beans, black beans, soybeans Good source of protein and fiber Best used in combination with other gluten-free flours to balance taste and texture Bean flours complement sorghum flour
Buckwheat	Strong, somewhat bitter flavor Best used in pancakes or yeast breads in combination with neutral gluten-free flours
Corn flour	Used in breads, waffles, and tortillas
Corn meal	Used in spoon breads and baking powder-leavened breads
Cornstarch	Works well in combination with tapioca starch
Flax	Ground flaxseeds increase nutritional value High in soluble fiber, which allows gel formation; retains moisture and gives spongy texture to baked goods Nutty, bold flavor Adds color to baked goods
Millet	Powdery consistency, color similar to cornmeal Delicate, sweet flavor Suitable for use in flatbreads and muffins
Nut flours	Almond, pecan, walnut, hazelnut, filbert, and chestnut Contribute flavor and nutrition to baked products Best used in combination with other gluten-free flours to balance taste and texture
Quinoa	Good source of protein Mild, slightly nutty flavor Suitable for cookies, cakes, and breads
Potato flour	Neutral flavor Blends well with stronger-flavored flours
Potato starch	Provides a light consistency to baked products Helps retain moisture, combines well with eggs Bland flavor, low in fiber and nutrients
Rice, Rice bran	Comes in brown, white, and sweet varieties Best used when combined with other gluten-free flours and binders or gums Neutral flavor Sweet rice flour is used in pie crusts and as a thickener (gummy)
Sorghum	Sweet, nutty flavor Best when used with other neutral gluten-free flours and gums
Teff	Taste similar to hazelnuts Gels well—a good thickener
Tapioca	Starchy, sweet flavor Adds chewy texture to breads Used in blends to improve color and crispiness of crusts

TABLE 9-3	Gluten-Free Baking Tips

TO INCREASE NUTRITION

- Use a variety of gluten-free flours in combination to maximize nutrition.
- Use whole grain or enriched, gluten-free flours (vitamins and minerals have been added).
- Substitute up to ¼ cup ground flaxseeds plus ¼ cup water for ¼ cup flour in a recipe (flax will absorb more moisture).

TO INCREASE MOISTURE

- Add gelatin, extra egg, or oil to the recipe.
- Honey can help retain moisture.
- Brown sugar often works better than white.

TO ENHANCE FLAVOR

- Add chocolate chips, nuts, or dried fruits.
- Double the amount of spices.
- Use high-quality extracts.

TO ENHANCE STRUCTURE

- Use a combination of gluten-free flours and mix together well before adding to other ingredients.
- Add dry milk solids or cottage cheese into recipe.
- Use evaporated milk in place of regular milk.
- To reduce grainy texture, mix rice flour or corn meal with liquid. Bring to a boil and cool before adding to recipe.
- Add extra egg or egg white if product is too crumbly.
- Do not overbeat; kneading time is shorter since there is no gluten to develop.

LEAVENING

- Starch flours need more leavening than wheat flours.
- Rule-of-thumb: Start with 2 teaspoons baking powder per cup of gluten-free flour.
- If baking soda and buttermilk are used to leaven, add 1⅛ teaspoon cream of tartar for each ½ teaspoon baking soda used to neutralize acid.
- For better rise, dissolve leavening in liquid before adding to other ingredients or add a little extra baking powder.

TEXTURE/LIGHTNESS

- Weigh dry ingredients.
- Sift flours and starches prior to measuring. Combine and sift again (together) after measuring to improve the texture of the product.
- Hold gluten-free dough at least ½ hour (up to overnight) in the refrigerator to soften and improve the final texture of the product.
- In products made with rice flour or corn meal, mix with the liquid called for in the recipe. Bring to a boil and cool before adding to recipe to help reduce grainy texture.

BAKING PANS AND UTENSILS

- Bake in smaller-than-usual portions at a lower temperature for a longer time (small loaf pans instead of standard size; use mini-muffins or English muffin tins instead of large muffin tins).
- Use dull or dark pans for better browning.

FRESHNESS

- Gluten-free baked goods can lose moisture and quality quickly. Wrap them tightly and store in the refrigerator or freezer in an airtight container to prevent dryness and staling.
- Refrigerate all flours for freshness and quality but bring to room temperature before measuring.

Adapted from "Gluten-Free Baking" by Colorado State University Cooperative Extension, 2009.

MENU IDEAS AND PRESENTATION

Following are menu ideas and tips for presenting soups:

- For appearance, strain soups such as broccoli, celery, cauliflower, and asparagus through a large-holed china cap to remove fibrous strands. Also, puree bean soups such as black bean and split pea to get a homogeneous product. Next, strain in a large-holed china cap to remove the skins. Add any vinegar of your choice or another acid such as lime juice to finish bean soups. Dark vinegars are best for darker soups such as black lentil and red bean. White or red wine vinegar, or cider vinegar, is best for lighter color bean soups. The acid will open up the flavor and bring out a more intense aroma that is pleasing to the palate.

- Garnish soups with an ingredient to accompany the soup whenever possible. For example, serve pieces of baked tortillas with a Mexican-style soup. The use of fresh vegetables, fruits salsas, or herbs as a garnish adds interest, color, and flavor. Also consider garnishing some soups, as appropriate, with a small dollop of whipped cream, roasted nuts, baked wheat croutons, low-fat seasoned sour cream, smoked chicken breast, avocado cilantro salad, seafood, or vegetable salad.

SUMMARY

1. Soups can spotlight vegetables, beans, lentils, split peas, and grains. Some of these ingredients, such as potatoes and rice, can act as thickeners.
2. Stocks, herbs, and spices are low-kcalorie ways to support flavor in soups.
3. If using concentrated bases, pick those listing beef, chicken, or fish extract as the first ingredient—not salt.
4. Steps are listed that explain how to fortify a base with flavor.
5. For appearance, strain soups such as broccoli or celery to remove fibrous strains. Puree bean soups to get a homogeneous product.
6. Garnish soups with an ingredient to accompany the soup whenever possible.

SALADS AND DRESSINGS

BALANCED INGREDIENTS AND PREPARATION

Components of salads go way beyond simple raw vegetables. Salads are a wonderful place to feature high-fiber components, lean proteins, and low-fat accompaniments. Consider using grains; beans; lentils; pasta; fresh fruits and juices; oven-dried, roasted, or grilled vegetables; dried fruits; poultry; meat; seafood; game; and herbs and spices such as ginger, Kaffir lime leaves, star anise, cardamom, curry, lavender, lemon balm, basil, oregano, thyme, sage, and cinnamon. As elsewhere in the kitchen, use fresh, high-quality ingredients for salads. When creating your dish, choose ingredients for compatibility of flavors, textures, and colors.

There are two categories of salads: simple and complex. The term *simple* refers to assorted greens and lettuces sometimes with the addition of toppings such as tomatoes, cucumbers, shredded carrots, avocados, or nuts. There are several guidelines to formulate a simple salad:

1. **Use fresh lettuces.** There are five distinct types of lettuce: leaf or loose-leaf lettuce, Cos or romaine, crisphead, butterhead, and stem. **Leaf lettuce** produces crisp leaves loosely arranged on a stalk. **Cos or romaine** forms an upright, elongated head. **Crisphead** varieties

Leaf lettuce A type of lettuce with crisp leaves loosely arranged on a stalk; the leaves are often colorful.

Cos or romaine A type of lettuce that forms an upright, elongated head with tall leaves.

Crisphead A type of lettuce, such as iceberg, that contain curved, overlapping leaves that form crispy, firm round heads.

include iceberg, which has little flavor and is not as rich in nutrients as many other lettuces. The **butterhead** varieties are generally small, loose types with tender leaves that have a delicate sweet flavor. **Stem lettuce**, also called Chinese lettuce, is grown mostly for its thick stem and used mainly as a vegetable. Lettuce varieties include red and green leaf, endives, escarole, Bibb, oak, limestone, dandelion, red and green romaine, and baby lettuces such as baby romaine, baby red oak leaf, and lola rossa (a red lettuce with ruffled leaves). Spicy greens include arugula or rocket, watercress, garden cress, mustard, amaranth, baby turnip greens, and mizuna (also known as Japanese mustard greens).

2. **Add sprouts. Sprouts**, young plants that have just emerged from their seeds before any leaves grow, include mung bean, alfalfa, broccoli, radish and mustard sprouts.

3. **Experiment with microgreens. Microgreens**, the first true leaves that develop after a seed sprouts, add visual interest and a host of flavors to salads. Examples include amaranth, red garnet amaranth, arugula, celery, cilantro, purple or pink orach, green or red mustard greens, dark red beet, and lemon basil. Depending on flavor, some work well in salads or simply as a garnish on salads or other menu items.

4. **Provide appealing combinations.** Color, texture, and flavors of ingredients should complement each other.

5. **Use high-quality vinegars and oils.** The dressing should represent the profile of the greens. Light vinaigrette works best with delicate lettuces, and thicker stronger dressings with heartier varieties of lettuces. Both may be enhanced with fresh milled black pepper.

Complex salad, by definition, means that it is more complicated and involves different ingredients, such as a variety of proteins, vegetables, fruits, grains, and legumes tossed with a flavorful dressing. There are four categories of complex salads:

1. Complex side salads that are accompaniments for sandwiches or entrées such as:
   - Black Bean Cake with Grilled Shrimp, Avocado, Tomato, and Corn Salad
   - Crabmeat and Shrimp Sandwich with Jicama, Mango, and Celery Root Salad

2. Bound protein salads, such as fresh tuna, chicken, or shrimp, may be tossed with nonfat yogurt dressing, green goddess, or a basic vinaigrette emulsified with Dijon mustard.

3. Complete entrée salads either with meat, starch, and vegetable or vegetarian style with legumes, grains, vegetables, and possibly dairy that provides the right balance for a nutritional meal. For example, legumes make wonderful salads. Black-eyed peas go well with flageolets and red adzuki beans. To add a little more color and develop the flavor, you might add chopped tomatoes, fresh cilantro, haricots verts, and roasted peppers with smoked chicken, seared tuna, salmon, or duck.

4. Fruit salads consisting of a variety of fresh and dried fruits and their juices with fresh mint, lavender, basil, or served fresh and pure.

There are a wide range of greens, ingredients, and dressing combinations you can use to create simple or complex salads. They will add healthy choices, variety, and creativity to your menus. The key is to make sure your ingredients blend well and complement each other for a well-received salad.

Dressings are used in much more than salads. They can often be used as an ingredient in entrées, appetizers, relishes, vegetables, and marinades. There are many categories of dressings. The best place to start is the basic vinaigrette, because it is a base component that blends well with ingredients such as herbs, spices, vegetables, and fruits to create many variations. The best ingredients to use include good-quality vinegars, first-pressed olive oil, infused oil and nut oils, or combinations with the addition of fresh herbs. Use the most flavorful oils in this process because you will need the strongest flavor with the least amount of fat. Other good-quality ingredients that add flavor without large amounts of kcalories or fat include Dijon mustard, shallots or garlic (which may be roasted for a robust flavor), a touch of honey, capers, hot sauces, reduced vinegars, lemon or lime juice, and fruit and vegetable juices.

Butterhead A type of lettuce with generally small, loose, and tender leaves with a delicate sweet flavor.

Stem lettuce A type of lettuce, also called Chinese lettuce, grown mostly for its thick stem.

Sprouts Young plants that have just emerged from their seeds before any leaves grow.

Microgreens The first true leaves that develop after a seed sprouts.

- Indian Spiced Chicken Breast, Griddled Sausage, Tomato Chutney, Braised Fava Beans, Fennel, and Cauliflower
- Pan-Seared Pork with Brussels Sprouts, Mushrooms, and Pearl Onions (Figure 10-6)
- Cedar-Planked Wild Striped Bass with Ratatouille and Butterscotch Bean Salad and Micro Arugula
- Cornmeal-Crusted Monkfish with Wheat Berries, Black Beans, Grilled Fennel, Green Beans, and Spicy Tomato Relish
- Everything Crusted Salmon, Glazed Carrots, Green Beans, Purple Pearl Onions, Lentils, Fava Beans, and Roasted Apple Mustard
- Coffee-Crusted Chicken with Spinach, Butternut Squash, Mashed Potatoes (Figure 10-7)

FIGURE 10-6 Seared Pork Roulade with Sautéed Brussels Sprouts, Caramelized Mushrooms, and Roasted Pearl Onions with Paprika Cream.

Photo by Peter Pioppo.

S U M M A R Y

1. For meat, poultry, and fish entrees, select about a 5-ounce raw portion of a lean cut (listed above).
2. Use smaller entrée portions with grains, beans, colorful seasonal vegetables, and lighter sauces.
3. It is essential to develop flavor in your protein without much salt. Use marinades, dry rubs, and cooking techniques such as grilling, searing, or poaching in flavorful liquids to develop flavor.
4. When using cheese in an entrée, use a small amount of a strong cheese and grate on top so the first taste will include the flavorful cheese.
5. Additional ideas for entrées are given.

FIGURE 10-7 Coffee-Crusted Chicken with Spinach, Butternut Squash, Mashed Potatoes.

Photo by Peter Pioppo.

BALANCED SAUCES

Up to this point, you may have noticed the lack of heavy "classical" sauces. Rather than foods covered with rich, heavy sauces, the emphasis in this book is on the natural integrity of the food itself. In sauce making, this has meant a great change in the techniques of thickening. Instead of roux to make sauces, this new style often thickens with purees or with reductions of fruits and vegetables or meat juices and stock.

Alternatives to classic sauces, many of which are high in fat, take many forms. Pureeing of vegetables or starchy foods is commonly done as a means to thicken soups, stews, and sauces without adding fat. Sauces made with vegetables purees are light and low in fat and kcalories. With the addition of herbs and spices, they make a colorful, tasty complement to entrées and side dishes. For example, you can puree roasted yellow or red peppers with ginger, garlic, and herbs to make a vibrant sauce for grilled swordfish or tuna.

Before puréeing, vegetables should be cooked until just tender, using methods such as steaming, roasting, grilling, or sautéing. Roasting caramelizes the vegetables' natural sugars, making them taste rich and sweet. Vegetables that roast well include peppers, eggplant, asparagus, broccoli, zucchini, cauliflower, and root vegetables. More delicate vegetables, such as mushrooms, can be sautéed. Other vegetables that make excellent purees are butternut acorn squash, artichokes, parsnips, peas, and carrots.

Coulis, a French term, refers to a thick puree or sauce made from vegetables or fruits. A vegetable coulis, such as bell pepper or tomato, can be served hot or cold to accompany entrées, appetizers, and side dishes. Recipe 10-37 is for Red Pepper Coulis. Fruit coulis is usually served cold as a dessert sauce.

A vegetable coulis is often made by cooking the main ingredient, such as tomatoes, with typical flavoring ingredients such as onions, garlic or shallots, and herbs in a liquid such as stock. The vegetables and flavorings are then pureed, and the consistency and flavoring of the product are adjusted. A vegetable coulis can also be made by cooking the vegetable with potato or rice (2 ounces per gallon) to give the sauce a silky texture and a smooth consistency when it is pureed and strained.

The texture of a vegetable or fruit coulis is quite variable, depending on the ingredients and how it is to be used. A typical coulis is about the consistency and texture of a thin tomato sauce. The color and flavor of the main ingredient should provide most of the eye appeal and flavor.

Salsas are versatile, colorful, low-fat, high-flavor sauces. Salsas are mixtures of vegetables and/or fruits and seasoning ingredients. Salsas are a staple of Mexican cuisine, typically made with tomatoes, chilies, onion, garlic, and fresh cilantro, but many types are found throughout Latin American. Most salsas consist of a chopped or almost pureed vegetable (most often tomatoes) and fruits, with the addition of a strong flavoring such as red, white, or green onions; garlic and lime juice; herbs and spices; and chilies. Chilies are frequently cooked to tame their flavor. Recipe 10-35 is for Papaya and Plantain Salsa.

Relishes are cooked, pickled foods, typically served as a cold condiment, and have recently exploded on menus as excellent sauce choices for meat, poultry, and seafood. Since salsas and relishes contain little or no fat, they rely on ingredients with an intense flavor, such as cilantro, thyme, dill, bay leaf, jalapeño peppers, chilies, lime and lemon juice, garlic, onions, pickling spice, coriander, and mustard. Figure 10-8 is an example of a relish.

Chutneys, such as tomato-papaya, mango-onion, tomato ginger, or tamarind mint, are made from fruits, vegetables, herbs, and spices, and are either hot or sweet. Chutney originated in India and the word chutney derives from the word *Chatni,* meaning strongly spiced. Chutney is a wonderful accompaniment to a wide variety of dishes adding a boost of flavor with little or no fat and moderate in kcalories. Recipe 10-34 explains how to make Papaya and White Raisin Chutney.

Coulis A sauce made of a puree of vegetables or fruits.

Salsa Mixtures of vegetables and/or fruits and flavor ingredients.

Relish Cooked, pickled foods typically served as a cold condiment.

Chutney A strongly spiced sauce from India, either hot or sweet, that is made with fruits, vegelables, and herbs.

FIGURE 10-8 Marinated Asparagus with Beans, Grains, and Roasted Vegetables.

Photo by Peter Pioppo.

Compote A preparation of fruit, fresh or dried, cooked in syrup flavored with spices such as vanilla, citrus peel, toasted spices, vanilla bean, or liqueur.

A **compote** is a preparation of fruit, fresh or dried, cooked in syrup flavored with spices such as vanilla, citrus peel, toasted spices, vanilla bean, or liqueur. It is often served as a dessert or as an accompaniment to a savory item. When using ripe fruits and creative seasoning, compote can be made with a reduced sugar content, yielding excellent results.

A *mojo* is a spicy sauce from the Caribbean and South America. It is traditionally a mixture of sour oranges and their juice, garlic, oil, and fresh herbs.

Glaze A stock reduced to a thick, gelatinous consistency with flavoring and seasonings.

Also, stock can be flavored and reduced to make a quality sauce that can be used in many dishes on the menu. **Glazes** are basic preparations in classical cookery. They are simply stocks reduced to a thick, gelatinous consistency with flavoring and seasonings. Meat glaze, or glace de viande, is made from brown stock. Glace de volaille is made from chicken stock, and glace de poisson is made from fish stock. To prepare a glaze, you reduce stock over moderate heat, frequently skimming off the scum and impurities that rise to the top. When the stock has reduced by about half, strain it through a fine mesh strainer into a smaller heavy pan. Place it over low heat and continue to reduce it until the glaze forms an even coating on a spoon. Cool, cover, and refrigerate or freeze.

Small amounts of glazes (remember, they are very concentrated) are used to enhance sauces and other items. They make sauces cleaner and fresher tasting than sauces made with heavy thickeners. They can also be added to soups to improve and intensify flavor. However, they cannot be used to re-create the stock from which they were made; the flavor is not the same after the prolonged cooking at higher temperatures.

Juices can be used as is for added flavor or can be reduced (boiled or simmered down to a smaller volume) to get a more intense flavor, a vibrant color, and a syrupy texture. Reduced juices make excellent sauces and flavorings. Use a commercially intended juicer or buy quality juices.

For example, orange juice can be reduced to orange syrup, which is excellent in salad dressings, marinades, and sauces. Also, freshly made beet juice can be used to enhance stocks, glazes, and sauces. Reduced beet juice with seasonings placed into a squirt bottle can be drizzled or composed on plates as a decoration and flavor, especially with salads, appetizers, and entrées. Other juices that are add flavor and color are carrot, fennel, celeriac, pomegranate, ginger, celery, asparagus, bell pepper (yellow, red, orange), herbs (watercress, cilantro, parsley, basil, chive), citrus, and tomato.

Flambé To light alcohol in preparation.

Wines, liqueurs, brandy, cognac, and other spirits are often added as flavorings at the end of cooking. Sherry, a popular flavor profile, is used in a variety of sauces. Brandy, rum and other liquors are often poured over a dish and **flambéed** (set afire) at the time of service. This adds some flavor but is done more in the dining room for show.

In preparations such as sauces, wines and spirits may be added during cooking to become part of the total flavor. They are then flavor builders rather than flavorings, playing one role in one dish and a different role in another dish.

With their many colors, flavors, and textures, each of these items is a contemporary, exciting and flavorful addition to a variety of preparations.

SUMMARY

1. Alternatives to traditional high fat sauces include coulis—a thick puree or sauce made from vegetables or fruits. A typical coulis is about the consistency and texture of a thin tomato sauce. The color and flavor of the main ingredient should provide most of the eye appeal and flavor.
2. Most salsas consist of diced or almost puréed vegetable and fruit, with the addition of strong flavorings.
3. Relishes are cooked, pickled foods and chutneys are made from fruits, vegetables, herbs, and spices, and are either hot or sweet.
4. A compote is a preparation of fruit, fresh or dried, cooked in syrup flavored with spices.
5. A mojo is a spicy sauce traditionally made of sour oranges, garlic, oil, and fresh herbs.
6. Reduced stock (glazes) and reduced juice can be used as ingredients in sauces.
7. Wines, liqueurs, brandy, cognac, and other spirits can be added as flavorings at the end of cooking.

SIDE DISHES

BALANCED INGREDIENTS AND PREPARATION

There is no end to the variety of side dishes for every entrée. The same entrée can take on a new face simply by changing the accompaniments. Besides the traditional side dish of vegetables and potatoes, consider using grains such as wheat berries and barley. Try legumes, such as heirloom beans and green lentils with the addition of seared tofu, roasted beets, root vegetables, citrus zest, dried fruits, and fresh herbs.

Consider techniques such as the following:

- **Puréeing.** For example, purée sweet potatoes, butternut squash, and carrots flavored with cinnamon, fresh grated nutmeg, honey, and thyme. You can also purée cauliflower with roasted garlic, skim milk and olive oil, or potatoes using the same preparation and similar ingredients yielding a delicious fluffy whipped purée.

- **Roasting.** Vegetables can be oven-roasted to bringing additional flavor, color, and texture to your preparation. The browning that occurs during roasting adds a deep rich character to vegetables. For example, potato wedges or slices can be seasoned and roasted. Vegetables such as cauliflower, broccoli, parsnips, carrot, asparagus, squashes, and eggplant can be seasoned and roasted to create a variety of side-dish options. This preparation doesn't need to be roasted on a rack. Season the vegetables in a pan with herbs, such as thyme stem, rosemary, oregano, granulated garlic, fresh onions, shallots, coriander, or pepper, and then add stock and a few sprays of a good flavorful oil. Vegetables can be roasted in the oven, at 375°F, which results in wonderful texture and flavor due largely to caramelization of the natural sugars. Larger, heartier vegetables, such as celery root, butternut, turnip, rutabagas, and sweet potatoes take longer to oven-roast than do vegetables such as peppers, radishes, and pattypan squash. A variety of fresh herbs and spices such as shallots, thyme, rosemary, oregano, cardamom, coriander, fennel seed, and chilies add flavor and uniqueness. For example, sautéed cipolinni onions with cinnamon, bay leaf, and thyme; deglazed with balsamic vinegar; and a touch of sugar goes well with pork cutlets or sliced filet of beef. Recipe 10-42 shows how to prepare Roasted Summer Vegetables, a colorful, well-balanced addition to a variety of proteins.

- **Grilling.** Grilled vegetables are a popular side dish and include favorites such as peppers, eggplant, mushrooms, zucchini, and summer squash, as well as others like asparagus, carrots, fennel, radicchio, and endive (Figure 8-8). For example, a grilled Portobello mushroom filled with artichoke and white bean purée or vegetable spelt salad, garnished with garden tomatoes and wilted arugula, is a satisfying side dish to fennel-dusted pork loin or center of the plate for a vegan entrée selection.

- **Steaming.** Steaming has been the traditional method that surrounds the product with vapor from a hot liquid in a covered vessel. Steamed foods stay moist, plump, and tender and retain most of their nutrients. This method is extremely well balanced with the addition of little or no additional kcalories, yet adding a blend of flavors through the flavored circulating steam.

- **Sautéing.** To sauté vegetables, simmer them with a liquid such as stock or juice in a non-stick pan, and add seasonings such as cardamom, coriander, and fennel. Cook over moderately high heat. As the vegetables cook, add more liquid and deglaze the pan. Once the vegetables are ready and the stock has reduced to a syrup consistency, add a little bit of butter or flavorful oil (about 1 teaspoon for four servings) to create a slightly thickened glaze and give the product a rich flavor and a shiny appearance. Once prepared, serve sautéed foods immediately.

Another method is to blanch vegetables in boiling water to the desired doneness and then shock them in ice water, drain, and lay out on towels before placing immediately in refrigerator. Dry sauté the blanched vegetables (remove any moisture) in a hot nonstick pan with stock, a touch of wine, fresh herbs, garlic, and chopped shallots, and finish with fresh black pepper and extra virgin olive oil, butter, or nut oil (1 teaspoon for four servings).

Grains, the edible seeds of various members of the grass family, add incredible flavor, texture, and appeal to dishes. Wheat and rice are the most important grains worldwide, but

Farro A grain that is the ancestor of modern wheat.

Kamut A grain that is an ancient relative to wheat.

Triticale A hybrid of wheat and rye.

there are many other grains available that are making their debut on menus across the globe. Other grains include wild rice (not actually a rice, but the seed of an unrelated grass in North America), **farro** (also called spelt), **kamut**, barley, oats, millet, quinoa (popular colors are ivory, red, and black), **triticale**, amaranth, and flaxseeds.

Legumes are plants that bear seed pods that split along two opposite sides when ripe, such as beans, lentils, and peas. Dried legumes have played an important role in the human diet for thousands of years, and are especially important today given the increased interest in healthy diets. Legumes are high in protein and therefore very important in vegetarian and vegan diets. Heirloom varieties such as black beluga lentils, anasazi, black calypso, butterscotch, peruano, cranberry, chestnut lima, Jacob's Cattle, and scarlet runner beans add color, flavor, nutrition and variety to many preparations.

Add grains and legumes to vegetable dishes, such as brown rice with quick-fried vegetables; wheat berries with roasted golden beets and dried cranberries; or chestnut lima beans with grilled zucchini, summer squash, peppers, and oven-dried tomatoes. See Recipe 10-24, which shows how to make Mixed-Grain Pilaf.

MENU IDEAS AND PRESENTATION

Following are menu ideas and tips for presenting side dishes.

- When using vegetables, you need to plan your menu selection by what is in season for maximum appearance and overall taste. Consider flavor, color, and whether the combination makes good culinary sense. This is where tradition and history become essential in the development of the dish you are creating. If you are preparing a roasted pork dish, this pairs well with Brussels sprouts, cabbage, stewed mushrooms, or hard squashes (see Figure 10-9). You could additionally crust it with oregano, basil, and roasted garlic and serve it with homemade vinegar peppers and oven-roasted plum tomatoes. The idea is to think your dishes through from start to finish. Furthermore, consider variety when planning menus. Vegetable selection doesn't mean switching from broccoli to cauliflower and then back to broccoli. There is a wide variety of vegetables, both seasonal and local, to choose from. Research, experiment with, and practice different vegetable cuts, cooking methods, and flavorings.

FIGURE 10-9 Slow-Cooked Pork Roast, Stuffed Cabbage Balls, and Roasted Summer Vegetables.

Photo by Peter Pioppo.

- Blanched vegetables should be reheated in a small amount of seasoned stock and then finished with an oil such as extra virgin olive oil, flavored nut oil, or sweet butter. These delicate coatings will be the first flavor your customers will taste, giving the dish a rich body and taste.

- Serve grains and beans. For example, rice and beans are a nutritious match that provides fiber and a complete protein. These versatile dishes using grains and legumes have endless possibilities; items such as Texmati brown rice with Butterscotch Calypso beans, or Colusari™ Red Rice with Gigante Beans and seasonings such as onions, garlic, fresh herbs spices, vegetables, olives, dried fruits, and citrus, creating a complete vegan meal or an attractive side dish.

- Green salads are often used as side dishes. There are so many varieties of greens on the market today, as well as a wide variety of garnishes to accent salads leaving room for creative options. With the addition of flavorful balanced dressings to accompany your salads, they become light alternatives as entrées, appetizers, or side dishes.

Additional examples of side dishes include these:

- Red Mashed Potatoes Roasted Garlic, Yogurt Cream, and Fresh Dill

- Couscous with Dried Fruit, Roasted Squash, Cucumber, and Mint

- Seven-Vegetable Stir-Fried Bamboo Rice

- Ratatouille Wrapped in Phyllo with Oven-Dried Tomatoes

- Oven-Baked French Fries and Sweet Caramelized Onions

- Sweet and Sour Savoy Cabbage with Hickory-Smoked Bacon

- Wild Mushroom Goulash

- Roasted Cauliflower and Garlic (Figure 10-10)

- Wheat Berry Risotto with Spring Onions and Fava Beans

- Portobello Pizza (Tomato, Basil, Wilted Spinach, and Low-Fat Mozzarella)

- Oven-Roasted Baby Beets, Sweet Onions and Pistachio Oil

FIGURE 10-10 Roasted Cauliflower with Lemon Thyme, Fresh Bay Leaf, and Roasted Garlic.

Photo by Peter Pioppo.

1. To cook vegetables, considering roasting, grilling, steaming, and sautéing.
2. To sauté vegetables, simmer them with stock or juice in a nonstick pan, and add seasonings such as cardamom and fennel. Cook over moderately high heat. As the vegetables cook, add more liquid and deglaze the pan. Once the vegetables are ready and the stock has reduced to a syrup consistency, add a little bit of butter or flavorful oil (about 1 teaspoon for four servings) to create a slightly thickened glaze and give the product a rich flavor and a shiny appearance.
3. Use grains and legumes in side dishes, such as wheat berries with roasted golden beets and dried cranberries.
4. When using vegetables, plan your menu by what is in season for maximum appearance and overall taste. There is a wide variety of vegetables, both seasonally and locally, to choose from.

 DESSERTS

BALANCED INGREDIENTS AND PREPARATION

Desserts have always been perceived as the rich, fattening, "bad for you" part of the meal that can never be part of a balanced diet. Newly composed sweet endings can find a place in a diet where moderation is the focus. They are not the heavy, oversweet, overportioned choices we are all accustomed to, but a delicate delight that will satisfy a person's sweet tooth while following a manageable lifestyle. Creativity and precise visual execution are the keys to your client's perception of these desserts. A great marketing tool in your establishment is to promote a three-course meal under 1,000 kcalories, which includes dessert and a glass of wine. This could be a way to introduce moderation, not elimination, on your balanced menus.

There are many different techniques to modify sugar, fat, and cholesterol in your dessert selections, while not compromising taste and satisfaction. Your focus needs to identify what will sell in your establishment, keeping in mind traditional, simple, approachable selections. For example, try a ricotta cheesecake with a crust made of roasted walnuts, spices, and Splenda; a buttery phyllo cylinder with maple pineapple cream and macerated berries; a banana polenta soufflé with dark chocolate sauce and glazed banana slices; Angel Food Savarin (Recipe 10-47), fresh fruits, and blackberry ice; or an old-fashioned strawberry shortcake with berry sauce and whipped cream. These desserts may sound too good to be true, but they are very much in line with the requirements of a balanced eating style.

You can also make a wide variety of desserts with fruits, either fresh cut or as a key ingredient in quick breads, cobblers, puddings, phyllo strudels, and even some cakes and cookies. The following recipes show how fruit can be used in many forms:

- Oatmeal-Crusted Peach Pie
- Icebox Cheesecake with Papaya, Kiwi, and Crystallized Ginger
- Apple Strudel with Orange Caramel Sauce
- Fresh Fruit Sorbets/Ices or Creams (see Recipe 10-46)
- Spiced Carrot Cake with Mango Custard Sauce
- Walnut–Dried Cranberry Shortcake, Berries, and Fresh Cream
- Warm Chocolate Torte with Raspberry Sauce and Blackberries (Figure 10-11)
- Poached Sickle Pears with Merlot Syrup and Almond Tuile
- Banana Polenta Soufflé (Recipe 10-50) (Figure 10-12)

FIGURE 10-11 Chocolate Torte with Crisp Phyllo and Blackberry Sauce.

Photo by Peter Pioppo.

MENU IDEAS AND PRESENTATION

Following are menu ideas and tips for presenting desserts:

- To make sorbet without sugar, puree and strain the fruits. Make sure the fruits are at the peak of ripeness. Churn in an ice cream machine or place in a freezer and hand stir every 10 minutes until frozen.

- Use angel food cake as a base to build a dessert. Serve it with fresh fruit sorbet, warm sautéed apples and cranberries, pear and ginger compote, mango, mint, papaya salsa, or additional delicious fruits of the season.

FIGURE 10-12 Banana Polenta Soufflé.

Photo by Peter Pioppo.

HOT TOPIC

SUSTAINABLE FOOD PURCHASING

Many restaurants and on-site foodservices are considering the environment, animal welfare, farm laborers, and other concerns when deciding which foods to purchase. Sustainability refers to meeting our needs, such as for food, without making it difficult for future generations to meet their own needs. Sustainable food is produced by farmers and ranchers who care about the health of their land and animals. In brief, sustainable food is good for the environment and the people who grow it.

Sustainable agriculture nourishes the long-term health of the soil through practices such as crop rotation and composting. Raising animals sustainably means providing an environment that reflects the needs of the animals—for example, letting chicken graze on grass daily and providing organic, all-vegetarian feed. Sustainable animal husbandry represents one of the strongest growth areas as consumer demand for organic meat, milk, and eggs increases.

Each foodservice needs to develop its own vision of sustainable food purchasing, ideally tied to the overall mission of the foodservice. For example, Kaiser Permanente (the largest non-profit healthcare provider in the United States) has a vision statement that it "aspires to improve the health of our members, employees, our communities, and the environment by increasing access to fresh, healthy food in and around KP facilities." It also states that Kaiser Permanente will purchase food in a way that is "environmentally sound, economically viable, and socially responsible." Your vision statement is simply a "big-picture" statement of your values and long-term goals.

The next step is to set up appropriate food-purchasing guidelines and specifications to match your goals. You will need to be familiar with the definitions of terms such as organic or certified humane (Table 10-1). Other

TABLE 10-1	Glossary for Sustainable Food Purchasing

Antibiotic claims—The USDA does not allow "antibiotic free" to be put on meat or poultry labels. It does allow "raised without antibiotics" or "no antibiotics administered" as long as no antibiotics were ever administered to the animal.

Cage free—This is a claim made by the producer that the poultry were raised without cages. There is usually no independent verification of "cage-free" claims. This does not guarantee that birds were raised on grass or had access to the outdoors. Birds may simply live in flocks in barns.

Certified Humane Raised & Handled®—The Certified Humane Raised & Handled® program is the only farm animal welfare and food labeling program in the United States dedicated to improving the welfare of farm animals in food production and include all stages of the animal's life including handling and slaughter. Humane Farm Animal Care's Animal Care Standards require that animals have ample space, shelter, and gentle handling to limit stress and additionally require that:

- The use of growth hormones and antibiotics is prohibited.
- Animals must be free to move and not be confined—Cages, crates, and tie stalls are prohibited. This means that chickens are able to flap their wings and dust bathe, and pigs have the space to move around and root.
- Livestock have access to sufficient, clean, and nutritious feed and water.
- They must have sufficient protection from weather elements and an environment that promotes well being.
- Managers and caretakers must be thoroughly trained, skilled, and competent in animal husbandry and welfare, and have good working knowledge of their system and the livestock in their care.
- Farmers and ranchers must comply with food safety and environmental regulations.

| TABLE 10-1 | Glossary for Sustainable Food Purchasing (*Continued*) |

Conventionally grown—Foods grown using chemical fertilizers, pesticides, and herbicides (in the case of farmed products) or hormones and antibiotics (in the case of animals).

Ecologically grown—An uncertified label that signifies that a crop is grown without using chemical fertilizers or herbicides. Products with this label can be grown using integrated pest management practices, which minimizes but does not rule out the use of chemical pesticides.

Fair trade certified—Fair trade goods, such as coffee, come from farmers, often around the world, who are justly compensated. Fair Trade USA helps farmers in developing countries build businesses that positively influence their communities. They encourage sustainable farming practices, and discourage the use of child labor and certain pesticide. Fair Trade USA is the third-party certifier of Fair Trade goods in the United States and one of the members of Fair Trade Labeling Organizations International.

Family farm—A farm managed by a family or individual who owns the animals or land, gets a good portion of their livelihood from the farm, and participates in the daily labor to work and manage the farm. The USDA defines a family farm as having less than $250,000 gross receipts annually.

Food Alliance certified—Food Alliance provides comprehensive third-party certification for social and environmental responsibility in agriculture and the food industry. Their products include meats, shellfish, eggs, dairy, grains, legumes, a variety of fruits and vegetables, and prepared products made with these certified ingredients. Their certification addresses these issues: safe and fair working conditions, humane treatment of animals, and environmental stewardship.

Free range—Free-range poultry have to be allowed access to the outside; however, it is not required that they be outside. The term is regulated by the USDA and allowed to be used only on poultry labels, not egg labels.

Genetically modified organism (GMO) claims—When a gene from one organism is purposely moved to improve or change another organism in a laboratory, the result is a genetically modified organism (GMO). Many genetically modified crops, such as corn, have been modified to resist insects or herbicides. Some companies label their foods as "GMO free" but there is normally no independent verification (unless the food is certified as Organic or Food Alliance).

Grassfed (American Grassfed Association Certified)—The American Grassfed Association defines grassfed products from ruminants, including cattle, bison, goats and sheep, that have eaten nothing but their mother's milk and fresh grass or grass-type hay all their lives. For grassfed nonruminants, including pigs and poultry, grass is a significant part of their diets, but not their entire diet, since these animals need to consume grains. To be American Grassfed certified, the animals must live on pasture and not receive hormones or antibiotics. In the USDA definition for grassfed, animals can receive hormones and antibiotics and also be confined.

Hormone claims—Hormones are not allowed in raising hogs or poultry. Therefore, the claim "no hormones added" cannot be used on the labels of pork or poultry unless it is followed by a statement that says "Federal regulations prohibit the use of hormones." The term "no hormones administered" may be approved for use on the label of beef products if sufficient documentation is provided to the USDA by the producer showing no hormones have been used in raising the animals. The terms "no antibiotics added" may be used on labels for meat or poultry products if producers provide sufficient documentation to the USDA.

Integrated pest management (IPM)—IPM is an effective and environmentally sensitive approach to pest management that relies on a combination of common-sense practices. IPM programs use current, comprehensive information on the lifecycles of pests and their interaction with the environment. This information, in combination with available pest control methods, is used to manage pest damage by the most economical means, and with the least possible hazard to people, property, and the environment. Pesticides are normally used as a last resort.

Local—How local is defined will vary by location. Foodservices often choose to define local as food grown within a specific distance from the point of consumer purchase, such as 200 miles.

Monterey Bay Aquarium Seafood Watch—This organization provides Seafood Watch regional guides that contain the latest information on sustainable seafood choices available in different regions of the United States. Its "Best Choices" are abundant, well managed, and fished or farmed in environmentally friendly ways. Their seafood to "avoid" are overfished and/or fished or farmed in ways that harm other marine life or the environment. View the guides online at their website.

Natural—When you are buying meat or poultry, natural products contain no artificial ingredient or added color and are only minimally processed. For other foods, natural means that there are no added colors, artificial flavors, or synthetic ingredients.

Organic—Organic food is produced by farmers who emphasize the use of renewable resources and the conservation of soil and water to enhance environmental quality for future generations. Organic meat, poultry, eggs, and dairy products come from animals that are given no antibiotics or growth hormones. Organic food is produced without using most conventional pesticides, fertilizers made with synthetic ingredients or sewage sludge, bioengineering, or ionizing radiation. Before a product can be labeled *organic*, a government-approved certifier inspects the farm where the food is grown to make sure the farmer is following all the rules necessary to meet USDA organic standards. Companies that handle or process organic food before it gets to your local supermarket or restaurant must be certified, too.

terms, such as buying "local," have no definition and you will have to determine what is appropriate for your location, such as buying foods grown within a 200-mile radius from your operation. For each category of food, you will also need to decide questions such as what percentage you want to buy organic (or other sustainable practice) and how much you want to buy locally. Another consideration is the ownership of the farms you want to deal with and their labor practices and policies. You need to consider to what extent your preference is to work with a family farm or a corporate farm (which tends to be much bigger).

Table 10-2 explains what to look for when purchasing sustainable foods in various categories. Keep in mind that you won't always be able to buy the "best" options. There are times when you will have to buy lesser options. Table 10-3 shows which fresh fruits and vegetables are the lowest and the highest in pesticide residue. This is an important consideration when you are picking which produce you want to purchase in organic form.

| TABLE 10-2 | Considerations for Purchasing Sustainable Foods |

FRUITS AND VEGETABLES

Best Choices	Better Choices	Avoid
• Product is certified organic. • Soil is fertilized through composting, composted manure, cover crops/mulching, copy rotation, and crop diversity. • Farm is local. • Farm is a family farm/small to mid-sized farm. • Employees receive livable wages, benefits, training, and safe conditions.	• Produce is grown using integrated pest management (using natural pest repellents and methods, only use pesticides when absolutely necessary). • Farm is regional or national.	• Nonorganic. • Chemical pesticides used regularly whether needed or not. • Farm is international. • Farm is a large corporate farm. • Farm pays below minimum wage and has unsafe working conditions.

ALSO NOTE:

• In areas with cold winters, you can have locally grown greens. Farmers can extend the growing season by using unheated greenhouses or hoophouses—structures covered with a thick layer of transparent greenhouse plastic that traps sunlight and warmth.

• Avoid hothouse tomatoes as heated greenhouses use much heat and electricity.

• Since bananas are imported, consider fair-trade bananas.

• Apples, peaches, and plums are hard to grow organically in the northeastern states for a variety of reasons. You may want to choose ecologically grown apples from the northeast rather than organic from the western states, which then are trucked long distances.

• See Table 10-3 for guidance on which produce contains the most pesticides as they are the best candidates to buy organic.

DAIRY

Best Choices	Better Choices	Avoid
• Cows are AGA grass-fed certified or pasture-raised (possibly some grain in winter months). • No growth hormones or antibiotics. • Cows are milked once or twice a day, seasonally. • Farm is local. • Farm is a family farm/small to mid-sized farm. • Employees receive livable wages, benefits, training, and safe conditions.	• Cows are USDA-certified grassfed (cows may still receive hormones and antibiotics). • Cows eat some grass. • Antibiotics are given when cows are sick. • Farm is located regionally or nationally.	• Animals raised in confined operations in which they are fed a grain diet. • Routine growth hormone administered. • Antibiotics are given. • Farm is a large corporate farm. • Farm pays below minimum wage and has unsafe working conditions.

EGGS

Best Choices	Better Choices	Avoid
• Chickens eat certified organic feed and they graze outside daily. • Farm is local. • Farm is a family farm/small to mid-sized farm. • Employees receive livable wages, benefits, training, and safe conditions.	• Chickens eat all-vegetarian feed with no antibiotics and they graze outside. • Farm is located regionally or nationally.	• Chickens are fed conventional grain feed containing antibiotics and animal byproducts. • Chicken are caged or in a crowded barn. • Farm is a large corporate farm. • Farm pays below minimum wage and has unsafe working conditions.

| TABLE 10-2 | Considerations for Purchasing Sustainable Foods (*Continued*) |

POULTRY

Best Choices

- Birds graze outside a good portion of the day in season.
- Birds eat certified organic feed.
- Birds do not receive antibiotics.
- Farm is local.
- Farm is a family farm/small to mid-sized farm.
- Employees receive livable wages, benefits, training, and safe conditions.

Better Choices

- Birds graze outside part of every day or have access to outdoors.
- Birds eat all-vegetarian feed with no antibiotics.
- Birds are given antibiotics only when flock is sick.
- Farm is located regionally or nationally.

Avoid

- Factory-farmed poultry raised in unsanitary and inhuman conditions, like small cages or overcrowded barns.
- Birds eat conventional grain feed, which sometimes contains animal byproducts or GMO grains, as well as antibiotics.
- Farm is a large corporate farm.
- Farm pays below minimum wage and has unsafe working conditions.

BEEF AND LAMB

Best Choices

- Animals are AGA certified grassfed, or eat pasture grasses and hay and grain in winter if there is no grass.
- Animals do not get any growth hormones or antibiotics.
- Farm is located locally.
- Farm is a family farm/small to mid-sized farm.
- Employees receive livable wages, benefits, training, and safe work conditions.

Better Choices

- Animals are USDA certified grass fed, or eat pasture grass with hay and grain—at least half of their diet comes from grazing on pasture.
- Antibiotics are given when an animal is sick.
- Farm is located regionally or nationally.

Avoid

- Conventionally grown animals are grain-fed on feedlots. Animals eat mostly grain and are routinely given growth hormones and antibiotics.
- Farm is a large corporate farm.
- Farm pays below minimum wage and has unsafe working conditions.

FISH

Monterey Bay Aquarium Seafood Watch Guide is a series of guides that contain the latest information on sustainable seafood choices available in different regions of the United States. Its "Best Choices" are abundant, well managed, and fished or farmed in environmentally friendly ways. Its seafood to "avoid" are overfished and/or fished or farmed in ways that harm other marine life or the environment. View the guides online at its website. www.montereybayaquarium.org/cr/cr_seafoodwatch/download.aspx

TABLE 10-3	Pesticides on Produce

MOST PESTICIDE RESIDUE	LEAST PESTICIDE RESIDUE
Apples	Onions
Celery	Corn
Sweet bell peppers	Pineapples
Peaches	Avocado
Strawberries	Cabbage
Nectarines (imported)	Sweet peas
Grapes	Asparagus
Spinach	Mangoes
Lettuce	Eggplant
Cucumbers	Kiwi
Blueberries (domestic)	Cantaloupe
Potatoes	Sweet potato
	Grapefruit
	Watermelon
	Mushrooms

Source: Environmental Working Group, 2012

RECIPES

The recipes noted below all can be found on the book's website: www.wiley.com/college/drummond.

Appetizers

10-1	Scallop and Shrimp Rolls in Rice Paper
10-2	Eggplant Rollatini with Spinach and Ricotta
10-3	Mussels Steamed in Saffron and White Wine
10-4	Roast Chicken and Shredded Mozzarella Tortellonis
10-5	Crab Cakes
10-6	Mahi-Mahi Eggroll
10-7	Eggplant and Sun-Dried Tomato Dip

Soups

10-8	Pasta e Fagioli
10-9	Butternut Squash Bisque

Salads

10-10	Wild Mushroom Salad
10-11	Heirloom Lettuces and Greens with Shaved Fennel and Orange Sections
10-12	Grilled Chicken and Greek Quinoa Salad

Dressings

10-13	Ginger Lime Dressing
10-14	Basic Herb Vinaigrette
10-15	Orange Vinaigrette
10-16	Green Goddess Dressing
10-17	Yogurt Dressing/Sauce
10-18	Cucumber Dill Dressing
10-19	Thai Dressing

Rubs and Marinades

10-20	Capistrano Spice Rub
10-21	Adobo Spice Rub
10-22	Cuisine Naturelle Marinade

Entrées

10-23	Grilled Pork Chop Adobo with Spicy Apple Chutney
10-24	Slates of Salmon
10-25	Braised Lamb
10-26	Meat Loaf
10-27	Classic Beef Stew
10-28	Vegetable Lasagna
10-29	Whole Wheat Pasta
10-30	Chicken Pot Pie
10-31	Grilled Chicken Burger
10-32	Grilled Turkey Burger
10-33	Chicken Quesadillas

Sauces

10-34	Papaya and White Raisin Chutney
10-35	Papaya and Plantain Salsa
10-36	Tomato Salsa
10-37	Red Pepper Coulis
10-38	Hot-and-Sour Sauce
10-39	Velouté Sauce
10-40	Carbonara Sauce
10-41	Tomato Sauce

Sides

10-42	Roasted Summer Vegetables
10-43	Mixed-Grain Pilaf
10-44	Potatoes Au Gratin

Desserts

10-45	Warm Chocolate Pudding Cake with Almond Cookie and Raspberry Sauce
10-46	Raspberry Creamed Ice
10-47	Angel Food Savarin
10-48	Fresh Berry Phyllo Cones
10-49	Carrot Cake
10-50	Banana Polenta Soufflé
10-51	Oatmeal Raisin Cookies

Breakfast

10-52	Stuffed French Toast Layered with Light Cream Cheese and Bananas
10-53	Glazed Grapefruit and Orange Slices with Maple Vanilla Sauce
10-54	Whole-Wheat Peach Chimichangas

Gluten-Free

10-55	Gluten-Free Bread
10-56	Gluten-Free Pizza Dough
10-57	Gluten-Free Black Bottom Cupcakes

PART

3

APPLIED NUTRITION

HANDLING CUSTOMERS' SPECIAL NUTRITION REQUESTS

CHAPTER 11

- Identify appropriate ingredients/menu items when customers request foods low in kcalories, fat and cholesterol, sugar, or sodium.

- Compare and contrast a food allergy with a food intolerance, and identify the most common food allergies.

- Describe how to set up a food allergy management plan in a restaurant, and identify foods to avoid for the most common food allergies.

- Give three examples of gluten-free foods from each food group, explain cross-contact, and give five examples of how to avoid cross-contact when preparing gluten-free menu items.

- Identify appropriate ingredients/menu items for a customer who is lactose intolerant.

- Identify complementary protein combinations, and use them along with vegetarian menu planning guidelines to plan a vegetarian menu that includes vegan options.

INTRODUCTION

Here are some examples of what your customers are asking for:

- "What menu items are low in sodium?"
- "What selections are vegetarian?"
- "What do you have that is low in fat?"

These types of requests are quite common, along with more requests for gluten-free foods. For example, a chef on a cruise ship had a woman who was gluten intolerant so she couldn't eat anything made with wheat flour. The menu for lunch one day was pan pizza. The server brought one to her and she said to her server, "I can't have that." And the server said yes you can, because the pastry chef made it with rice flour. The woman started crying. She hasn't had pizza in 30 years and she just started crying. She was just so touched that the server and chef would go out of their way to make something special for her, which she really enjoyed eating.

The days of foodservice establishments serving plain steamed vegetables as a healthy alternative are long gone. Unless you have a customer with a very restricted diet, you will have to be much more creative to accommodate special requests. Whole grains, carbohydrates with fiber, fresh vegetables, fresh fruits, lean proteins, and the use of healthy oils are not just a fad; they are a way of life for many people seeking a healthy lifestyle that we know can prevent diseases such as heart disease. With the increasing number of people eating out, the need for balance in foodservice establishments is essential to survival.

As operator/chef, you have incorporated a variety of choices into your menu so the ability to mix and match for special requests can be easily remedied. Vegetarian, low fat, and other requests are becoming fairly common. Part of your daily regime as a chef is to be prepared for these special needs without creating kitchen havoc. Dishes for a special need must show the same skill, taste, and presentation of your original menu items so your reputation goes untarnished and your customers are happy. The best approach when designing your menus is to have choices that follow the basic dietary guidelines discussed in this book. With more people seeking healthier foods in restaurants, the need for menu selections to be better balanced and limited in rich ingredients is greater than ever. Many diners are not out for that once-in-a-blue-moon special occasion; many people eat or take out several times a week.

To respond to special guest requests, keep in mind these basic preparations:

- When marinating meats, there are many no-salt, no-sugar rubs and seasonings that can be used. The addition of salt to your proteins can be done at the time of cooking (or not at all), so a request for no salt is easy to accommodate.

- Blanched vegetables should be reheated in a small amount of seasoned stock, then finished with a moderate amount of whole butter, an extra-virgin olive oil, or flavored nut oil. These delicate oils will be the first flavor your customers will taste, giving a rich body and the taste fulfillment up front with fewer fats than before.

- Dips and chips are and always will be an American favorite for appetizers. This can still be a great selection to your menu with a new twist. Hummus, baba ghanoush, white bean and roasted garlic, artichoke, and goat cheese are well-accepted favorites dips that can be accompanied by baked whole-wheat tortilla chips, melba toast, baked multigrain croutons, or a variety of different vegetables.

- Create a well-balanced dressing that is low in fat and made with extra-virgin olive oil and good vinegars as we have suggested, and finished with fresh herbs and spices. Use this as one of your house dressings so it is available to prepare a number of different choices.

- Keep a flavorful stock or clear broth for reheating vegetables, because some new eating habits recommend starting your meal with a cup of clear broth. This helps to curb the appetite. This can be a kitchen staple as well as a quick-serve dual purpose inexpensive starter on your menu, increasing sales for an appetizer that may not have been purchased otherwise.

- Desserts, desserts, desserts. Creating balanced desserts that your guests can enjoy should not cause stress. There are more than enough ways to add limited sugar to a menu item that can appeal to the majority of eating styles today. A ricotta cheesecake with a roasted walnut and spiced crust; a flourless chocolate cake with fresh fruit garnish; a toasted oatmeal, chocolate, banana, pecan pudding with a kiwi or mixed fruit salsa; a buttery phyllo cylinder with maple cream pineapple chutney and berries in season; a banana polenta soufflé with chocolate sauce and glazed banana slices; or an old-fashioned berry shortcake with fruit sauce berries and rich whipped cream. Sounds too good to be true, but these options are very much within the guidelines for a balanced eating style.

The following discussion will coach you into looking at menus in a different light so that you can accommodate any special request from your diners. With the changing habits of your customers, there are some basic menu items that you should consider having so that you can more easily accommodate special requests.

SUMMARY

1. Vegetarian, low fat, and other requests are becoming fairly common. Part of your daily regime as a chef is to be prepared for these special needs without creating kitchen havoc. The best approach when designing your menus is to have choices that follow the basic dietary guidelines discussed in this book. With more people seeking healthier foods in restaurants, the need for menu selections to be better balanced and limited in rich ingredients is greater than ever.

2. Have these basic preparations ready to respond to almost any special request: marinate meats without using salt; reheat blanched vegetables in a small amount of seasoned stock, then finished with a moderate amount of whole butter, an extra-virgin olive oil, or flavored nut oil; whole-grain chips with balanced dips such as hummus work well for an appetizer; have a well-balanced dressing for salads and stock for soups and sauces; and consider balanced desserts featuring fruits.

LOW KCALORIE

You can control the number of kcalories in a dish in many ways, but you will probably wind up either adjusting the portion size and/or the amount of fat in the menu item. Here are some simple ways to offer menu items lower in kcalories:

1. Adjust the portion size smaller.

2. Offer a menu item in different sizes. Perhaps you want to offer a salad in small and large sizes, or entrées in a half-size.

FIGURE 11-1 Phyllo Bavarian with Berries. Fruit-based desserts work well for low kcalorie menu selections.

Photo by Peter Pioppo.

3. Allow customers to "share" an entrée or other menu items.

4. Offer to put sauces and salad dressings on the side.

5. If customers have a choice of sides, make sure at least one is balanced, such as sliced apples or baked sweet potato fries with a burger.

6. Choose cooking methods such as broiling, roasting, or steaming, and make sure the menu mentions the cooking method so the customer knows.

7. Have balanced sauces and dressings available.

8. For dessert, use fruit as the primary ingredient (Figure 11-1).

Following are lower kcalorie options for each section of the menu (Table 11-1).

TABLE 11-1	Menu Choices for Low Kcalorie Diet	
LOW KCALORIE DIET	**GOOD CHOICES**	**USE IN MODERATION**
Appetizer and Soup	Raw vegetables and fruits	Fatty proteins
	Dips made with nonfat yogurt/sour cream	Regular cheeses
	Juice	Cream soups
	Clear soups with vegetables, lean proteins	Nuts
		No fried foods
Salads	Fresh vegetables and fruits	Regular cheeses
	Lean proteins such as grilled chicken, egg whites, fish, beans, and chickpeas	Nuts
		Avocado
	Vinaigrette, pureed, and other dressings using moderate amounts of oil, or fat-free dressings, served on the side	Bacon
		Raisins
		Balanced (unsaturated) oils
Breads	Whole-grain breads and rolls	Biscuits
	Light or fat-free margarine or butter	Cornbread
		Muffins made with refined flours and solid fat
		Croissants
Entrées	Lean proteins such as chicken (no skin) and fish	High fat meat cuts such as 80% ground beef
	Pasta with vegetables	Cheese—use shredded cheese so you can use less and choose varieties that pack the greatest flavor punch
	Sauces with small amounts of fat, such as salsa and chutney	
	Small quesadilla or soft taco stuffed with lean protein and vegetables	Nuts
		No fried foods

LOW KCALORIE DIET	GOOD CHOICES	USE IN MODERATION
Side Dishes	Roasted and steamed vegetables with sauces etc. on the side Whole grains such as wild rice or quinoa	Fats and oils in sauces and preparation
Desserts	Fruits and fruit-based desserts Puddings made with nonfat milk	Regular cakes, pies, cookies, ice cream
Beverages	Coffee, tea Drinks with no or little sweetening Diet drinks Nonfat milk Unsweetened 100% juice	Sweetened drinks such as regular soda and sweet tea
Breakfast	Whole-grain cereals Eggs Whole-grain breads and muffins 100% juice	Bacon and sausage No fried foods

SUMMARY

1. You can control the number of kcalories in a dish in many ways, but you will probably wind up either adjusting the portion size and/or the amount of fat in the menu item. Also, choose cooking methods such as broiling or steaming that don't add kcalories.
2. Use Table 11-1 to pick appropriate menu choices low in kcalories.

LOW FAT AND CHOLESTEROL

The biggest sources of fat and saturated fat in the American diet are animal foods such as:

- Fat and eggs used to make grains-based desserts (such as cookies)
- Cheese
- Whole milk
- Beef and pork (sausage, franks, and bacon)
- French fries (fried foods)
- Dairy desserts such as ice cream

Saturated fat is also found in eggs, poultry skin, and butter. Animal fat often contains at least 50 percent saturated fat. Most vegetable oils are rich in unsaturated fats, but don't forget there are ones high in saturated fat: coconut, palm kernel, and palm oils, often called tropical oils. They are used in some processed foods, such as commercial baked goods and frozen whipped nondairy toppings.

For a low-fat and low-cholesterol diet, you need to really limit the foods just mentioned, or use lower-in-fat versions. Better yet, switch to lean proteins, nonfat dairy, fruits, vegetables, balanced oils, and balanced sauces, as discussed in "Balanced Sauces" in Chapter 10, and whole grains. For example, grilled salmon with spring vegetables is an excellent choice for a low-fat and low-cholesterol diet (Figure 11-2). Following are lower fat and cholesterol options for each section of the menu (Table 11-2).

FIGURE 11-2 Everything Salmon with Lentil Salad. Grilled salmon is an excellent choice for a low-fat and low-cholesterol diet.

Photo by Peter Pioppo.

TABLE 11-2	Menu Choices for a Low-Fat and Low-Cholesterol Diet	
LOW FAT AND CHOLESTEROL	**GOOD CHOICES**	**USE IN MODERATION**
Appetizer and Soup	Raw vegetables and fruits	Balanced (unsaturated) oils for sautéing, etc.
	Baked pita wedges (or other whole-grain baked chip) with vegetable-based dip or hummus	Soups made with cream or roux-based thickeners
	Clear soups with vegetables and lean protein	
Salads	Raw vegetables and fruits	Balanced (unsaturated) oils in dressings
	Vinaigrette, pureed, and other dressings using moderate amounts of oil, or fat-free dressings, served on the side	Cheese
		Regular mayonnaise
		Nuts and seeds
		No bacon
Breads	Whole-grain breads and rolls	Olive oil with or on bread
	Light or fat-free margarine	Biscuits
		Cornbread
		Muffins, etc. made with solid fats
Entrées	Lean proteins such as chicken (no skin) and fish	Regular cheeses
	Pasta with vegetables	Nuts and seeds
	Meatless entrées using beans, whole grains, and/or vegetables	Balanced (unsaturated) oils
	Whole-grain pasta	No fatty cuts of beef, pork lamb, regular ground beef, spare ribs
	Tofu, tempeh, low-fat meat analogs	No processed meats such as sausage, bacon, bologna, salami, and hot dogs
		No fried foods
Side Dishes	Roasted and steamed vegetables	Balanced (unsaturated) oils in sauces and preparation
	Whole-grains such as wild rice or quinoa	No fried foods

LOW FAT AND CHOLESTEROL	GOOD CHOICES	USE IN MODERATION
Desserts	Fruits and fruit-based desserts Low-fat and nonfat frozen yogurt, sorbet Meringue cookies	Regular cakes, pies, ice cream
Beverages	Coffee, tea Nonfat or 1% milk Unsweetened 100% juice	2% Milk
Breakfast	Whole-grain cereals Egg whites or egg substitutes in omelets, etc. cooked using vegetable oil cooking spray Whole-grain breads and muffins made with balanced (unsaturated) oils Pancakes and waffles using vegetable oil spray to cook	Whole eggs Butter Cheese No fried foods No doughnuts, Danish, croissants, biscuits

SUMMARY

1. For a diet low in fat, saturated fat, and cholesterol, switch from fried foods and high fat dairy, meats, and desserts to lean proteins (especially poultry and seafood), fruits, vegetables, whole grains, nonfat dairy, balanced oils, and balanced sauces.
2. Table 11-2 lists appropriate menu choices for a diet low in fat and cholesterol.

LOW SUGAR

Most customers looking for low-sugar menu items are trying to avoid added sugars. Added sugars, like granulated white sugar or high fructose corn syrup, are added to a food for sweetening and flavor (such as soda and cookies) or added to foods or beverages at the table (such as sugar added to coffee). Currently, sugar-sweetened beverages are the primary source of added sugars in the American diet, followed by grain-based desserts (cakes, cookies, pies, etc.), sugar-sweetened fruit drinks, dairy-based desserts (ice cream, etc.), and candy. The menu categories with the most issues with sugar tend to be desserts and beverages, as seen in Table 11-3.

Many customers asking for low-sugar foods have diabetes. If you have diabetes, your body cannot make or properly use insulin. This leads to high blood glucose, or sugar, levels in your blood. Healthy eating helps keep your blood sugar in your target range. Treatment for diabetes is individualized and includes a balanced diet that supports a healthy weight, physical activity, as well as insulin or other medications as needed. Diet is a critical part of managing diabetes, because controlling your blood sugar can prevent the complications of diabetes.

TABLE 11-3   Menu Choices for a Low-Sugar Diet

LOW-SUGAR DIET	GOOD CHOICES	USE IN MODERATION
Appetizer and Soup	Most appetizers and soups are low in sugar	
Salads	Most salads are low in sugar	Avoid gelatin salads unless artificially sweetened
Breads	Whole-grain breads and rolls	
Entrées	Most entrées are low in sugar	
Side Dishes	Most side dishes are low in sugar	
Desserts	Fruits and fruit-based desserts with little sugar added, such as fruit crisps Puddings or gelatins made with sugar substitute Egg custard Low sugar or sugar-free cakes, pies, ice cream, and gelato	Cakes, pies, cookies Ice cream, frozen yogurt
Beverages	Coffee, tea Unsweetened iced tea, waters Diet sodas and drinks 100% fruit juices	No soft drinks, juice drinks, iced tea, or lemonade with sugar added
Breakfast	Whole-grain cereals with less than 4 grams sugar/serving such as corn flakes Eggs Whole-grain breads Muffins made with small amounts of sugar Pancakes and waffles Fresh fruit Crushed fresh fruit to top toast and pancakes/waffles or sugar-free syrup	Pancake syrup

FIGURE 11-3 Fruit Salad. Fruits contain natural sugars so they work well in a low-sugar diet.

Photo by Peter Pioppo.

If you have diabetes, a registered dietitian can help make an eating plan just for you. It should take into account your weight, medicines, lifestyle, and other health problems you have. Healthy diabetic eating includes:

- Limiting foods that are high in sugar
- Spreading your meals and snacks over the day
- Being careful about when and how many carbohydrates you eat
- Eating a variety of whole-grain foods, fruits (Figure 11-3), and vegetables every day
- Eating less fat
- Limiting your use of alcohol

SUMMARY

1. Currently, sugar-sweetened beverages are the primary source of added sugars in the American diet, followed by grains-based desserts (cakes, cookies, pies, etc.), sugar-sweetened fruit drinks, dairy-based desserts (ice cream, etc.), and candy.
2. Table 11-3 lists appropriate menu choices for a low-sugar diet.
3. If you have diabetes, your body cannot make or properly use insulin. This leads to high blood glucose, or sugar, levels in your blood. Healthy eating helps keep your blood sugar in your target range. Treatment for diabetes is individualized and includes a balanced diet that supports a healthy weight, physical activity, as well as insulin or other medications as needed. Diet is a critical part of managing diabetes, and involves paying attention to when and how many carbohydrates you eat.

LOW SODIUM

Virtually all Americans consume more sodium than they need. The estimated average intake of sodium for all Americans ages 2 years and older is approximately 3,300 mg per day. The sodium AI for individuals ages 9 to 50 years is 1,500 mg per day. Lower sodium AIs were set for children and older adults because their kcalorie requirements are lower. For everyone 14 years and older, the Tolerable Upper Intake Level (UL) is set at 2,300 mg per day. As you can see, Americans are taking in more sodium than the UL.

On average, the higher your sodium intake, the higher your blood pressure will be. Research shows in adults that as sodium intake decreases, so does blood pressure. Keeping blood pressure in the normal range is very important because it reduces your risk of cardiovascular disease, stroke, and kidney disease.

The major source of sodium in the diet is salt, which contains 2,300 milligrams (a little more than 2 grams) of sodium. Sodium is used to enhance the flavor of foods; it also enhances the texture of food and serves as a preservative. High amounts of salt and sodium are found in processed foods and restaurant food, and it is estimated that these foods provide 80 percent of the sodium in most people's diets. Very little of the sodium in foods is naturally occurring.

When you are looking for menu items that are lower in sodium, there are two places to look. If a menu item is high in sodium, it is most likely that you are either purchasing a processed, high-sodium main ingredient and/or you are using a lot of salt/high-sodium condiments in the recipe. Don't forget that unprocessed meats, poultry, and fish, fresh fruits and vegetables, as well as beans and peas (as long as they are not canned) are all low in sodium (Figure 11-4). Use the following guide to have low-sodium options available for your diners (Table 11-4). The Hot Topic in this chapter discusses how to decrease sodium in a recipe and continue to have exceptional flavor.

FIGURE 11-4 Vegetable Caper Relish Salad. Fresh vegetables and fruits with a balanced dressing are always good choices on a low-sodium diet.

Photo by Peter Pioppo.

TABLE 11-4 | Menu Choices for a Low-Sodium Diet

LOW-SODIUM DIET	GOOD CHOICES	USE IN MODERATION
In all menu categories, limit high-sodium ingredients such as salt, soy sauce, MSG, seasoned salts, Worcestershire sauce, ketchup, barbecue sauce, most Asian sauces, and any ingredient that is pickled, brined, smoked, or cured.		
Appetizer and Soup	Raw vegetables and fruits with dips low in sodium such as hummus	No regular soup bases, mixes, or bouillon cubes
	Fresh seafood with low-sodium accompaniments	
	Soups made with small amount of salt	
Salads	Raw fruits and vegetables	No commercial salad dressings unless lower than 100 mg sodium/2 tablespoons
	Salad dressings made from scratch with little or no sodium	
Breads	Whole-grain breads and rolls	No breads with salt on top
Entrées	Proteins such as fresh beef, chicken, fish, and beans (not processed)	Convenience breaded poultry or fish
	Pasta with low-sodium sauce made from scratch and vegetables	No cured and/or smoked meats and fish such as bacon, salt pork, ham, bologna, corned beef, frankfurters, luncheon meats, and canned tuna unless low-sodium versions
	Meatless entrées using beans, whole grains, and/or vegetables made from scratch with low sodium	No canned beans unless low-sodium version
Side Dishes	Roasted and steamed fresh or frozen vegetables	No canned vegetables or tomato products unless low-sodium versions
	Whole grains such as wild rice or quinoa	
Desserts	Fruits and fruit-based desserts	Commercial baked goods
	Sorbets and fruit ices	
	Desserts with less than 200 mg sodium/serving (often made on premises)	
Beverages	Coffee, tea, milk, soda	No canned tomato or vegetable juices unless low-sodium versions
	Unsweetened 100% juice	
Breakfast	Whole-grain cereals with less than 300 mg sodium/serving such as shredded wheat	Pancake and waffle mix
	Whole-grain breads	
	Muffins	
	Pancakes and waffles (made from scratch)	
	Low sodium bacon, sausage	
	Fresh fruit	

S U M M A R Y

1. If a menu item is high in sodium, it is most likely that you are either purchasing a processed, high-sodium main ingredient and/or you are using a lot of salt/high-sodium condiments in the recipe. Don't forget that unprocessed meats, poultry, and fish, fresh fruits and vegetables, as well as beans and peas (as long as they are not canned) are all low in sodium.
2. Table 11-4 lists appropriate menu choices for a low-sodium diet.

FOOD ALLERGIES

Do you start itching whenever you eat peanuts? Does seafood cause your stomach to churn? Symptoms like those cause millions of Americans to suspect they have a food allergy, when indeed most probably have a food intolerance. True food allergies affect a relatively small percentage of people. Experts estimate that only 2 percent of adults and from 4 to 8 percent of children are truly allergic to certain foods. So what's the difference between a food intolerance and a food allergy? A **food allergy** involves an abnormal immune-system response. If the response doesn't involve the immune system, it is called a **food intolerance**. Symptoms of food intolerance may include gas, bloating, constipation, dizziness, and difficulty sleeping.

Food allergy symptoms are quite specific. Food allergens, the food components that cause allergic reactions, are usually proteins. These food protein fragments are not broken down by cooking or by stomach acids or enzymes that digest food. When an allergen passes from the mouth into the stomach, the body recognizes it as a foreign substance and produces antibodies to halt the invasion. As the body fights off the invasion, symptoms begin to appear throughout the body. The most common sites (Figure 11-5) are the mouth (swelling of the lips or tongue, itching lips), the digestive tract (stomach cramps, vomiting, diarrhea), the skin (hives, rashes, or eczema), and the airways (wheezing or breathing problems). Allergic reactions to foods usually begin within minutes to a few hours after eating.

Food intolerance may produce symptoms similar to those of food allergies, such as abdominal cramping. But whereas people with true food allergies must avoid the offending foods altogether, people with food intolerance can often eat small amounts of the offending foods without experiencing symptoms. The most common food intolerance is lactose intolerance, discussed later in this chapter and also in Chapter 3. Lactose is found in milk and dairy products and causes gas, bloating, and diarrhea in people with lactose intolerance.

Food allergies are much more common in infants and young children, who may outgrow them later. In general, the more severe the first allergic reaction is, the longer it takes to outgrow it. Adults also have food allergies—especially to fish and shellfish. The following foods account for 90 percent of all food allergies.

- Milk
- Eggs
- Peanuts
- Tree nuts (such as walnuts, pecans, almonds, cashews, pistachio nuts)
- Fish
- Shellfish (such as crab or shrimp)
- Soy
- Wheat

Peanuts are the leading cause of death from food allergies. As little as 1/5 to 1/5,000 of a teaspoon of the offending food has caused death.

Most cases of allergic reactions to foods are mild, but some are violent and life-threatening and can be caused by simply a trace amount of the offending food. The greatest danger in food allergy comes from *anaphylaxis*, also known as anaphylactic shock, a rare allergic reaction involving a number of body parts simultaneously. Like less serious allergic reactions, anaphylaxis usually occurs after a person is exposed to an allergen to which he or she was sensitized by previous exposure. That is, it does not usually occur the first time a person eats a particular food. Anaphylaxis can produce severe symptoms in as little as 2 to 15 minutes. Signs of such a reaction include difficulty breathing, swelling of the mouth and throat, a drop in blood pressure, and loss of consciousness. The sooner anaphylaxis is treated, the greater the person's chance of surviving is.

There is no specific test to predict the likelihood of anaphylaxis, although allergy testing may help determine which foods a person is allergic to and provide some guidance as to the severity of the allergy. Experts advise people who are prone to anaphylaxis to carry medication—usually injectable epinephrine—with them at all times and to check the medicine's expiration date regularly.

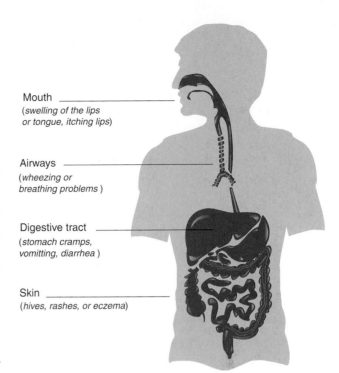

Mouth
(*swelling of the lips or tongue, itching lips*)

Airways
(*wheezing or breathing problems*)

Digestive tract
(*stomach cramps, vomitting, diarrhea*)

Skin
(*hives, rashes, or eczema*)

FIGURE 11-5 Common sites for allergic reactions.

Food allergy An abnormal response to a food triggered by your body's immune system. Allergic reactions to food can sometimes cause serious illness and death.

Food intolerance An unpleasant reaction to food that, unlike a food allergy, does not involve an immune system response. Symptoms can include gas, bloating, constipation, or dizziness.

Diagnosing a food allergy begins with a thorough medical history to identify the suspected food, the amount that must be eaten to cause a reaction, the amount of time between food consumption and the development of symptoms, how often the reaction occurs, and other detailed information. A complete physical examination and selected laboratory tests are conducted to rule out underlying medical conditions not related to food allergy. Several tests, such as skin testing and blood tests, are available to determine whether a person's immune system is sensitized to a specific food.

Once diagnosed, most food allergies are treated by avoidance of the food allergen. Avoiding allergens is relatively easy to do at home, where you can read food labels. Manufacturers must list each allergen by its recognizable name, such as "milk" instead of "casein." If "milk," for example, is not identified in the ingredient list, the label must state "Contains milk."

As a chef or foodservice professional, you can do the following to create a food allergy management plan to assist customers with food allergies:

1. *Who will answer guests' questions?* Individuals with food allergies need restaurants to provide them with accurate information about ingredients so they can make informed decisions about what to order from the menu. Recipes with complete ingredient information should be available for customers to look at. Some restaurants keep a book with specific dishes that are free of peanuts and other common allergens to suggest to customers. It is best to designate a person, such as a manager, on each shift who can discuss allergy concerns with the customer, check with the chef to ensure the item is prepared correctly, and hand-carry the plate separately from the rest of the table's order directly to the customer. High-risk menu choices for customers with allergies include the following:

   - Fried foods, because cooking oil is used for many foods and cross-contact can occur
   - Sauces in entrées or desserts
   - Combination foods, such as stews, contain many ingredients
   - Foods served on a buffet/cafeteria station because of the possibility of cross-contact

2. *Who will be responsible for keeping recipe information up-to-date as well as check ingredients used in menu items?* A manufacturer may change the ingredients in salsa, for example, and these changes are important to note. It is essential to carefully read the food labels of all ingredients. Table 11-5 lists ingredients to avoid for different food allergies. Keep on the alert for allergens in strange places, such as anchovies or sardines (both fish) in Worcestershire sauce or milk protein in imitation butter flavor.

Cross-contact This occurs when an allergen-containing food, such as peanuts, comes in contact with a safe food and contaminates it with the allergen.

3. *What steps will the kitchen and waitstaff take to avoid cross-contact?* **Cross-contact** occurs when an allergen-containing food comes in contact with a safe food. As a result, each food then contains small amounts of the other food, often invisible to us. Such contact may be either direct (such as putting cheese on a burger) or indirect via hands or utensils. Something as simple as picking up a muffin containing nuts, setting it down, and then picking up a nut-free muffin can cause cross-contact. It is also possible for cross-contact to occur if a customer's food is carried on a tray with other items. You have to ensure that preparation and cooking equipment, as well as serving trays and utensils, are thoroughly cleaned before using them to prepare and serve an allergen-free dish. Remember that even trace amounts of the offending food, sometimes even a strong smell of the food, can cause anaphylactic reactions.

4. *How should staff handle an allergic reaction?* Staff should immediately dial 911 and give the location of the restaurant, including cross-streets. Keep the guest seated, as sometimes standing up causes more problems.

All staff need training in the nature of food allergies, the foods commonly involved in anaphylactic shock, the restaurant's procedure for identifying and handling customers with food allergies, and emergency procedures.

The prevalence of food allergies is growing and probably will continue to grow. Take it seriously when a customer asks whether there are walnuts in your Waldorf salad. Some customers who ask such questions are probably just trying to avoid an upset stomach, but for some it's a much more serious, and possibly life-threatening, matter. Since you don't know which customer really has food allergies, take every customer seriously. Also consider putting a note on your menu to ask customers to talk with a manager about their food allergies before ordering. Unfortunately, some allergic reactions in restaurants have occurred when the customer did not mention their food allergies to employees.

TABLE 11-5   Foods to Avoid for Different Food Allergies

Egg allergies usually start in childhood. But it's an allergy that is frequently outgrown by the time the child grows up to be an adult. In baking, for leavening, you can use the following to substitute per egg for up to 3 eggs: 1½ tablespoons water, 1½ tablespoons oil, plus 1 teaspoon baking powder. As a binder, you can substitute 1/4 cup applesauce per egg.

EGG ALLERGIES

Ingredients to Avoid	Foods to Avoid	Foods That MAY Contain Egg
Albumin (also spelled albumen)	Whole eggs, egg whites, egg yolks	Baked goods and desserts such as cakes, cookies, ice cream, and sherbet
Dried or powdered egg	Egg substitutes	Foods prepared using breading or batters
Egg white	Custard	Pasta such as egg noodles or macaroni
Egg yolk	All baked goods made with eggs such as angel food cake, meringue, cream pies, etc.	Stock clarified with eggs
Egg solids		Meatloaf, meatballs
Globulin	Pancakes, waffles, and French toast	Pretzels
Lecithin	Hollandaise sauce, Bearnaise sauce, Newburgh sauce, and mayonnaise	Salad dressings
Livetin	Soufflés	Tartar sauce
Lysozyme	Nougat	Fudge and cream-filled chocolates
Ovalbumin or other words starting with "ova" or "ovo"	Surimi	Marshmallows, marzipan
Simplesse		Divinity and fondant
Vitellin		
Meringue		

Peanut allergies can trigger a severe reaction if the individual is very sensitive. Some unexpected sources of peanuts include some Asian and Mexican dishes, some sauces such as pesto and gravy, and specialty pizzas. Peanuts go by many names, such as beer nuts, ground nuts, or monkey nuts. Some individuals with peanut allergies can safely eat peanut oil.

PEANUT ALLERGIES

Ingredients to Avoid	Foods to Avoid	Foods That MAY Contain Peanuts
Peanuts	Peanuts	Baked goods with unknown ingredients, particularly cookies, carrot cake, pumpkin cake or pie, and fruit and nut rolls
Peanut oil—cold pressed, expeller pressed or extruded	Peanut oil	Breakfast foods such as cereal, muesli, and breakfast/granola bars
Peanut flour	Nuts roasted in peanut oil	Asian foods—especially Chinese, Indian, Indonesian, Thai, and Vietnamese (like satay, pad Thai, egg rolls)
Peanut protein hydrolysate	Peanut butter	
Arachis bouillon	Artificial nuts	
Hydrolyzed vegetable protein lecithins (soy lecithin is OK)	Beer nuts	African and Mexican dishes
Mandelonas (peanuts soaked in almond flavoring)	Goobers	Some vegetarian foods
	Ground or mixed nuts	Chili
	Monkey nuts	Salad dressings and condiments like barbecue sauce and Worcestershire sauce
	Nut meat or pieces	Sauces such as gravy, pesto, prepared sauces, and enchilada sauce
		Candies, marzipan, pralines, and nougat

Tree nuts include the nuts listed below, such as almonds, cashews, walnuts, and pistachios. Peanuts are not included in this group, because peanuts are legumes and grow underground. Some unexpected sources of tree nuts include pasta, honey, breading, salads and dressings, and meat-free burgers. Many experts advise patients allergic to tree nuts to avoid peanuts and other tree nuts because of the high likelihood of cross-contact at processing facilities, which process tree nuts and peanuts on the same equipment. Furthermore, a person with an allergy to one type of tree nut has a higher chance of being allergic to other types.

| TABLE 11-5 | **Foods to Avoid for Different Food Allergies** *(Continued)* |

TREE NUT ALLERGIES

Ingredients to Avoid	Foods to Avoid	Foods That MAY Contain Tree Nuts
Almond	Foods containing nuts and nut product listed on left	Meat-free burgers
Artificial nuts	Nut butters	Breading for chicken
Beechnut	Nut oils	Fish dishes
Brazil nut	Almond paste	Pasta
Butternut	Nut paste, Nutella®	Mortadella (may contain pistachios)
Cashew	Mandelonas (peanuts soaked in almond flavoring)	Barbecue sauce
Chestnut	Nougat, marzipan	Salads and salad dressing
Coconut	Natural flavorings and extracts such as almond extract	Honey
Filbert/hazelnut		Baked goods, pie crusts, and crackers
Ginkgo nut		Chocolates
Hickory nut		Some cereals
Litchi/lychee nut		
Macadamia nut		
Marzipan/almond paste		
Natural nut extract		
Pecan		
Pesto		
Pine nut		
Pistachio		
Praline		
Walnut		

Milk allergy is most common in young children, who usually outgrow it as they get older. Some hidden sources of milk include luncheon meats that are sliced on slicers also used to slice cheese.

MILK ALLERGIES

Ingredients to Avoid	Foods to Avoid	Foods That MAY Contain Milk
Any milk or cream in all forms	All dairy products including any type of milk or cream, ice cream, yogurt, cheese, cheese food, butter	Artificial butter flavor
Any cheese, yogurt, or ice cream	Pudding, custard	Baked goods—bread, cakes, cookies, crackers, etc.
Butter	Milk chocolate	Luncheon meats, frankfurters, sausages
Casein or casein hydrolysate	All baked goods made with dairy	Salad dressings
Caseinates *(in all forms)*	All foods made with dairy/cheese such as cream sauces, cheese sauces, puddings, and custards	Soups
Ghee		Vegetarian cheese
Half-and-half		Tofu
Lactalbumin		Margarine
Lactose		Nondairy products
Lactulose		Cocoa mixes
Milk protein hydrolysate		Chocolate, caramel candies, and nougat
Tagatose		
Whey (in all forms)		

Soybean allergy is one of the more common food allergies, especially in infants and children. Although Americans don't eat a lot of soybeans, soybeans appear in a number of processed foods and therefore can be hard to avoid. Soy products are found in baked goods, crackers, cereals, soups, and sauces, as well as other foods.

SOY ALLERGIES

Ingredients to Avoid	Foods to Avoid	Foods That MAY Contain Soy
All ingredients that include the word "soy"	All ingredients that include the word "soy" such as soy milk, soybean, soy protein, soy nuts, soy sauce	Meat or cheese substitutes
Hydrolyzed soy or vegetable protein	Edamame	Milk substitutes
Protein extender	Tofu	Canned soups, commercial entrées
Textured vegetable protein (TVP)	Miso	Breading/stuffing mixes
Soy sauce and shoyu sauce	Natto	Baked goods, crackers, desserts
	Tempeh	Cereals and granola bars
	Tamari	Low-fat peanut butter
	Shoyu sauce	Salad dressings, mayonnaise
	Soy pasta	Sauces and gravies
	Imitation bacon bits made with soy	Instant coffee, hot cocoa mixes, malt beverages
	Malt vinegars	Most individuals allergic to soy can eat soy lecithin as well as soy oil that has been highly refined (NOT cold pressed, expeller pressed, or extruded soybean oil).

Wheat allergy is not the same as having celiac disease that requires a gluten-free diet (discussed next). Children are the ones most likely to have a wheat allergy. Wheat can be hard to avoid and you need to read food labels carefully. Wheat can be found in unlikely places such as luncheon meats or french fries.

WHEAT-FREE DIET

Ingredients to Avoid	Foods to Avoid	Foods That MAY Contain Wheat
Any ingredient with the word "wheat" or parts of wheat (germ, bran, gluten, starch, malt)	Bulgur	Bread, crackers, pretzels
Bread crumbs	Couscous	All bakery goods
Any wheat flour including all-purpose, bread, cake, durum, enriched, graham, high gluten, pastry, self-rising, stone ground, etc.	Cracker meal	Breakfast cereals and bars
	Durum	Pasta
	Einkorn	Soups
	Emmer	Luncheon meats and frankfurters
Gelatinized starch	Farina	Sauces and gravies
Modified food starch	Any food made with wheat flour, including graham or high-gluten flour	French fries
Malt	Hydrolyzed wheat protein	Pancakes and waffles
	Kamut®	Ice cream
	Matzoh	
	Pasta made from wheat or durum wheat	
	Semolina	
	Spelt	
	Triticale	
	Whole-wheat berries	

Being allergic to shellfish is not something that is normally outgrown. Shellfish allergies are usually lifelong. Some people are allergic to only one type of shellfish, but can eat others. However, some people must avoid all shellfish.

SHELLFISH FREE DIET

Ingredients to Avoid	Foods to Avoid	Foods That MAY Contain ShellFish
Fish stock	Crabs, crayfish, lobster, languistines, prawns, shrimp (crevette, scampi), mollusks, krill	Appetizers, soups, salads, entrées containing ingredient or foods to left
Seafood flavoring such as crab or clam extract	Clams (cherrystone, littleneck, pismo, quahog), oysters, mussels, scallops	Any food served in a seafood restaurant may contain shellfish protein due to cross-contact
	Abalone, snails (escargot), limpets, periwinkles	
	Squid, octopus, cuttlefish	
	Surimi	

SUMMARY

1. A food allergy involves an abnormal immune-system response. The most common sites (Figure 11-5) are the mouth (swelling of the lips or tongue, itching lips), the digestive tract (stomach cramps, vomiting, diarrhea), the skin (hives, rashes, or eczema), and the airways (wheezing or breathing problems).

2. If the response doesn't involve the immune system, it is called a food intolerance. Symptoms of food intolerance may include gas, bloating, constipation, dizziness, and difficulty sleeping. Food intolerances are much more common than food allergies.

3. Food allergies are much more common in infants and young children, who may outgrow them later. Adults also have food allergies—especially to fish and shellfish. The following foods account for 90 percent of all food allergies: milk, eggs, peanuts, tree nuts, fish, shellfish, wheat, and soy.

4. Most cases of allergic reactions to foods are mild, but some are violent and life-threatening and can be caused by simply a trace amount of the offending food.

5. Whereas people with true food allergies must avoid the offending foods altogether, people with food intolerances can often eat small amounts of the offending foods without experiencing symptoms. The most common food intolerance is lactose intolerance.

6. As a chef or foodservice professional, you can do the following to create a food allergy management plan to assist customers with food allergies: Determine who will answer guest questions, determine who will be responsible for keeping recipe information up-to-date as well as check ingredients used in menu items, determine what steps the kitchen and waitstaff will take to avoid cross-contact, and determine what to do in an emergency.

7. Table 11-5 lists foods to avoid for different food allergies.

GLUTEN FREE

Celiac disease An inherited autoimmune disease. When a person with celiac disease consumes any food, beverage, or medication containing gluten (found in wheat, barley, and rye—or oats that are contaminated by contact with these other grains), his or her immune system is "triggered" and responds by damaging the lining of the intestinal tract. Symptoms include abdominal pain, diarrhea, and/or a severe skin rash.

Gluten sensitivity (also called non-celiac gluten intolerance) A condition that is not a wheat allergy or an autoimmune disease that does not seem to damage the intestines. Common symptoms of gluten sensitivity include abdominal pain, diarrhea, fatigue, and headaches.

Celiac disease is an inherited autoimmune disease that usually affects several organs in the body before it is diagnosed and treated. When a person with celiac disease consumes any food, beverage, or medication containing gluten (found in wheat, barley, and rye, or oats that are contaminated with these grains), his or her immune system is "triggered" and responds by damaging the lining of the intestinal tract. As a result, symptoms include recurrent abdominal pain, bloating, diarrhea, weight loss, lactose intolerance, and malnutrition, often accompanied by nonintestinal symptoms such as anemia and fatigue. Some people with celiac disease develop a severe skin rash. Treatment for celiac disease involves a gluten-free diet for life. Even tiny amounts of gluten can be harmful.

Other people without celiac disease also have gastrointestinal symptoms after eating gluten. They have a condition called "non-celiac gluten intolerance" or simply **gluten sensitivity**. Gluten sensitivity is NOT a wheat allergy or an autoimmune disease like celiac and it does not seem to damage the intestines. Common symptoms of gluten sensitivity include abdominal pain, diarrhea, fatigue, and headaches. A gluten-free diet is helpful for many people with gluten sensitivity but some can tolerate small to moderate amounts of gluten without symptoms.

It is the gluten in wheat, barley, and rye that causes celiac disease (Figure 11-6). Gluten is found in breads and other baked goods, where it provides strength, elasticity, and the structure needed to hold dough together and seals in the gases produced during fermentation, allowing the dough to rise. To avoid gluten you need to check food labels for these ingredients that must be avoided:

- Wheat—including durum wheat and flour, enriched flour, farina, graham flour, plain flour, white flour, wheat bran, wheat germ, semolina, kamut, and spelt wheat
- Barley
- Rye
- Triticale (a cross between wheat and rye)

FIGURE 11-6 The gluten in wheat, rye, and barley is essential for bread to rise, but people with celiac disease must avoid gluten.

Courtesy of the US Department of Agriculture.

- Oats (unless labeled gluten-free)
- Malt as in barley malt extract or flavoring or malt vinegar
- Brewer's yeast
- Dextrin
- Modified food starch

Oats do not contain gluten but they may be contaminated with wheat, barley, or rye through harvesting, processing, or travel. Therefore, oats are generally not recommended unless labeled "gluten-free." The majority of individuals with celiac disease can tolerate a daily intake of a limited amount (e.g., 50 grams) of oats that are free of gluten from wheat, rye, barley or their crossbred hybrids. Oats add variety, taste, satiety, dietary fiber, and other essential nutrients to the diet of individuals with celiac disease and may make their diet more appealing.

So what grains, starches, and flours can someone who is avoiding gluten eat? There actually are quite a few:

- Brown rice, wild rice, white rice
- Corn and cornmeal
- Amaranth
- Quinoa
- Millet
- Teff
- Sorghum
- Buckwheat groats—buckwheat is not technically a grain and it does not contain gluten
- Flax
- Starches/thickeners such as potato starch, tapioca starch, arrowroot, and corn starch

You may notice that several of the ancient grains—such as quinoa, amaranth, millet, and teff—are all gluten-free. Ancient grains have not only been around for thousands of years, but they have remained unchanged as well. By contrast, corn and rice have been bred selectively to look and taste much different than their ancestors.

Gluten-free flours (Figure 9-6) lack the structure and elasticity-providing properties of gluten, so baked goods made with them require additional ingredients to stabilize their shape and consistency. Good stabilizers include natural gums such as guar gum and xanthan gum, egg white powder, and fresh egg whites. These substitutions have little effect on taste, but they do change the mixing and baking methods.

TABLE 11-6   **Gluten-free Diet**

FOOD GROUP	FOODS TO ALLOW	POTENTIAL GLUTEN SOURCE (FOODS TO QUESTION)	FOODS TO AVOID
Dairy	Milk, buttermilk, half-and-half Cream Most ice cream Plain yogurt Cheese Processed cheese Cottage cheese Cream cheese	Flavored yogurt, frozen yogurt Chocolate milk Milk shakes Flavored shredded cheese or cheese blends Cheese spreads Cheese sauces	Malted milk Malted milk powders
Grains	Breads, bakery products, mixes, and cereals made with amaranth, arrowroot, corn bran, corn flour, cornmeal, cornstarch, flax, legume flours, millet, nut flours, potato flour, potato starch, gluten-free oat products such as oat flour or oatmeal, quinoa, rice bran, rice flours, sorghum flour, soy flour, sweet potato flour, tapioca, taro, teff Pasta and noodles made from beans, corn, lentils, peas, potato, quinoa, rice, soy, or any grains/starches listed above Corn tacos, corn tortillas, rice tortillas Plain rice such as basmati, brown, white, wild	Buckwheat flour (pure buckwheat flour is gluten-free but sometimes it is mixed with wheat flour) Rice and corn cereals (may contain barley malt, barley malt extract or flavoring) Seasoned or flavored rice mixes Rice crackers, rice cakes, corn cakes	Breads, bakery products, mixes, pasta, and cereals made with wheat (including bulgur, couscous, durum flour, einkorn, emmer, farro, farina, graham flour, kamut, semolina, splet, wheat bran, wheat germ), rye, triticale, or barley (including malt, malt flavoring, and malt extract) Matzoh, matzoh meal Modified food starch Tabouli
Fruits and Vegetables	Any fresh, frozen, or canned fruits or vegetables 100% fruit or vegetable juices	Dried fruits (may be dusted with flour) Fruits or vegetables with sauces	French fries cooked in oil used for other products containing gluten Scalloped potatoes Battered fried vegetables Fruit pie fillings Commercial salad dressings made with ingredients not allowed
Protein	Plain meat, poultry, fish, or shellfish Dried beans and peas Nuts and seeds Eggs Tofu	Luncheon meats, frankfurters, sausages, meat and sandwich spread, frozen burgers, meatloaf, ham, dried meats, imitation fish products, meat substitutes (may contain fillers made from wheat) Flavored egg products (liquid or frozen) Baked beans Seasoned or dry-roasted nuts Nut butters (some may contain wheat germ) Flavored tofu Tempeh, miso	Products breaded with ingredients not allowed Products made with hydrolyzed vegetable protein, marinades, or soy sauce Products prepared with cream sauces or gravies Frozen chicken or turkey containing chicken broth or hydrolyzed wheat protein Seitan (wheat gluten), fu (Japanese wheat gluten)

| TABLE 11-6 | Gluten-free Diet (Continued) |

FOOD GROUP	FOODS TO ALLOW	POTENTIAL GLUTEN SOURCE (FOODS TO QUESTION)	FOODS TO AVOID
Desserts/ Snacks	Plain popcorn Nuts Most potato chips and corn chips Cakes, cookies, and pies made with gluten-free flours and allowed ingredients Most ice cream, sherbet, sorbet, and Italian ices Custard and gelatin desserts Honey, jam, jelly, maple syrup Brown and white sugar, confectioner's sugar	Milk pudding Hard candies Chocolate bars Seasoned potato chips, corn chips, or nuts Cake icings or frostings (may contain wheat flour or wheat starch)	Bakery products or donuts made with wheat, rye, or barley Potato chips with ingredients not allowed Puddings made with wheat flour Ice cream or sherbet that contain gluten stabilizers or made with cookie dough, etc. Ice cream cones Marshmallows Licorice
Beverages	Coffee, tea, cocoa made with pure cocoa powder Soft drinks Distilled alcoholic beverages and wine	Flavored and herbal teas, flavored coffees, hot chocolate mixes Instant tea or coffee	Cereal and malt-based beverages Beer, ale, and lager Beverages that contain nondairy cream substitutes or barley malt extract/flavoring
Condiments	Catsup, relish, plain prepared mustard Herbs, spices, salt, pepper Olives, plain pickles Vinegar (except malt vinegar) Gluten-free soy sauce, gluten-free teriyaki sauce	Specialty prepared mustards Worcestershire sauce Salsa Seasoning mixes	Soy sauce made from wheat Malt vinegar Any sauce made with wheat flour and/or hydrolyzed wheat protein

Staples of the gluten-free diet include:

- Fruits and vegetables—All fresh fruit and vegetables are free of gluten. Most processed fruits and vegetables are free of gluten, too—but it is necessary to check labels for gluten-containing ingredients. Some dried fruit are dusted with oat flour, which is unlikely to be gluten-free. Frozen vegetables prepared with seasonings or sauces may contain gluten.

- Protein—Plain red meat, poultry, fish, dried beans and peas, nuts, seeds, and eggs do not contain gluten. When proteins such as chicken are breaded, sauced, or seasoned, the chances increase that they will contain gluten.

- Dairy—Most milk, yogurt, and cheese, as well as soy milk, do not contain gluten.

- Grains—As discussed above, you can use grains such as all forms of rice, amaranth, quinoa, and others as side dishes. For hot breakfast, cook kasha, teff, or gluten-free oats, or make buckwheat pancakes. Amaranth and buckwheat groats work well in soups, stews, and chili.

Table 11-6 outlines foods that are appropriate for a gluten-free diet, as well as food that are potential sources of gluten.

There are many processed gluten-free breads, pastas, breakfast cereals, and baking mixes available. Two considerations when buying mixes and processed gluten-free products are that many are low in fiber and/or low in nutrients because they are not enriched or fortified. Therefore, when reading the food label, look for products that list a gluten-free whole grain as the

b. Many ethnic cuisines use a variety of herbs, spices, and other flavorings instead of relying on sodium. See "Toasting Spices and Using Spice Blends" in Chapter 8 for ingredients in common ethnic blends such as Indian or Italian.

c. Use lower-sodium versions of soy sauce, teriyaki sauce, barbecue sauces, and other high-sodium condiments.

d. Use colorful and flavorful salsas, coulis, vegetable purees, relishes, chutneys, compotes, and mojos to top or otherwise accompany entrées.

e. Use cooking techniques and methods that are flavorful. For example, grill vegetables, then marinate or roast for maximum taste. Braising, roasting, grilling, and steaming bring out natural flavors.

f. Use moderate amounts of oils high in monounsaturated and/or polyunsaturated fats.

7. **Use umami to boost flavor.**
Umami, the fifth basic taste, differs from the traditional sweet, sour, salty, and bitter tastes by providing a savory, sometimes meaty, sensation. The umami taste receptor is very sensitive to glutamate, an amino acid found in protein that occurs in foods such as meat, fish, and milk. Despite the frequent description of umami as meaty, many foods, including mushrooms, tomatoes, and Parmesan cheese, have a higher level of glutamate than an equal amount of beef or pork. This explains why foods that are cooked with mushrooms or tomatoes seem to have a fuller, rounder taste than when cooked alone. When incorporating umami ingredients such as Parmesan cheese and tomato products into recipes, chefs can reduce the salt content of foods without sacrificing flavor. Chefs can also build umami flavor through cooking techniques. Any process that breaks down protein—such as drying, aging, curing, and slow cooking—increases umami because glutamate is released from protein.

8. **Draining and rinsing canned foods,** such as canned beans, can dramatically cut their sodium levels.

Restaurants around the country are working on offering lower-sodium foods. With the menu labeling requirement, restaurants with 20 or more locations have to provide sodium information for standard menu items in writing upon request. Restaurant chains have cut sodium from certain menu items. Taco Bell has cut 20 percent of the sodium from its menu across the board.

Another example of restaurants lowering sodium is evident in the National Salt Reduction Initiative (NSRI). NSRI is a partnership that includes 19 national health organizations, 9 national and regional health associations, and 44 cities, states, and related entities. New York City is coordinating the effort. The partnership is working with the restaurant industry and food manufacturers to lower salt levels in commonly consumed products. The goal is to reduce Americans' salt intake by 20 percent over five years.

As a first step, NSRI worked with industry to set salt-reduction targets for 25 restaurant food categories and 62 packaged foods. A number of major manufacturers and restaurant chains have agreed to pursue salt reduction targets for one or more food categories. For restaurants, having a set maximum level of sodium per serving of a given food is a requirement. The next step is to assess the impact on people's salt intake. The NSRI is a voluntary initiative, not a regulatory measure. It is modeled after a program developed in the United Kingdom.

WEIGHT MANAGEMENT

CHAPTER 12

- Explain how you gain or lose weight, and discuss at least two factors that play a role in the development of obesity.

- Define overweight and obesity, and determine how much you should weigh.

- Recognize risks of being obese.

- Describe how to use the following components of a weight loss program to lose weight: eating plan, exercise, behavior and attitude modification, and support.

INTRODUCTION TO WEIGHT MANAGEMENT

As you probably know, overweight and obesity are among the most pressing health challenges in the United States. Just take a look at these statistics.

	In the 1970s	In 2010
Children ages 2–5	5% were obese	12% were obese
Children ages 6–11	4% were obese	18% were obese
Children 12–19	6% were obese	18% were obese
Adults	15% were obese	36% were obese

The prevalence of obesity in the United States increased during the last decades of the twentieth century. More recently, there appears to have been a slowing of the rate of increase or even a leveling off. However, over the last decade there has been an increase in the prevalence of obesity among men and boys, but not among women and girls (Centers for Disease Control and Prevention, 2012). In 2009, obesity and overweight cost the US economy $270 billion because of increased need for medical care and loss of economic productivity.

Keeping a healthy body weight across the lifespan is vital to maintaining good health and quality of life. **Kcalorie balance** over time is the key to weight management. Kcalorie balance refers to the relationship between the kcalories you take in from foods and beverages and the kcalories you use through physical activity and your basal metabolism. Basal metabolism, discussed in Chapter 1, refers to the energy your body uses to meet its basic needs. You cannot control your metabolic rate (except by building more muscles), but you can control what you eat and drink, as well as how many kcalories you use in physical activity.

Kcalorie balance The balance between kcalories consumed in foods and beverages and kcalories used up through physical activity and basal metabolism.

Kcalories consumed must equal kcalories used for a person to maintain the same body weight. Consuming more kcalories than expended will result in weight gain. Conversely, eating fewer kcalories than you use will result in weight loss. This can be achieved over time by eating fewer kcalories, being more physically active, or, best of all, a combination of the two.

A variety of factors play a role in obesity: behavioral, cultural, socioeconomic, genetic, and environmental. The choices you make in eating and physical activity can contribute to overweight and obesity. A diet of energy-dense foods rich in fats and sugars, along with a sedentary lifestyle, are major contributors to weight gain. Your food choices as well as your level of activity are likely influenced by your cultural and socioeconomic background. For example, people in poorer neighborhoods are at a higher risk of becoming too heavy because they may not have access to grocery stores that are stocked with healthy foods, such as fresh fruits and vegetables, and they often don't have safe places to be physically active. Genetics play a role in obesity; however genes do not always predict future health. Genes and behavior may both be needed for a person to be overweight. Also, some illnesses may lead to obesity or weight gain.

The overall environment in which many Americans now live, work, learn, and play, which for many can be very stressful, has contributed to the obesity epidemic. Ultimately, individuals choose the type and amount of food they eat and how physically active they are. However, choices are often limited by what is available in a person's environment, including stores, restaurants, schools, and worksites. Environment affects both sides of the kcalorie balance equation—it can promote overconsumption of kcalories and discourage physical activity.

The food supply has changed dramatically over the past 40 years. Foods available for consumption increased in all major food categories from 1970 to 2008. Average daily kcalories available per person in the marketplace increased approximately 600 kcalories, with the greatest increases in the availability of added fats and oils, grains, milk and milk products, and caloric sweeteners.

Portion sizes have also increased. Research has shown that when larger portion sizes are served, people tend to consume more kcalories. Restaurant portion sizes have grown over the past 20 to 30 years. Bags of snack foods or soft drinks in the grocery store or vending machines are offered in larger and larger sizes that contain multiple servings, while a 1-ounce bag of snack food or an 8-ounce soft drink, which are the recommended single serving sizes, can, at times, be difficult to find. Americans are surrounded by larger portion sizes at relatively low prices, appealing to the consumer's economic sensibilities. However, the cost to America's health may be higher than most people realize.

SUMMARY

1. Kcalorie balance is the key to weight management. Kcalories consumed must equal kcalories expended for a person to maintain the same body weight. Eating fewer kcalories than you use will result in weight loss.
2. Many factors play a role in obesity: behavioral, cultural, socioeconomic, genetic, and environmental. A diet of energy-dense foods rich in fats and sugar, along with a sedentary lifestyle, are major contributors to weight gain. Large portion sizes and the ready availability of inexpensive food everywhere has contributed to obesity levels.

HOW MUCH SHOULD I WEIGH?

The best way to determine if you have a healthy weight is to use the **body mass index (BMI)** chart in Table 12-1. BMI is a direct calculation based on height and weight, and it applies to both men and women. Obesity in children is defined as a BMI greater than or equal to the age- and sex-specific 95th percentiles of the 2000 CDC growth charts (see the growth charts in Appendix G). BMI does not directly measure the percentage of body fat, but it is more accurate at approximating body fat than is measuring body weight alone. BMI is found by dividing a person's weight in kilograms by that person's height in meters squared.

The National Institutes of Health defines **overweight** for adults as a BMI of 25 to 29.9. A BMI of 30 or greater is considered **obese**. These cutoff points were chosen because studies show that as BMI rises above 25, blood pressure and blood cholesterol rise, HDL (good cholesterol) levels fall, and maintaining normal blood sugar levels becomes more difficult. Thus, individuals with a BMI of 25 or higher run a greater risk of heart disease, heart attacks, high blood pressure, stroke, and type 2 diabetes.

Using BMI is simple, quick, and inexpensive, but it does have limitations. One problem with using BMI as a measurement tool is that very muscular people may fall into the "overweight" category when they are actually healthy and fit. Another problem is that people who have lost muscle mass, such as the elderly, may be in the healthy weight category when they actually have reduced nutritional reserves. Further evaluation of a person's weight and health status is necessary.

Because BMI doesn't tell you how much of your excess weight is fat and where that fat is located, the National Institutes of Health has asked physicians to measure patients' waistlines. Studies show that excessive abdominal fat is more health-threatening than is fat in the hips or thighs. A woman whose waist measures more than 35 inches (88 cm) and a man whose waist measures more than 40 inches (100 cm) may be at particular risk for developing health problems. Studies indicate that increased abdominal fat is related to the risk of developing heart disease, high blood pressure, and stroke. Body fat concentrated in the lower body (around the hips, for example) seems to be less harmful.

Body mass index (BMI) A measure of weight in kilograms relative to height in meters squared. BMI is considered a reasonably reliable indicator of total body fat, which is related to the risk of disease and death. BMI categories include underweight, healthy weight, overweight, and obese.

Overweight Having a body mass index of 25 to 29.9.

Obese Having a body mass index of 30 or greater.

TABLE 12-1   **Body Mass Index Chart**

Locate the height of interest in the left-most column and read across the row for that height to the weight of interest. Follow the column of the weight up to the top row that lists the BMI. BMI of 19–24 is the healthy weight range, BMI of 25–29 is the overweight range, and BMI of 30 and above is in the obese range.

BMI	19	20	21	22	23	24	25	26	27	28	29	30	31	32	33	34	35
HEIGHT							**WEIGHT IN POUNDS**										
4'10"	91	96	100	105	110	115	119	124	129	134	138	143	148	153	158	162	167
4'11"	94	99	104	109	114	119	124	128	133	138	143	148	153	158	163	168	173
5'	97	102	107	112	118	123	128	133	138	143	148	153	158	163	168	174	179
5'1"	100	106	111	116	122	127	132	137	143	148	153	158	164	169	174	180	185
5'2"	104	109	115	120	126	131	136	142	147	153	158	164	169	175	180	186	191
5'3"	107	113	118	124	130	135	141	146	152	158	163	169	175	180	186	191	197
5'4"	110	116	122	128	134	140	145	151	157	163	169	174	180	186	192	197	204
5'5"	114	120	126	132	138	144	150	156	162	168	174	180	186	192	198	204	210
5'6"	118	124	130	136	142	148	155	161	167	173	179	186	192	198	204	210	216
5'7"	121	127	134	140	146	153	159	166	172	178	185	191	198	204	211	217	223
5'8"	125	131	138	144	151	158	164	171	177	184	190	197	203	210	216	223	230
5'9"	128	135	142	149	155	162	169	176	182	189	196	203	209	216	223	230	236
5'10"	132	139	146	153	160	167	174	181	188	195	202	209	216	222	229	236	243
5'11"	136	143	150	157	165	172	179	186	193	200	208	215	222	229	236	243	250
6'	140	147	154	162	169	177	184	191	199	206	213	221	228	235	242	250	258
6'1"	144	151	159	166	174	182	189	197	204	212	219	227	235	242	250	257	265
6'2"	148	155	163	171	179	186	194	202	210	218	225	253	241	249	256	264	272
6'3"	152	160	168	176	184	192	200	208	216	224	232	240	248	256	264	272	279
			HEALTHY WEIGHT						**OVERWEIGHT**					**OBESE**			

Source: Evidence Report of Clinical Guidelines on the Identification, Evaluation and Treatment of Overweight and Obesity in Adults, 1998. NIH/National Heart, Lung and Blood Institute (NHLBI).

Total body fat measurements also give an indication of obesity. For men, a desirable percentage of body fat is under 25 percent (22 percent if under 40 years of age); for women, under 35 percent (32 percent if under 40 years of age). Body fat is most often measured by using special calipers to measure the skinfold thickness of the triceps and other parts of the body. Because half of all your fat is under the skin, this method is quite accurate when performed by an experienced professional.

SUMMARY

1. Overweight for adults is having a BMI of 25 to 29.9. A BMI of 30 or greater is considered obese. Individuals with a BMI of 25 or higher run a greater risk of heart disease, high blood pressure, stroke, and type 2 diabetes.
2. Because BMI doesn't tell you how much of your excess weight is fat, physicians measure patients' waistlines. A woman whose waist measures more than 35 inches (88 cm) and a man whose waist measures more than 40 inches (100 cm) may be at particular risk for developing cardiovascular health problems. Excessive stomach fat is more health-threatening than is fat in the hips or thighs.

HOW OBESITY AFFECTS YOUR HEALTH

An obese individual is at increased risk for:

- Type 2 diabetes
- High blood cholesterol levels
- Hypertension (high blood pressure)
- Heart disease
- Stroke
- Certain types of cancer (esophagus, pancreas, colon and rectum, breast after menopause, endometrium, kidney, thyroid, and gallbladder)
- Liver and gallbladder disease
- Sleep apnea (interrupted breathing during sleeping) and respiratory problems
- Osteoarthritis (a degeneration of cartilage and its underlying bone within a joint)

In addition, surgery is riskier for obese individuals. Losing weight often decreases blood pressure and blood cholesterol levels and brings diabetes under better control. Although obesity does not cause these medical conditions, losing weight can help reduce some of their negative effects.

Obesity creates a psychological burden that in terms of suffering may be its greatest adverse effect. In American and other Westernized societies, there are powerful messages that people, especially women, should be thin, and that to be fat is a sign of poor self-control. Negative attitudes about the obese have been reported in children and adults, health-care professionals, and the overweight themselves. People's negative attitudes toward the obese often translate into discrimination in employment opportunities, job earnings, and other areas.

SUMMARY

An obese individual is at increased risk for type 2 diabetes, high blood cholesterol, high blood pressure, heart disease, stroke, certain types of cancer, and other conditions. Obesity also creates a psychological burden.

HOW TO LOSE WEIGHT

Weight loss can be achieved by eating and drinking fewer kcalories and/or by burning more kcalories in physical activity. The people with the greatest long-term success are doing BOTH—eating less and being more active. For example, walking 30 minutes each day OR drinking one less soda each day are steps you can take that can have a big impact on your weight over time.

First, set a realistic weight loss goal—perhaps ½ to 1 pound/week. Reducing kcalories by 500 each day amounts to about 1 pound lost in a week. In any case, kcalories should not be restricted below 1,200 without medical supervision because getting adequate nutrients is impossible below that level.

The good news is that no matter what your weight loss goal is, even a modest weight loss, such as 5 to 10 percent of your total body weight, is likely to produce health benefits, such as improvements in blood pressure, blood cholesterol, and blood sugars. For example, if you weigh 200 pounds, a 5 percent weight loss equals 10 pounds, bringing your weight down to 190 pounds. While this weight might still be in the overweight or obese range, this modest weight loss can decrease your risk factors for heart disease, stroke, and diabetes. So even if the overall goal seems large, see it as a journey rather than just a final destination. You'll learn new eating and physical activity habits that will help you live a healthier lifestyle. These habits will

TABLE 12-2 Nutrient Composition of a Low-Kcalorie Diet

NUTRIENT	RECOMMENDED INTAKE
Kcalories	Approximately 500 to 1,000 kcal/day reduction from usual intake
Total fat	30% or less of total kcalories
Saturated fatty acids	8–10% of total kcalories
Monounsaturated fatty acids	Up to 15% of total kcalories
Polyunsaturated fatty acids	Up to 10% of total kcalories
Cholesterol	Less than 300 mg/day
Protein	Approximately 15% of total kcalories
Carbohydrate	55% or more of total kcalories
Sodium	No more than 2.3 g of sodium
Calcium	1,000–1,500 mg/day
Fiber	20–30 g/day

Source: Adapted from "The Practical Guide: Identification, Evaluation, and Treatment of Overweight and Obesity in Adults," National Institutes of Health, 2000.

help you maintain your weight loss over time. To lose weight, we will now look at these weight loss program components: eating plan, exercise, behavioral and attitude modification, social support, and support to maintain weight loss.

EATING PLAN

The foundation for weight loss is a diet that restricts kcalories. A diet that creates a deficit of 250 to 1,000 kcal/day will theoretically lead to a weight loss of ½ to 2 pounds per week. A moderate reduction in kcaloric intake is designed to achieve a slow, but progressive, weight loss. It's natural for anyone trying to lose weight to want to lose it very quickly. But evidence shows that people who lose weight gradually and steadily (about 1 to 2 pounds per week) are more successful at keeping weight off. Healthy weight loss isn't just about a "diet" or "program." It's about an ongoing lifestyle that includes long-term changes in daily eating and exercise habits.

Table 12-2 shows the nutrient composition of a low-kcalorie diet. This type of diet will decrease other risk factors such as high blood cholesterol and hypertension. In general, your food choices should:

- Emphasize fruits, vegetables, whole grains, and fat-free or low-fat milk and milk products
- Include lean meats, poultry, fish, bean, eggs, and nuts
- Be low in saturated fats, *trans* fats, cholesterol, sodium, and added sugars

In this way, your diet will be nutritionally adequate.

There are many diets out there for the millions of Americans who struggle with weight. There is a growing body of evidence that one diet is no better than the next when it comes to weight loss. So if you prefer to eat a high-protein, low-carbohydrate diet or the Weight Watchers diet, it likely doesn't matter where your kcalories come from, as long as you're eating fewer kcalories. Unfortunately, there are still a number of weight loss diets that don't help you learn new eating habits, nor do they provide adequate nutrition. Beware of these types of diets that often also guarantee quick results.

Whatever diet you are on, be sure to get in adequate amounts of fruits, vegetables, whole grains, legumes, fat-free or low-fat dairy, and lean proteins to ensure you get enough vitamins and minerals. The water and fiber found in fruits, vegetables, whole grains, and legumes, as well as the protein in meat, poultry, and fish, all help you feel "full" so they are useful when you want to lose weight. Eat plenty of lean protein, the nutrient proven to have the strongest influence on satiety (feeling full).

While it would be nice to take off pounds quicker, a dieter who severely restricts kcalories will actually lose more muscle tissue than is desirable, and wind up with a lower basal metabolic rate. Muscles burn more kcalories than fat, so maintaining as much muscle as possible during weight loss is important. Less muscle tissue means a lower basal metabolic rate and this just makes it easier to regain weight.

Weigh yourself frequently—at least weekly. For some dieters, daily weigh-ins are helpful and keep them honest and motivated. At the minimum you should weigh yourself once a week (Figure 12-1).

Here are some eating tips:

- No foods should be forbidden because that only makes them more attractive. Instead, allow yourself small portions of favorite foods as appropriate.

- In order to cut kcalories without eating less and feeling hungry, you need to replace some higher kcalorie foods with foods that are lower in kcalories and fat and will fill you up. In general, this means foods with lots of water and fiber in them such as fruits and vegetables, broth-based soups such as vegetable soup, whole grains, and legumes (Table 12-3). For example, use more vegetables in a casserole or pasta dish to increase the water and fiber content and decrease the kcalories. Both water and fiber reduce kcalorie intake and promote satiety—or feeling full.

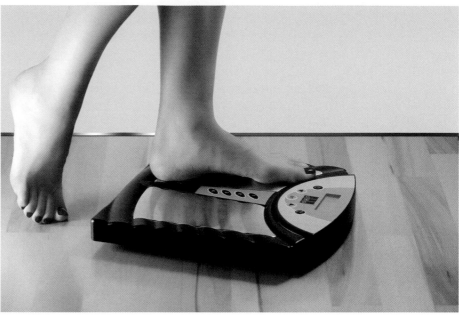

FIGURE 12-1 Weigh yourself frequently—at least weekly.

Courtesy of O.G.F./Shutterstock.

TABLE 12-3	Lower-Kcalorie Alternatives	
	INSTEAD OF . . .	**REPLACE WITH . . .**
DAIRY PRODUCTS 	Evaporated whole milk	Evaporated fat-free (skim) or reduced fat (2%) milk
	Whole milk	Low-fat (1%), reduced fat (2%), or fat-free (skim) milk
	Ice cream	Sorbet, sherbet, low-fat or fat-free frozen yogurt, or ice milk (check label for calorie content)
	Whipping cream	Imitation whipped cream (made with fat-free [skim] milk) or low-fat vanilla yogurt
	Sour cream	Plain low-fat yogurt
	Cream cheese	Neufchatel or "light" cream cheese or fat-free cream cheese
	Cheese (cheddar, American, Swiss, jack)	Reduced-calorie cheese, low-calorie processed cheeses, etc.; fat-free cheese
	Regular (4%) cottage cheese	Low-fat (1%) or reduced fat (2%) cottage cheese
	Whole-milk mozzarella cheese	Part skim milk, low-moisture, mozzarella cheese
	Whole-milk ricotta cheese	Part skim milk ricotta cheese
	Coffee cream (half and half) or non-dairy creamer (liquid, powder)	Low-fat (1%) or reduced fat (2%) milk or fat-free dry milk powder
CEREALS, GVRAINS, AND PASTA 	Ramen noodles	Rice or noodles (spaghetti, macaroni, etc.)
	Pasta with cream sauce (alfredo)	Pasta with red sauce (marinara)
	Pasta with cheese sauce	Pasta with vegetables (primavera)
	Granola	Bran flakes, crispy rice, etc.
		Cooked grits or oatmeal
		Whole grains (e.g., couscous, barley, bulgur, etc.); reduced-fat granola

TABLE 12-3   **Lower-Kcalorie Alternatives** (*Continued*)

MEAT, FISH, AND POULTRY

Cold cuts or lunch meats (bologna, salami, liverwurst, etc.)	Low-fat cold cuts (95% to 97% fat-free lunch meats, low-fat pressed meats)
Hot dogs (regular)	Lower-fat hot dogs
Bacon or sausage	Canadian bacon or lean ham
Regular ground beef	Extra-lean ground beef such as ground round or ground turkey (read labels)
Chicken or turkey with skin, duck, or goose	Chicken or turkey without skin (white meat)
Oil-packed tuna	Water-packed tuna (rinse to reduce sodium content)
Beef (chuck, rib, brisket)	Beef (round loin) trimmed of external fat (choose select grades)
Pork (spareribs, untrimmed loin)	Pork tenderloin or trimmed, lean smoked ham
Frozen breaded fish or fried fish (homemade or commercial)	Fish or shellfish, unbreaded (fresh, frozen, canned in water)
Whole eggs	Egg whites or egg substitutes
Frozen dinners (containing more than 13 grams of fat per serving)	Frozen dinners (containing less than 13 grams of fat per serving and lower in sodium)
Chorizo sausage	Turkey sausage, drained well (read label)

BAKED GOODS

Croissants, brioches, etc.	Hard French rolls or soft "brown 'n serve" rolls
Donuts, sweet rolls, muffins, scones, or pastries	English muffins, bagels, reduced fat or fat-free muffins or scones
Crackers	Low-fat crackers (choose lower in sodium) Saltine or soda crackers (choose lower in sodium)
Cake (pound, chocolate, yellow)	Cake (angel food, white, gingerbread)
Cookies	Reduced fat or fat-free cookies (graham crackers, ginger snaps, fig bars) (compare calorie level)

SNACKS AND SWEETS

Nuts	Popcorn (air-popped or light microwave), fruits, vegetables
Ice cream, e.g., cones or bars	Frozen yogurt, frozen fruit, or chocolate pudding bars
Custards or puddings (made with whole milk)	Puddings (made with skim milk)

FATS, OILS, AND SALAD DRESSINGS

Regular margarine or butter	Light-spread margarines, diet margarine, or whipped butter, tub or squeeze bottle
Regular mayonnaise	Light or diet mayonnaise or mustard
Regular salad dressings	Reduced calorie or fat-free salad dressings, lemon juice, or plain, herb-flavored, or wine vinegar
Butter or margarine on toast or bread	Jelly, jam, or honey on bread or toast
Oils, shortening, or lard	Nonstick cooking spray for stir-frying or sautéing
	As a substitute for oil or butter, use applesauce or prune puree in baked goods.

MISCELLANEOUS

Canned cream soups	Canned broth-based soups
Canned beans and franks	Canned baked beans in tomato sauce
Gravy (homemade with fat and/or milk)	Gravy mixes made with water or homemade with the fat skimmed off and fat-free milk included
Fudge sauce	Chocolate syrup
Avocado on sandwiches	Cucumber slices or lettuce leaves
Guacamole dip or refried beans with lard	Salsa

Source: Aim for a Healthy Weight. NIH Publ. No. 05-5213, National Institutes of Health.

FIGURE 12-2 Eating breakfast is very important if you want to lose weight.

Courtesy of Dusan Zidar/ Shutterstock.

- Another way to cut kcalories is to eat smaller portions. Strong evidence shows that portion size is associated with body weight, and eating smaller portions is associated with weight loss. Unfortunately, the more food on your plate, the more you are likely to eat. Pay more attention to how hungry you are, rather than how much food is on the table. Also slowly learn to eat less—perhaps a smaller amount of margarine on vegetables or less peanut butter on a sandwich. The kcalories you save will add up.

- Studies show that people who skip breakfast and eat fewer times during the day tend to be heavier than people who eat a healthy breakfast and eat four or five times a day (Figure 12-2). This may be because people who skip meals tend to feel hungrier later on, and eat more than they normally would. It may also be that eating many small meals throughout the day helps people control their appetite.

- Low-fat and fat-free foods are *not* necessarily low kcalorie. Read the nutrition label. Many low-fat and fat-free foods, such as desserts, are no lower in kcalories than their regular counterparts.

- Avoid fried foods, especially French fries and anything battered.

- Choose your carbohydrates carefully. Your best carbohydrate-containing foods are nutrient-packed foods such as fruits, vegetables, whole grains, and dairy. Fiber-containing foods such as whole grains help provide a feeling of fullness with fewer kcalories. Limit carbohydrate foods with added sugars—such as regular soda. Also limit foods containing primarily white flour and sugar: cakes, cookies, and doughnuts, to name a few. It's okay to drink diet soda and other beverages using artificial sweeteners.

- No foods can burn fat. Some foods with caffeine may speed up your metabolism (the way your body uses energy or kcalories) for a short time, but they do not cause weight loss.

- Drink plenty of water, and foods high in water such as broth-based soups, because they make you feel full and reduce hunger.

- Watch out for empty kcalories in alcoholic beverages and mixers, as well as fancy coffee drinks such as flavored lattes.

Diet analysis programs, such as SuperTracker (at the MyPlate website), or a diet analysis available from a website or application such as SparkPeople, help you identify high-kcalorie foods and monitor how many kcalories you are eating.

A **very-low-kcalorie diet (VLCD)** is a doctor-supervised diet that typically uses commercially prepared formulas to promote rapid weight loss in patients who are obese. These formulas, usually liquid shakes or bars, replace all food intake for several weeks or months.

Very-low-kcalorie diet A doctor-supervised diet that usually uses commercially prepared formulas to promote rapid weight loss in obese patients. These formulas, usually liquid shakes or bars, replace all food intake for several weeks or months.

FIGURE 12-3 Increasing exercise helps you lose weight and improves your mood.

Courtesy of bikeriderlondon/Shutterstock.

VLCD formulas need to contain appropriate levels of micronutrients to ensure that patients meet their nutritional requirements. Some physicians also prescribe VLCDs made up almost entirely of lean protein foods, such as fish and chicken. People on a VLCD consume about 800 kcalories per day or less.

VLCD formulas are not the same as the meal replacements you can find at grocery stores or pharmacies, which are meant to substitute for one or two meals a day. Over-the-counter meal replacements such as bars, entrees, or shakes, should account for only part of one's daily kcalories.

When used under proper medical supervision, VLCDs may produce significant short-term weight loss in patients who are moderately to extremely obese. VLCDs should be part of comprehensive weight-loss treatment programs that include exercise, behavioral therapy, nutrition counseling, and/or drug treatment.

EXERCISE

Regular physical activity is important for good health, and it's especially important if you're trying to lose weight or to maintain a healthy weight (Figure 12-3). When losing weight, more physical activity increases the number of kcalories your body uses for energy. You burn more kcalories not only during exercise but also for a few hours after vigorous prolonged activity. The burning of kcalories through physical activity, combined with reducing the number of kcalories you eat, creates a "kcalorie deficit" that results in weight loss.

Dieters who exercise also tend to lose more fat and retain more muscle than dieters who simply cut back on kcalories. They are also more likely to follow their diet plans and keep the weight off, possibly in part because exercising makes you feel better physically and mentally.

Much weight loss occurs because of decreased caloric intake. However, evidence shows exercise is crucial for maintaining weight loss. Regular exercise builds and tones muscles, which, in turn, raises your basal metabolic rate, so you end up burning more kcalories every day.

Physical activity reduces risks of cardiovascular disease and diabetes beyond that produced by weight reduction alone. Regular physical activity also helps to do the following:

- Reduce high blood pressure
- Reduce risk for type 2 diabetes, heart attack, stroke, and several forms of cancer
- Reduce arthritis pain and associated disability
- Reduce risk for osteoporosis and falls
- Reduce symptoms of depression and anxiety
- Improve psychological well-being
- Increase stamina and resistance to fatigue

When it comes to weight management, people vary greatly in how much physical activity they need. Here are some guidelines to follow:

- **To maintain your weight:** Work your way up to 150 minutes (2½ hours) of moderate-intensity aerobic activity or 75 minutes of vigorous-intensity aerobic activity, or an equivalent mix of the two each week. For example, walk briskly 30 minutes a day for five days each week. Strong scientific evidence shows that physical activity can help you maintain your weight over time. However, the exact amount of physical activity needed to do this is not clear since it varies greatly from person to person. It's possible that you may need to do more than the equivalent of 150 minutes of moderate-intensity activity a week to maintain your weight.

- **To lose weight and keep it off:** To lose weight, you should increase your weekly minutes of aerobic physical activity gradually over time (while eating fewer kcalories) to about 250 to 300 minutes per week of moderate-intensity activity. Some adults who need to lose weight may need to do more than five hours per week of moderate-intensity activity to meet weight loss goals. This may sound like a lot. However, your weight is a balance of the number of kcalories you eat and drink and the physical activity you do.

What do moderate- and vigorous-intensity mean? During activity, if your breathing and heart rate is noticeably faster but you can still carry on a conversation—it's probably moderately intense. Examples include:

- Walking briskly (a 15-minute mile)
- Light yard work (raking/bagging leaves or using a lawn mower)
- Light snow shoveling
- Actively playing with children
- Biking at a casual pace

If your heart rate is increased substantially and you are breathing too hard and fast to have a conversation, it's probably vigorously intense. Examples include:

- Jogging/running
- Swimming laps
- Rollerblading/inline skating at a brisk pace
- Cross-country skiing
- Most competitive sports (football, basketball, or soccer)
- Jumping rope

Table 12-4 shows kcalories used in common physical activities at both moderate and vigorous levels.

TABLE 12-4	Kcalories per Hour Expended in Common Physical Activities	
MODERATE PHYSICAL ACTIVITIES	IN 1 HOUR	IN 30 MINUTES
Hiking	370	185
Light gardening/yard work	330	165
Dancing	330	165
Golf (walking and carrying clubs)	330	165
Bicycling (less than 10 miles per hour)	290	145
Walking (3½ miles per hour)	280	140
Weight training (general light workout)	220	110
Stretching	180	90
VIGOROUS PHYSICAL ACTIVITIES	IN 1 HOUR	IN 30 MINUTES
Running/jogging (5 miles per hour)	590	295
Bicycling (more than 10 miles per hour)	590	295
Swimming (slow freestyle laps)	510	255
Aerobics	480	240
Walking (4½ miles per hour)	460	230
Heavy yard work (chopping wood)	440	220
Weight lifting (vigorous effort)	440	220
Basketball (vigorous)	440	220

Source: www.choosemyplate.gov 2012.

TABLE 12-5	Types of Physical Activity for Optimum Fitness
AEROBIC ACTIVITIES	Aerobic exercise increases your heart rate, works your muscles, and raises your breathing rate. These activities move large muscles in your arms, legs, and hips over and over again. Examples include walking, jogging, bicycling, swimming, and tennis.
STRENGTH-TRAINING ACTIVITIES	These activities increase the strength and endurance of your muscles. Examples of strength-training activities include working out with weight machines, free weights, and resistance bands. (A resistance band looks like a giant rubber band.) Push-ups and sit-ups are examples of strength-training activities you can do without any equipment.
	Aim to do strength-training activities at least twice a week. In each strength-training session, you should do 8 to 10 different activities using the different muscle groups throughout your body, such as the muscles in your abdomen, chest, arms, and legs. Repeat each activity 8 to 12 times, using a weight or resistance that will make you feel tired. When you do strength-training activities, slowly increase the amount of weight or resistance that you use. Also, allow one day in between sessions to avoid excess strain on your muscles and joints.
STRETCHING ACTIVITIES	While strong muscles are important for movement, flexibility allows you to move your muscles and joints through the full range of motion. Being flexible also helps reduce the chance of muscle injury.
	You should do stretching activities after your muscles are warmed up—for example, after strength training. Stretching your muscles before they are warmed up may cause injury.

The optimal fitness routine should include aerobic training (such as walking or biking), strength-training activities, and stretching activities (Table 12-5). Find activities that you enjoy. You may want to consider buying a pedometer. Health experts recommend taking at least 10,000 steps a day, which is roughly four to five miles depending on the length of your stride.

BEHAVIOR AND ATTITUDE MODIFICATION

Behavior modification Use of demonstrated behavior change techniques to increase or decrease the frequency of behaviors, such as restricting eating to the kitchen or dining room table, to eat fewer kcalories.

Behavior modification deals with using demonstrated behavior change techniques to increase or decrease the frequency of behaviors that, in this case, affect weight gain. For example, if you eat a lot in response to seeing foods such as chips and cookies laying on the kitchen table or counters, you can put foods away or at least cover them so you don't see them. To change behaviors, you will need to set specific, achievable goals.

Most people trying to lose weight focus on just that one goal: weight loss. However, the most productive areas to focus on are the eating and exercise changes that will lead to that long-term weight change. Effective goals are:

1. Specific and measurable
2. Attainable and realistic
3. Forgiving—less than perfect

"Exercise more" is a commendable ideal, but it's not specific. "Walk five miles every day" is specific and measurable, but is it realistic if you are just starting out? "Walk 30 minutes every day" is more attainable, but what happens if you're held up at work one day and there's a thunderstorm during your walking time another day? "Walk 30 minutes, five days each week" is specific, attainable, and forgiving. In short, a great goal!

Additional elements of behavior and attitude modification can be grouped into several categories: self-monitoring, stimulus or cue control, eating behaviors, reinforcement or self-reward, self-control, and attitude modification. Let's take a look at each category.

Self-monitoring involves keeping a food diary or daily record of types and amounts of foods and beverages consumed, as well as time and place of eating, mood at the time, and

TABLE 12-6	Food Diary					
TIME	WHAT WAS EATEN?	WHERE?	HOW MUCH?	HUNGRY?	WITH WHOM?	MOOD?

degree of hunger felt (Table 12-6). Its purpose is to increase awareness of what is actually being eaten and whether the eating is in response to hunger or other stimuli. Hunger is a physiological need for food, whereas appetite is a psychological need. Eating should be in response to hunger, not to appetite. Through self-monitoring of your eating behaviors, cues or stimuli to overeating can be identified. For example, passing a bakery may be a cue to stop and buy a dozen cookies. Examples of behavioral modification techniques for cue or stimulus control are in Table 12-7.

TABLE 12-7	Behavioral Modification Techniques for Cue or Stimulus Control

Food Purchasing, Storage, and Cooking

1. Plan meals a week or more ahead.
2. Make a shopping list.
3. Do food shopping after eating, on a full stomach.
4. Do not shop for food with someone who will pressure you to buy foods you do not need.
5. If you feel you must buy high-kcalorie foods for someone else in the family who can afford the kcalories, buy something you do not like or let that person buy and store that particular food.
6. Store food out of sight and limit storage to the kitchen.
7. Keep low-kcalorie snacks on hand and ready to eat.
8. When cooking, keep a small spoon such as a half-teaspoon measure on hand to use if you must taste while cooking.

Mealtime

1. Do not serve food at the table or leave serving dishes on the table.
2. Leave the table immediately after eating.
3. Brush your teeth after eating to replace food cravings with a fresh taste.

Holidays and Parties

1. Eat and drink something before you go to a party.
2. Drink fewer alcoholic beverages.
3. Bring a low-kcalorie food to the party.
4. Decide what you will eat before the meal.
5. Stay away from the food as much as possible.
6. Concentrate on socializing.
7. Be polite but firm and persistent when refusing another portion or drink.

Eating behaviors need to be modified to discourage overeating. First, one or two eating areas, such as the kitchen and dining room tables, need to be set up so that eating occurs only in these designated locations. Eat only while sitting at the table, and make the environment as attractive as possible. Do not read, watch television, or do anything else while you eat, because you can easily form associations between certain activities and food, such as television and snacking. Do not eat while standing at cabinets or refrigerators.

Second, plan three meals and at least two snacks daily, preferably at specified times of the day. Third, eat slowly by putting your fork down between bites, talking to others at the table, eating a high-fiber food that requires time to chew, and drinking a no-kcalorie beverage to help you fill up. In addition, take smaller bites, savor each bite, use a smaller plate to make the food look bigger, and leave a bite or two on the plate. If you clean your plate, you are responding to the sight of food, not to real hunger.

Reward yourself for positive steps taken to lose weight but do not use food as a reward. An effective reward is something that is desirable and timely. The reward might be tangible, such as a book or a payment toward buying a more costly item, or intangible, such as an afternoon off from watching the kids or getting a massage. Numerous small rewards, delivered for meeting smaller goals, are more effective than bigger rewards after a long, difficult effort.

Overeating sometimes occurs in reaction to stressful situations, emotions, or cravings. The food diary is very useful in identifying these situations. Then you can handle these situations in new ways. You can exercise or use relaxation techniques to relieve stress, or you can switch to a new activity, such as taking a walk, knitting, reading, or engaging in a hobby to help take your mind off food. If you allow yourself five minutes before getting something to eat, you often will go on to something else and forget about the food. Positive self-talk is important for good self-control. Instead of repeating a negative statement such as "I cannot resist that cookie," say, "I will resist that cookie."

The most common attitude problem obese people have is thinking of themselves as either on or off a diet. Being on a diet implies that at some point the diet will be over, resulting in weight gain if old habits are resumed. Dieting should not be so restrictive or have such unrealistic goals that the person cannot wait to get off the diet. When combined with exercise, behavior and attitude modification, social support, and a maintenance plan, dieting is really a plan of sensible eating that allows for periodic indulgences.

Another attitude that needs modification revolves around using words such as always, never, and every. The following are examples of unrealistic statements using these terms.

- I will always control my desire for chocolate.

- I will never eat more than 1,500 kcalories each day.

- I will exercise every day.

Goals stated in this manner decrease the likelihood that you will ever accomplish them and thus result in discouragement and feelings of failure.

Even with reasonable goals, occasional lapses in behavior occur. This is when having a constructive attitude is critical. After a person eats and drinks too much at a party one night, for example, feelings of guilt and failure are common. However, those feelings do nothing to help people get back on their feet. Instead, the dieter must stay calm, realize that what is done is done, and understand that no one is perfect.

SUPPORT

You are more likely to lose weight and keep it off if you have support from family and friends. Having someone who is easy to talk to and understands what you are going through can help you reach your goals. Supporters can model good eating habits and give praise and encouragement. Commercial programs that rely on group support and discussions about diet and exercise work well for many dieters.

Support is also available using the Internet and technology such as mobile exercise or weight reduction applications (Figure 12-4). The following websites offer a variety of online

FIGURE 12-4 Mobile exercise and weight reduction applications are useful when trying to lose weight.

Courtesy of Robert Kneschke/Shutterstock.

tools and mobile applications that will allow you to count kcalories, plan menus, and find recipes and social support:

- CalorieCount
- LoseIt
- MealSnap
- MyFitnessPal
- SparkPeople
- WebMd
- WeightWatchers

MAINTENANCE SUPPORT

Keeping weight off requires persistent effort and attention. The following tips are some of the common characteristics among people who have successfully lost weight and maintained that loss over time:

1. Continue to follow a healthy eating pattern lower in kcalories when compared to your diet prior to weight loss.
2. Keep eating patterns consistent regardless of changes in your routine. Plan ahead for weekends, vacations, and special occasions.
3. Eat breakfast every day.
4. Keep monitoring how many kcalories you eat and how much physical activity you get.
5. Exercise an average of 30 to 60 minutes/day.
6. Watch less than 10 hours of television per week.
7. Keep monitoring your weight at least weekly.
8. Continue to get support from family, friends, and others.

Support is great to stay on course and get you over any bumps.

SUMMARY

1. To lose weight, and keep it off, you need to eat and drink fewer kcalories and burn more kcalories in physical activity. Even a modest weight loss, such as 5 to 10 percent of your total body weight, is likely to produce health benefits.

2. The foundation for weight loss is a diet that restricts kcalories. From 250 to 1,000 kcalories per day will theoretically lead to a weight loss of ½ to 2 pounds per week. People who lose weight gradually and steadily—about 1–2 pounds per week—are more successful at keeping weight off.

3. In many cases, one diet is no better than the next when it comes to weight loss, but beware of diets that don't help you learn new eating habits or provide enough nutrients.

4. In order to cut kcalories without eating less and feeling hungry, you need to replace some higher kcalorie foods with foods that are lower in kcalories and fat and will fill you up. In general, this means foods with lots of water and fiber in them such as fruits and vegetables, broth-based soups such as vegetable soup, whole grains, and legumes. Both water and fiber reduce kcalorie intake and promote satiety—or feeling full. Additional eating tips are given.

5. Dieters who exercise tend to lose more fat and keep more muscle tissue. Regular exercise raises your metabolic rate and is crucial to maintain weight loss. Regular physical activity reduces risk of cardiovascular disease and diabetes, reduces blood pressure, improves psychological well-being, and other benefits.

6. To lose weight, you should do about 250 to 300 minutes per week of moderate-intensity activity. To maintain your weight, you should do about 150 minutes per week of moderate-intensity activity such as walking briskly. The optimal fitness routine should include aerobic training, strength-training, and stretching.

7. Behavior modification involves using behavior change techniques to increase or decrease the frequency of behaviors that, in this case, affect weight gain. Elements of behavior and attitude modification include realistic goal-setting, self-monitoring (food diary), stimulus or cue control, eating behaviors, self-reward, self-control, and attitude modification.

8. You are more likely to lose weight and keep it off it you have support from family and friends. Support is also available using the Internet and mobile applications.

9. Tips are given to help dieters keep weight off.

CHECK-OUT QUIZ

1. To lose weight, you need to eat fewer kcalories than you burn or use each day.
   a. True
   b. False

2. If your BMI is 27, you are obese.
   a. True
   b. False

3. An obese person has an increased risk for heart disease, stroke, and type 2 diabetes.
   a. True
   b. False

4. When losing weight, it is desirable to severely restrict kcalories and lose the weight quickly.
   a. True
   b. False

5. So you don't get discouraged, just weigh yourself every two or three weeks.
   a. True
   b. False

6. To lose weight, replace higher kcalorie foods with foods high in water and fiber such as fruits and vegetables, legumes, and broth-based soups.
   a. True
   b. False

7. Dieters who exercise tend to lose more fat and retain more muscle.
   a. True
   b. False

8. To lose weight and keep it off, you need at least 150 to 200 minutes per week of moderate-intensity activity.
   a. True
   b. False

9. List ten tips for losing weight including at least two from each of these categories: eating, exercising, behavior and attitude modification, and support.

10. Which kind of diet works best for losing weight: high protein, low fat, or high carbohydrate?

NUTRITION WEB EXPLORER

For a complete list of websites for the following activities, please visit the companion book page at www.wiley.com/college/drummond.

WEIGHT-CONTROL INFORMATION NETWORK (WIN)

On the WIN homepage, click on "For the Public" under Publications. Read one of the publications and write a one paragraph summary of what you read.

NATIONAL WEIGHT CONTROL REGISTRY

The National Weight Control Registry tracks the habits of people who have lost significant amounts of weight and kept it off. Under "Research Findings" and "Success Stories," find out how they were successful.

COOKING LIGHT—LOW CALORIE RECIPES

At *Cooking Light* magazine, look at any category of low-kcalorie recipes—such as "Appetizers Under 100 Calories," "250-Calorie Entrees," or "Treats Under 150 Calories." Write up a couple of paragraphs explaining how the kcalories were kept low in some of the recipes you looked at.

SHAPE UP AMERICA!

At Shape Up America!, click on "Shape Up & Drop 10 ™" under "Adults." Choose one of the ten steps to read. Write a summary of what you learned.

SPARKPEOPLE

Register at this website and learn about all the tools available, including social support, to anyone who wants to lose weight.

Low-Kcalorie Meals

You will be preparing a low-kcalorie meal to include an appetizer, entrée, side dish, salad, and dessert.

Your instructor will hand out a worksheet that you will use to describe why your menu item is lower in kcalories than its traditional counterpart. You will also evaluate each item on appearance and taste.

HOT TOPIC

WEIGHT LOSS MYTHS

Myth: Fad diets work for permanent weight loss.

Fact: Fad diets are not the best way to lose weight and keep it off. Fad diets often promise quick weight loss or tell you to cut certain foods out of your diet. You may lose weight at first on one of these diets. But diets that strictly limit kcalories or food choices are hard to follow. Most people quickly get tired of them and regain any lost weight.

Fad diets may be unhealthy because they may not provide all of the nutrients your body needs. Also, losing weight at a very rapid rate (more than three pounds a week after the first couple of weeks) may increase your risk for developing gallstones (clusters of solid material in the gallbladder that can be painful).

Myth: Starches are fattening and should be limited when trying to lose weight.

Fact: Many foods high in starch, like whole-grain bread, rice, pasta, cereals, beans, fruits, and some vegetables (like potatoes and yams) are low in fat and kcalories. They become high in fat and kcalories when eaten in large portion sizes or when covered with high-fat toppings like butter, sour cream, or mayonnaise. Foods high in starch are an important source of energy for your body.

Myth: Certain foods, like grapefruit, celery, or cabbage soup, can burn fat and make you lose weight.

Fact: No foods can burn fat. Some foods with caffeine may speed up your metabolism (the way your body uses energy, or kcalories) for a short time,

but they do not cause weight loss. The best way to lose weight is to cut back on the number of kcalories you eat and be more physically active.

Myth: A high-protein diet is better for losing weight than a low-fat diet.

Fact: Reduced-kcalorie diets result in weight loss regardless of which macronutrient they emphasize. A high-protein diet, or a low-fat diet, will both help you lose weight as long as you are eating fewer kcalories than you are burning.

Myth: Natural or herbal weight-loss products are always safe and effective.

Fact: A weight-loss product that claims to be "natural" or "herbal" is not necessarily safe. These products are not usually scientifically tested to prove that they are safe or that they do everything the label says they do. For example, herbal products containing ephedra (now banned by the US government) have caused serious health problems and even death. Newer products that claim to be ephedra-free are not necessarily danger-free, because they may contain ingredients similar to ephedra.

Myth: Low-fat or fat-free means no kcalories.

Fact: A low-fat or fat-free food may be lower in kcalories than the same size portion of the full-fat product. But many processed low-fat or fat-free foods have just as many kcalories as the full-fat versions of the same foods—or even more kcalories. They may contain added sugar, flour, or starch thickeners to improve flavor and texture after fat is removed. These ingredients add kcalories. Read the Nutrition Facts on a food package to find

out how many kcalories are in a serving. Check the serving size, too—it may be less than you are used to eating.

Myth: Fast foods are always an unhealthy choice and you should not eat them when dieting.

Fact: Fast foods can be part of a healthy weight-loss program with a little bit of know-how. Avoid supersized combo meals, or split one with a friend. Sip on water or fat-free milk instead of soda. Choose salads and grilled foods, like a grilled chicken breast sandwich or small hamburger. Try a "fresco" taco (with salsa instead of cheese or sauce) at taco restaurants. Fried foods, like French fries and fried chicken, are high in fat and kcalories, so order them only once in a while, order a small portion, or split an order with a friend. Also, use only small amounts of high-fat, high-kcalorie toppings, like regular mayonnaise, salad dressings, bacon, and cheese.

Myth: Skipping meals is a good way to lose weight.

Fact: Studies show that people who skip breakfast and eat fewer times during the day tend to be heavier than people who eat a healthy breakfast and eat four or five times a day. This may be because people who skip meals tend to feel hungrier later on, and eat more than they normally would. It may also be that eating many small meals throughout the day helps people control their appetites.

Myth: Eating after 8 P.M. causes weight gain.

Fact: It does not matter what time of day you eat. It is what and how much you eat and how much physical activity you do during the whole day that determines whether you gain, lose, or maintain your weight. No matter when you eat, your body will store extra kcalories as fat.

Myth: Lifting weights is not good to do if you want to lose weight, because it will make you "bulk up."

Fact: Lifting weights or doing strengthening activities like push-ups and crunches on a regular basis can actually help you maintain or lose weight. These activities can help you build muscle, and muscle burns more kcalories than body fat. So if you have more muscle, you burn more kcalories—even sitting still. Doing strengthening activities two or three days a week will not "bulk you up." Only intense strength training can build very large muscles.

Myth: Nuts are fattening and you should not eat them if you want to lose weight.

Fact: In small amounts, nuts can be part of a healthy weight-loss program. Nuts are high in kcalories and fat. However, most nuts contain healthy fats that do not clog arteries. Nuts are also good sources of protein, dietary fiber, and minerals including magnesium and copper.

Myth: Eating red meat is bad for your health and makes it harder to lose weight.

Fact: Eating lean meat in small amounts can be part of a healthy weight-loss plan. Red meat, pork, chicken, and fish contain some cholesterol and saturated fat (the least healthy kind of fat). They also contain healthy nutrients like protein, iron, and zinc.

Choose cuts of meat that are lower in fat and trim all visible fat. Lower-fat meats include pork tenderloin and beef round steak, flank steak, and extra lean ground beef. Also, pay attention to portion size. Three ounces of meat or poultry is the size of a deck of cards.

Myth: Dairy products are fattening and unhealthy.

Fact: Low-fat and fat-free milk, yogurt, and cheese are just as nutritious as whole-milk dairy products, but they are lower in fat and kcalories. Dairy products have many nutrients your body needs. They offer protein to build muscles and calcium to strengthen bones. Most milk and some yogurt are fortified with vitamin D to help your body use calcium.

NUTRITION FOR ALL AGES

CHAPTER 13

- Describe how to ensure enjoyable mealtimes with young children and teach them good eating habits.

- Plan menus for preschool and school-age children and identify the nutrients that children are most likely to be lacking, their food sources, and why they are important.

- Identify three nutrients that are very important for adolescents and why they are important.

- Plan menus for adolescents.

- Describe factors that influence the nutritional status of older adults, and identify nutrients of concern for older adults, their food sources, and why they are important.

- Plan menus for healthy older adults.

- Describe signs and treatment of eating disorders and who is most likely to have an eating disorder.

- Plan nutritious menus for athletes.

This chapter looks at nutrition for all ages, as well as how to plan nutritious menus for children, adolescents, and older adults. Eating disorders and nutrition for the athlete are also discussed at the end of this chapter.

Whether you are 5 or 65, or somewhere in between, there is a growing consensus among nutrition experts about what is a healthy diet:

- High in healthy carbohydrates such as fruits, vegetables, whole grains, legumes, and soy products

- Low in refined carbohydrates such as sugar, high-fructose corn syrup, and white flour (or foods containing sugars and white flour)

- High in good fats such as vegetable oils high in unsaturated and omega-3 fatty acids (also found in fish and nuts)

- Low in bad fats such as saturated fats, trans fats, and hydrogenated fats

- Little or no red meat

- Lean protein (skinless poultry, fish, legumes, etc.)

This diet may be tweaked a little if you have diabetes, heart disease, or other diseases, but all in all, it does represent a healthy way of eating and a very different diet than most Americans of various ages are eating (Figure 13-1). This chapter starts by discussing nutrition for children starting at one year old.

NUTRITION AND MENU PLANNING FOR CHILDREN

Although much of the nutrition advice we read on the Internet or hear from the television is for adults—indeed, the first part of this book is mostly for adults—other groups have their own special nutrition needs and concerns. Children and teenagers are all growing, and growth requires more kcalories and nutrients. This section looks at nutrition for children.

NUTRITION FOR CHILDREN

Table 13-1 shows the estimated energy requirement (EER) for kcalories and the RDA for protein for children. A two-year-old needs about 1,000 kcalories a day. By age six, a child needs closer to 1,700 kcalories daily. Energy needs of children of similar age, sex, and size can vary due to differing BMRs (basal metabolic rates), growth rates, and activity levels. Energy and protein needs decline gradually per pound of body weight.

For children two to three years of age, fat intake should be 30 to 40 percent of total kcalories. Children and adolescents from 4 to 18 years of age should keep fat intake between 25 and 35 percent of kcalories, with most fats coming from sources of polyunsaturated and monounsaturated fatty acids, such as vegetable oils, nuts, and fish. Children over two years old should limit consumption of saturated and trans fats to 10 percent or less of total kcalories as do adults.

Problem nutrients for children include fiber, iron, calcium, and vitamin D. For a variety of reasons, children are not eating enough foods rich in these nutrients. Fiber is abundant in plants, so fruits, vegetables, whole grains, legumes (dried beans, peas, and lentils), and nuts provide fiber. Fiber is *not* found in meat, poultry, fish, dairy products, or eggs.

Lack of iron can cause fatigue and affect behavior, mood, and attention span. A balanced diet with adequate consumption of iron-rich foods such as lean meat (ground meat is easier for younger children to chew), whole grain and enriched breads and cereals, eggs, and legumes is important to get enough iron. A source of vitamin C, such as citrus fruits, increases the amount of iron absorbed.

The RDA for calcium increases from 700 mg for one- to three-year-olds to 1,000 mg for four- to eight-year-olds. Children from two to three years old should drink two cups per day of skim or low-fat milk or equivalent. It is recommended that children from four to eight years old should drink 2½ cups per day. Milk also supplies vitamin D, which may also be low in some children's diets if children don't get enough vitamin-D-fortified milk and fortified cereals.

During **growth spurts**, the requirements for kcalories and nutrients are greatly increased. Appetite fluctuates tremendously, with a good appetite during growth spurts (periods of rapid growth) and a seemingly terrible appetite during periods of slow growth. Parents may worry and force a child to eat more than

FIGURE 13-1 Based on national survey data, grain-based desserts account for a greater proportion of daily kcalories than any other food group in people ages one and older.

Courtesy of the US Department of Agriculture, Agricultural Research Center.

Growth spurts Periods of rapid growth.

TABLE 13-1	Energy and Macronutrients for Children			

ESTIMATED ENERGY REQUIREMENT (EER) AND RECOMMENDED DIETARY ALLOWANCE (RDA) FOR PROTEIN

Gender and Age	Height	Weight	EER*	Protein
Male 1–3 years	34 inches	27 pounds	1,046 kcal	1.1 grams/kilogram body weight
Female 1–3 years	34	27	992	1.1
Male 4–8 years	45	44	1,742	0.95
Female 4–8 years	45	44	1,642	0.95

ACCEPTABLE MACRONUTRIENT DISTRIBUTION RANGES

Age	Carbohydrate	Fat	Protein
1–3 years	45–65%	30–40%	5–20%
4–18 years	45–65%	25–35%	10–30%
Over 18 years	45–65%	20–35%	10–35%

*The EER is for healthy moderately active Americans and Canadians.

Source: Adapted with permission from the Dietary References Intakes for Energy, Carbohydrates, Fiber, Fat, Protein, and Amino Acids (Macronutrients). © 2002 by the National Academy of Sciences. Courtesy of the National Academy Press, Washington, DC.

needed at such times, when the child appears to be "living on air." A decreased appetite at times in childhood is perfectly normal. As long as the child is choosing nutrient-dense kcalories, nutritional problems are unlikely. In preparation for the adolescent growth spurt, children accumulate stores of nutrients, such as calcium, that will be drawn on later, as intake cannot meet all the demands of this intensive growth spurt.

Preschoolers exhibit some food-related behaviors that drive their parents crazy, such as **food jags**, eating mostly one food for a period of time. Food jags usually don't last long enough to cause any harm. Preschoolers often pick at foods or refuse to eat vegetables or drink milk. Lack of variety, erratic appetites, and food jags are typical of this age group. Toddlers (ages one to three) tend to be pickier eaters than older preschoolers (ages four to five). Toddlers are just starting to assert their independence and love to say no to parental requests. They may wage

Food jags A habit of young children in which they eat mostly one food for a period of time.

TABLE 13-2	Age-Appropriate Cooking Activities

2½–3-YEAR-OLDS

- Wash fruits and vegetables
- Peel bananas
- Stir batters
- Slice soft foods with table knife (cooked potatoes, bananas)
- Pour
- Fetch cans from low cabinets
- Spread with a knife (soft onto firm)
- Use rotary egg beater (for a short time)
- Measure (e.g., chocolate chips into 1-cup measure)
- Make cracker crumbs

4–5-YEAR-OLDS

- Grease pans
- Open packages
- Peel carrots
- Set table (with instruction)
- Shape dough for cookies/hamburger patties*
- Snip fresh herbs for salads or cooking
- Wash and tear lettuce for salad, separate broccoli, cauliflower
- Place toppings on pizza or snacks

6–8-YEAR-OLDS

- Take part in planning part of or entire meal
- Set table (with less supervision)
- Make a salad
- Find ingredients in cabinet or spice rack
- Shred cheese or vegetables
- Garnish food
- Use microwave, blender, or toaster oven (with previous instruction)
- Measure ingredients
- Present prepared food to family at table
- Roll and shape cookies

9–12-YEAR-OLDS

- Depending on previous experience, plan and prepare an entire meal

*Children should not put their hands in their mouths while handling raw hamburger meat or dough with eggs. It can carry harmful bacteria. They should wash hands before and after handling food, especially raw meat.

a control war, and parents need to set limits without being too controlling or rigid. Here are several ways to make mealtimes less stressful and teach good eating habits for preschoolers:

1. Make mealtimes as relaxing and enjoyable as possible.

2. Don't nag, bribe, force, or cajole a child to eat. Stay calm. Pushing or prodding children almost always backfires. Children learn to hate the foods they are encouraged to eat and to desire the foods used as rewards, such as cake, ice cream, and candy. Once children know that you won't allow eating to be made into an issue of control, they will eat when they're hungry and stop when they're full. Your child is the best judge of when he or she is full.

3. Allow children to choose what they will eat from two or more healthy choices. You are responsible for choosing which foods are offered, and the child is responsible for deciding how much he or she wants of those foods.

4. Let children participate in food selection and preparation. Table 13-2 lists cooking activities for children of various ages.

5. Respect your child's preferences when planning meals but don't make your child a quick peanut butter sandwich, for instance, if he or she rejects your dinner.

6. Make sure your child has appropriate-size utensils and can reach the table comfortably.

7. Preschoolers love rituals, so start them early with the habit of eating three meals plus one to two snacks each day at fairly regular times. Also, eat with your preschooler and model good eating habits.

8. Serve small portions.

9. Do not use desserts as a reward for eating meals. Make dessert a normal part of the meal and make it nutritious.

10. Ask children to try new foods (just a little bite) and praise them when they try something different. Encourage them by telling them about someone who really likes the food or relating the food to something they think is fun. Realize, though, that some children are less likely to try new things, including new foods.

11. Expect preschoolers to reject new foods at least once, if not many times (Figure 13-2). Simply continue presenting the new food, perhaps prepared differently, and one day they will try it (usually after 12 to 15 exposures).

12. Be a good role model.

13. Be consistent at mealtimes.

14. If all else fails, keep in mind that children under six have more taste buds (which may explain why youngsters are such picky eaters) and that this, too, will pass.

FIGURE 13-2 Expect pre-schoolers to reject new foods many times before they will try it.

Courtesy of Levranii/Shutterstock.

FIGURE 13-3 School lunches provide children with important nutrients at low or no cost.

Courtesy of the US Department of Agriculture.

Luckily, school-age children are much better eaters. Although they generally have better appetites and will eat a wider range of foods, they may dislike vegetables and casserole dishes. Both preschoolers and school-age children learn about eating by watching others: their parents, siblings, friends, and teachers. Parents, siblings, and friends provide role models for children and influence children's developing food patterns. Parents' interactions with their children also influence what foods they will or will not accept.

When children go to school, their peers influence their eating behaviors as well as what they eat for lunch. Breakfast and/or lunch for school-age children often consist of a school meal (Figure 13-3) or a packed meal from home.

Having breakfast makes a difference in how children perform at school. Breakfast also makes a significant contribution to the child's intake of kcalories and nutrients for the day. Children who skip breakfast usually don't make up for calories at other meals.

Preschoolers and school-age children also learn about food by watching television. Research shows that children who watch a lot of television are more apt to be overweight. Television, as well as the Internet, not only keeps them away from more robust activities but also exposes them to commercials for sugared cereals, candy, and other empty-calorie foods. Both obesity and inactivity are a serious health problem among school-age children.

So what can parents do to make sure their children eat nutritious diets and get exercise? Be a good role model by eating a well-balanced and varied diet. Have nutritious food choices readily available at home and serve a regular, nutritious breakfast. Maintain regular family meals as much as possible. Family meals are an appropriate time to model healthy eating habits and try out new foods. It is important to get moving, by limiting television watching and encouraging physical activity. Children should participate in at least 60 minutes of moderate intensity physical activity most days of the week, preferably daily. Also limit the time children watch television, play video games, or surf the web to no more than two hours per day. Eating behaviors learned during childhood (and adolescence) are maintained into adulthood.

MENU PLANNING FOR CHILDREN

By the time children are four years old, they can eat amounts that count as regular servings eaten by older family members—that is, ½ cup of fruit or vegetable, 1 slice of bread, and 2 to 3 ounces of cooked lean meat, poultry, or fish. Children two to three years of age need the same variety of foods as four- to six-year-olds but may need slightly smaller portions (except for milk). Two- to three-year-old children need 2 servings from the dairy group each day, such as 2 cups of milk. Four- to eight-year-old children need 2-1/2 servings.

Menu Planning Guidelines for Preschoolers

1. Offer simply prepared foods and avoid casseroles or any foods that are mixed together, as children need to be able to identify what they are eating. Some toddlers and preschoolers may not want their foods to touch on the plate and may prefer a plate with divided sections.

2. Children learn to like new foods by being presented with them repeatedly, as many as 12 or more times. Put a small amount of a new food along with a meal and don't require the child to eat it if he or she doesn't want to. The more times you do this, the more likely it is that the child will eventually eat the food.

3. Offer at least one colorful food, such as carrot sticks.

4. Preschoolers like nutritious foods in all food groups but may be reluctant to eat vegetables. Part of this concern may be due to the difficulty involved in getting them onto a spoon or fork. Vegetables are more likely to be accepted if served raw and cut up as finger foods. However, when serving celery, be sure to take off the strings. Serve cooked vegetables somewhat undercooked so that they are a little crunchy. Brightly colored, mild-flavored vegetables such as peas and corn are more popular with children. Use the tips in Figure 13-4 to make fruits and vegetables children will eat.

5. Provide at least one soft or moist food that is easy to chew at each meal. A crisp or chewy food is important, too, to develop chewing skills.

6. Avoid strong-flavored and highly salted foods. Children have more taste buds than adults, and so these foods taste too strong to them.

7. Preschoolers love carbohydrate foods, including cereals, breads, and crackers, as they are easy to hold and chew. Try whole grains.

8. Smooth-textured foods such as pea soup and mashed potatoes should not have any lumps—children find this unusual.

9. Before age four, when food-cutting skills start to develop, a child needs to have food served in bite-size pieces that are eaten as finger foods or with utensils. For example, cut meat into strips or use ground meat, cut fruit into wedges or slices, and serve pieces of raw vegetables instead of a mixed salad. Other good finger foods include cheese sticks, wedges of hard-boiled egg, dry ready-to-eat cereal, fish sticks, arrowroot biscuits, and graham crackers.

10. Serve foods warm, not hot; a child's mouth is more sensitive to hot and cold than an adult's is. Also, little children need little plates, utensils, and cups, as well as seats that allow them to reach the table comfortably.

11. Cut-up fruits and vegetables make good snacks. Let preschoolers spread peanut butter on crackers or use a spoon to eat yogurt. Snacks are important to preschoolers because they need to eat more often than adults.

12. Serve good sources of the nutrients children are most likely to be lacking: fiber, iron, and fiber, iron, calcium, and vitamin D.

13. To minimize choking hazards for children under age four:
    - Avoid large chunks of any food.
    - Slice hot dogs in quarters lengthwise.
    - Shred hard raw vegetables and fruits.
    - Remove pits from apples, cherries, plums, peaches, and other fruits.
    - Cut grapes and cherry tomatoes in quarters.
    - Spread peanut butter thin.
    - Chop nuts and seeds fine.
    - Check to make sure fish is boneless.
    - Avoid popcorn and hard or gummy candies.

kid-friendly veggies and fruits

10 tips for making healthy foods more fun for children

ChooseMyPlate.gov

Encourage children to eat vegetables and fruits by making it fun. Provide healthy ingredients and let kids help with preparation, based on their age and skills. Kids may try foods they avoided in the past if they helped make them.

1 smoothie creations
Blend fat-free or low-fat yogurt or milk with fruit pieces and crushed ice. Use fresh, frozen, canned, and even overripe fruits. Try bananas, berries, peaches, and/or pineapple. If you freeze the fruit first, you can even skip the ice!

2 delicious dippers
Kids love to dip their foods. Whip up a quick dip for veggies with yogurt and seasonings such as herbs or garlic. Serve with raw vegetables like broccoli, carrots, or cauliflower. Fruit chunks go great with a yogurt and cinnamon or vanilla dip.

3 caterpillar kabobs
Assemble chunks of melon, apple, orange, and pear on skewers for a fruity kabob. For a raw veggie version, use vegetables like zucchini, cucumber, squash, sweet peppers, or tomatoes.

4 personalized pizzas
Set up a pizza-making station in the kitchen. Use whole-wheat English muffins, bagels, or pita bread as the crust. Have tomato sauce, low-fat cheese, and cut-up vegetables or fruits for toppings. Let kids choose their own favorites. Then pop the pizzas into the oven to warm.

5 fruity peanut butterfly
Start with carrot sticks or celery for the body. Attach wings made of thinly sliced apples with peanut butter and decorate with halved grapes or dried fruit.

6 frosty fruits
Frozen treats are bound to be popular in the warm months. Just put fresh fruits such as melon chunks in the freezer (rinse first). Make "popsicles" by inserting sticks into peeled bananas and freezing.

7 bugs on a log
Use celery, cucumber, or carrot sticks as the log and add peanut butter. Top with dried fruit such as raisins, cranberries, or cherries, depending on what bugs you want!

8 homemade trail mix
Skip the pre-made trail mix and make your own. Use your favorite nuts and dried fruits, such as unsalted peanuts, cashews, walnuts, or sunflower seeds mixed with dried apples, pineapple, cherries, apricots, or raisins. Add whole-grain cereals to the mix, too.

9 potato person
Decorate half a baked potato. Use sliced cherry tomatoes, peas, and low-fat cheese on the potato to make a funny face.

10 put kids in charge
Ask your child to name new veggie or fruit creations. Let them arrange raw veggies or fruits into a fun shape or design.

Center for Nutrition Policy and Promotion

Go to www.ChooseMyPlate.gov for more information.

DG TipSheet No. 11
June 2011
USDA is an equal opportunity provider and employer.

FIGURE 13-4 Kid-friendly veggies and fruits.
Courtesy of the US Department of Agriculture.

Menu Planning Guidelines for School-Age Children

1. Serve a wide variety of foods, including children's favorites: tuna fish, pizza (use vegetable toppings), macaroni and cheese, hamburgers (use lean beef combined with ground turkey breast), and peanut butter.

2. Good snack choices are important, as children do not always have the desire or the time to sit down and eat. Snacks can include fresh fruits and vegetables, dried fruits, unsweetened fruit juices, whole-grain bread and cereals, popcorn (without excessive fat), pretzels, muffins, milk, yogurt, cheese, pudding, sliced lean meats and poultry, and peanut butter.

3. Balance menu items that are higher in fat with those containing less fat.

4. Pay attention to serving sizes.

5. Offer iron-rich foods such as meat in hamburgers, peanut butter, baked beans, chili, dried fruits, and fortified dry cereals.

6. As children grow, they need to eat more high-fiber foods such as fruits, vegetables, beans and peas, and whole-grain foods.

7. Serve good sources of calcium.

8. Breakfast is a very important meal before school. The first meal of the day should be rich in protein and good carbohydrates—the whole-grain variety that will sustain you for a long spell rather than the sugary kind that will push your blood sugar up, then let it fall.

9. Get kids involved in planning and making meals—they can help taste test foods, give feedback, and suggest foods.

All children up to age ten need to eat every four to six hours to keep their blood glucose at a desirable level. Therefore, snacking is necessary between meals. Nutritious snack choices for both preschoolers and school-age children are noted above.

SUMMARY

1. For children two to three years of age, fat intake should be 30 to 40 percent of total kcalories. Children and adolescents from 4 to 18 years of age should keep fat intake between 25 and 35 percent of kcalories, with most fats coming from sources of polyunsaturated and monounsaturated fatty acids. Children over two years old should limit consumption of saturated and trans fats to 10 percent or less of total kcalories.

2. Problem nutrients for children include fiber, iron, calcium, and vitamin D. Fiber is abundant in legumes, fruits and vegetables, and whole grains. Iron is rich in lean meat, enriched breads and cereals, and legumes. Dairy products are sources of calcium and vitamin D. Vitamin D is also found in fortified cereals.

3. During growth spurts, children's appetites are good; otherwise, their appetites may seem poor.

4. Preschoolers can be fussy eaters, often have food jags, and can take the pleasure out of mealtimes.

5. Guidelines for eating with preschoolers are detailed in this section.

6. School-age children are much better eaters than preschoolers.

7. Children's eating habits are influenced by family, friends, teachers, availability of school breakfast and lunch programs, television, and the Internet.

8. Parents need to be good role models for children, have nutritious foods choices available, serve regular meals including breakfast, and encourage one hour of physical activity most days of the week.

9. Menu planning guidelines for preschoolers and school-age children are detailed in this section. By the time children are four years old, they can eat regular serving sizes such as ½ cup of fruits or vegetables.

NUTRITION AND MENU PLANNING
FOR ADOLESCENTS

Adolescence is the period of life between about 11 and 21 years of age. The beginning of adolescence is marked by puberty, the process of physically developing from a child to an adult. Puberty starts at about age 10 or 11 for girls and 12 or 13 for boys. The growth spurt is intense for 2 to 2½ years, and then there are a few more years of growth at a slower place. The timing of puberty and rates of growth show much individual variation.

Whereas before puberty the proportion of fat and muscle was similar in males and females, males now put on twice as much muscle as females, and females gain proportionately more fat. In adolescent girls, an increasing amount of fat is being stored under the skin, particularly in the abdominal area. The male also experiences a greater increase in bone mass than does the female.

NUTRITION FOR ADOLESCENTS

Table 13-3 compares the need for kcalories and protein for adolescents with what adults need. Energy requirements for adolescents vary greatly depending on gender, body composition, level of physical activity, and current rate of growth. The highest levels of nutrients are for individuals growing at the fastest rate.

Adolescent males now need more kcalories, protein, magnesium, and zinc for muscle and bone development than do females; however, females need increased iron due to the onset of menstruation (Figure 13-5). Owing to their big appetites and high kcalorie needs, teenage boys are more likely than girls to get sufficient nutrients. Females have to pack more nutrients into fewer kcalories, and this can become difficult if they decrease their food intake to lose weight.

Problem nutrients for adolescents include iron, calcium, vitamin D, and potassium. Inadequate iron intake increases the incidence of iron-deficiency anemia, especially among those adolescents at highest risk, such as pregnant adolescents, vegetarians, and competitive athletes.

Calcium and vitamin D are very important for adolescents because half of their peak bone mass is built during this time. The greatest capacity to absorb calcium occurs in early adolescence, when the body can absorb almost four times more calcium than it can during young adulthood. The RDA for calcium increases from 1,000 mg at age 8 to 1,300 mg from ages 9 to 18 for both males and females. Adolescents need four glasses of milk (or milk equivalents)

TABLE 13-3		Energy and Macronutrients for Adolescents		

ESTIMATED ENERGY REQUIREMENT (EER) AND RECOMMENDED DIETARY ALLOWANCE (RDA) FOR PROTEIN

Gender and Age	Height	Weight	EER*	Protein
Male 9–13 years	57 inches	79 pounds	2,279 kcal	0.95 grams/kilogram body weight
Female 9–13 years	57	81	2,071	0.95
Male 14–18 years	68	134	3,152	0.85
Female 14–18 years	64	119	2,368	0.85
Male 19–30 years	70	154	3,067	0.80
Female 19–30 years	64	126	2,403	0.80

ACCEPTABLE MACRONUTRIENT DISTRIBUTION RANGES

Age	Carbohydrate	Fat	Protein
4–18 years	45–65%	25–35%	10–30%
Over 18 years	45–65%	20–35%	10–35%

*The EER is for healthy moderately active Americans and Canadians.

Source: Adapted with permission from the Dietary References Intakes for Energy, Carbohydrates, Fiber, Fat, Protein, and Amino Acids (Macronutrients). © 2002 by the National Academy of Sciences. Courtesy of the National Academy Press, Washington, DC.

to meet their calcium and vitamin D needs, but very few adolescents consume that much.

Few Americans, including all age-gender groups, consume potassium in amounts equal to or greater than the AI. In view of the health benefits of adequate potassium intake, increased intake of good sources of potassium is necessary. Unprocessed whole foods such as fruits and vegetables (especially winter squash, potatoes, oranges, grapefruits, and bananas), milk, yogurt, legumes, and meats are excellent sources of potassium.

With their increased independence, adolescents assume responsibility for their own eating habits. Teenagers are not fed; they make most of their own food choices. They eat more meals away from home, such as at fast-food restaurants, and skip more meals than they did previously. Irregular meals and snacking are common due to busy social lives and after-school activities and jobs. Adolescents are more likely to skip breakfast than either younger children or adults. Teenagers will tell you that they lack the time or discipline to eat right, although many are pretty well informed about good nutrition practices. Ready-to-eat foods such as cookies, chips, and soft drinks are readily available, and teenagers pick them up as snack foods.

FIGURE 13-5 Adolescent males need more kcalories, protein, magnesium, and zinc for muscle and bone develop than females do; however females need increased iron due to the start of menstruation.

Courtesy of Monkey Business Images/Shutterstock.

The media have a powerful influence on adolescents' eating patterns and behaviors. Advertising for not-so-nutritious foods and fast foods permeates television, radio, and billboards. In addition, questionable eating habits are portrayed on television shows. Studies show that the prevalence of obesity increases as the number of hours of watching television increases.

A typical meal at a fast-food restaurant—a 4-ounce hamburger, french fries, and a regular soft drink—is high in kcalories, fat, and sodium. However, more nutritious choices are available at fast-food restaurants, such as grilled chicken, salads, and low-fat milk.

Parents can positively influence adolescents' eating habits by being good role models and by having nutritious breakfast and snack foods available at home, and having dinner together when possible. Adolescents can become involved in food purchasing and preparation. Parents can also influence their children's fitness level by limiting their sedentary activities and encouraging exercise.

Both adolescent boys and girls are influenced by their body images. Adolescent boys may take nutrition supplements and fill up on protein in hopes of becoming more muscular. Adolescent girls who feel they are overweight may skip meals and modify their food choices in hopes of losing weight. Whether overweight or not, teens need regular exercise.

MENU PLANNING FOR ADOLESCENTS

Here are some tips about planning menus for adolescents:

1. Emphasize carbohydrates containing fiber, such as whole-grain breads and cereals, fruits, vegetables, brown rice, and dried beans and peas. These foods supply kcalories along with needed nutrients. Whole-grain products contain more fiber and other nutrients than do their refined counterparts.

2. Offer well-trimmed lean beef, poultry, and fish. Don't think that just because adolescents need more calories, fatty meats are in order. Their fat calories should come mostly from foods that are moderate or low in saturated fat.

3. Low-fat milk and nonfat milk need to be offered at all meals. Girls are more likely to need to select nonfat milk than are boys. Other forms of calcium also need to be available, such as pizza, macaroni and cheese, and other entrées using cheese, along with yogurt, frozen yogurt, ice milk, and puddings made with nonfat or low-fat milk.

4. Have nutritious choices available for hungry on-the-run adolescents looking for a snack. Nutritious snack choices include fresh fruit, whole-grain muffins and other quick breads, crackers or rolls with low-fat cheese or peanut butter, vegetable-stuffed pita pockets, yogurt or cottage cheese with fruit, and fig bars.

5. Emphasize quick and nutritious breakfasts, such as whole-grain pancakes or waffles with fruit, juices, whole-grain toast or muffins with low-fat cheese, cereal topped with fresh fruit, and bagels with peanut butter.

6. Serve good sources of the nutrients adolescents are most likely to be lacking: iron, calcium, vitamin D, and potassium.

SUMMARY

1. Adolescence is the period from 11 to 21 years of age. Puberty starts at about age 10 or 11 for girls and 12 or 13 for boys. The growth spurt is intense for 2 to 2½ years, and then there are a few more years of growth at a slower pace.

2. Males now put on twice as much muscle as females do, and females gain proportionately more fat.

3. Adolescent males now need more kcalories, protein, magnesium, and zinc for muscle and bone development than females do; however, females need increased iron due to the onset of menstruation.

4. Owing to their big appetites and high kcalorie needs, teenage boys are more likely than girls to get sufficient nutrients. Females have to pack more nutrients into fewer kcalories.

5. Problem nutrients for adolescents include iron, calcium, vitamin D, and potassium. Adolescents need four glasses of milk (or milk equivalents) to meet their calcium and vitamin D needs, but very few adolescents consume that much.

6. Teenagers make most of their own food choices. They eat more meals away from home and skip more meals than previously.

7. Parents can positively influence adolescents' eating habits by being good role models, having nutritious breakfast and snack foods available, and eating dinner together when possible.

8. Menu-planning guidelines for adolescents are given in this section.

NUTRITION OVER THE LIFESPAN: OLDER ADULTS

The number of older adults—persons age 65 years or older—has been steadily increasing in the United States. The older population will grow rapidly between the years 2010 and 2030 as the "baby boom" generation reaches age 65. People 65+ represented 12.9 percent of the population in the year 2009 but are expected to grow to be 19.3 percent of the population by 2030. Person reaching age 65 have an average life expectancy of an additional 20 years for females and 17.3 years for males. The 85+ population continues to grow and is projected to increase 15 percent from 2010 to 2020.

THE AGING PROCESS

Before looking at nutrition during aging, let's take a look at what happens to our bodies as we age. Sensitivity to taste and smell declines slowly with age. The taste buds become less sensitive, and the nasal nerves that register aromas need extra stimulation to detect smells. That's why older adults may find ordinarily seasoned foods too bland. Medications also may alter the ability to taste, and dental problems may affect the ability to chew. The movement of food through the gastrointestinal tract slows down over time, causing problems such as constipation and heartburn. **Heartburn**, a burning sensation in the area of the throat, has nothing to do with the heart. It occurs when acidic stomach contents are pushed into the lower part of the esophagus or throat.

The basal metabolic rate declines between 8 and 10 percent from age 30 to age 70. This change is largely the result of loss of muscle mass as we age. From the mid-twenties to the

Heartburn A painful burning feeling in your chest or throat that happens when stomach acid backs up into your esophagus, the tube that carries food from your mouth to your stomach.

mid-seventies, people slowly lose about 24 pounds of muscle, which is replaced by 22 pounds of fat. Combined with a general decrease in activity level, these factors clearly indicate a need for decreased kcalorie intake, which generally does take place during aging. But the elderly need not lose all that muscle mass. By doing regular weight-training exercises, the elderly can increase muscular strength, increase basal metabolism, improve appetite, and improve blood flow to the brain (Figure 13-6). Table 13-4 gives the recommendations for physical activity for older Americans.

After age 35, the functional abilities of almost every organ system start to decline. The functioning of the cardiovascular system declines with age. The heart does not pump as hard as before. Blood pressure increases normally with age. Lung capacity decreases but does not restrict the normal activity of healthy older persons. However, it might limit vigorous exercise. Kidney function deteriorates over time, and adequate fluid intake is important, especially since older adults have a decreased thirst sensation. Loss of bones occurs normally during aging, and osteoporosis is common.

FIGURE 13-6 For older adults, regular physical activity is one of the most important things they can do for their health.

Courtesy of Robert Kneschke/ Shutterstock.

TABLE 13-4	Recommendations for Physical Activity for Older Adults

Older Adults Need at Least:

2 hours and 30 minutes (150 minutes) of <u>moderate-intensity aerobic activity</u> (i.e., brisk walking) every week **and**

<u>muscle-strengthening activities</u> on two or more days a week that work all major muscle groups (legs, hips, back, abdomen, chest, shoulders, and arms).

OR

1 hour and 15 minutes (75 minutes) of <u>vigorous-intensity aerobic activity</u> (i.e., jogging or running) every week **and**

<u>muscle-strengthening activities</u> on two or more days a week that work all major muscle groups (legs, hips, back, abdomen, chest, shoulders, and arms).

OR

An equivalent mix of moderate- and vigorous-intensity <u>aerobic activity</u> **and**

<u>muscle-strengthening activities</u> on two or more days a week that work all major muscle groups (legs, hips, back, abdomen, chest, shoulders, and arms).

FOR EVEN GREATER HEALTH BENEFITS:

Older Adults Should Increase Their Activity to:

5 hours (300 minutes) each week of <u>moderate-intensity aerobic activity</u> **and**

<u>muscle-strengthening activities</u> on two or more days a week that work all major muscle groups (legs, hips, back, abdomen, chest, shoulders, and arms).

OR

2 hours and 30 minutes (150 minutes) each week of <u>vigorous-intensity aerobic activity</u> **and**

<u>muscle-strengthening activities</u> on two or more days a week that work all major muscle groups (legs, hips, back, abdomen, chest, shoulders, and arms).

OR

An equivalent mix of moderate- and vigorous-intensity <u>aerobic activity</u> **and**

<u>muscle-strengthening activities</u> on two or more days a week that work all major muscle groups (legs, hips, back, abdomen, chest, shoulders, and arms).

Source: Division of Nutrition, Physical Activity, and Obesity, National Center for Chronic Disease Prevention and Health Promotion, 2011.

NUTRITION FOR OLDER ADULTS

Older adults do eat fewer kcalories as they age. Between decreased physical activity and a decreased basal metabolic rate, they need about 20 percent fewer kcalories to maintain their weight. This means that there is less room in the diet for empty-kcalorie foods such as sweets, alcohol, and fatty foods. At a time when good nutrition is so important to good health, there are many obstacles to fitting more nutrients into fewer kcalories, such as medical conditions, dentures, and medications.

Nutrients of concern to the elderly include the following:

1. **Water.** Due to decreased thirst sensation and other factors, fluid intake is very important for older adults. Water is also important to prevent constipation.

2. **Calcium.** Current intakes for calcium are below the recommendations for people over 70 years old (1,200 mg). To meet this recommendation, an individual would need to eat four servings of dairy products or other calcium-rich food daily.

3. **Vitamin B$_{12}$.** On average, Americans ages 50 years and older consume adequate vitamin B$_{12}$. Nonetheless, many people 50 years and older have reduced ability to absorb naturally occurring vitamin B$_{12}$. However, crystalline vitamin B$_{12}$, the type of vitamin B$_{12}$ used in supplements and in fortified foods, is much more easily absorbed. Therefore, people 50 years and older are encouraged to include foods fortified with vitamin B$_{12}$, such as fortified cereals, or take dietary supplements.

4. **Vitamin D.** Vitamin D is unique in that sunlight on the skin enables the body to make vitamin D. Older people tend to spend less time outside and they have less of the vitamin D precursor in the skin that is needed to make vitamin D, so they may not be getting enough vitamin D. In the United States, most dietary vitamin D is obtained from fortified foods, especially fluid milk and some yogurts. Some other foods and beverages, such as breakfast cereals, margarine, orange juice, and soy beverages, also are commonly fortified with this nutrient. It is also available as a dietary supplement.

Fiber and potassium can also be problem nutrients. Both are found in fruits, vegetables, beans, and peas. Table 13-5 gives tips for healthy eating for older adults.

MENU PLANNING FOR OLDER ADULTS

Figure 13-7 shows how the number of servings (using MyPlate) generally decrease as adults age. Half the plate still contains fruits and vegetables in a variety of colors to ensure variety, fiber, vitamins, minerals, and adequate intake of phytochemicals. Frozen or canned fruits and vegetables are nutritious alternatives to fresh produce that are often more affordable, easier to prepare, and don't spoil as readily.

TABLE 13-5	Tips for Healthy Eating for Older Adults

1. Eat with other people. This usually makes mealtime more enjoyable and stimulates the appetite. Taking a walk before eating also stimulates the appetite.

2. Prepare larger amounts of food and freeze some for heating up at a later time. This saves cooking time and is helpful for someone who is reluctant to cook.

3. If big meals are too much, eat small amounts more frequently during the day. Eat regular meals.

4. If getting to the supermarket is a bother, go at a time when it is not busy or engage a delivery service (this is more expensive, though).

5. Use unit pricing, sales, and coupons to cut back on the amount of money spent on food.

6. Take advantage of community meal programs for the elderly such as Meals on Wheels and meals at senior centers.

7. To perk up a sluggish appetite, increase use of herbs, spices, lemon juice, vinegar, and garlic.

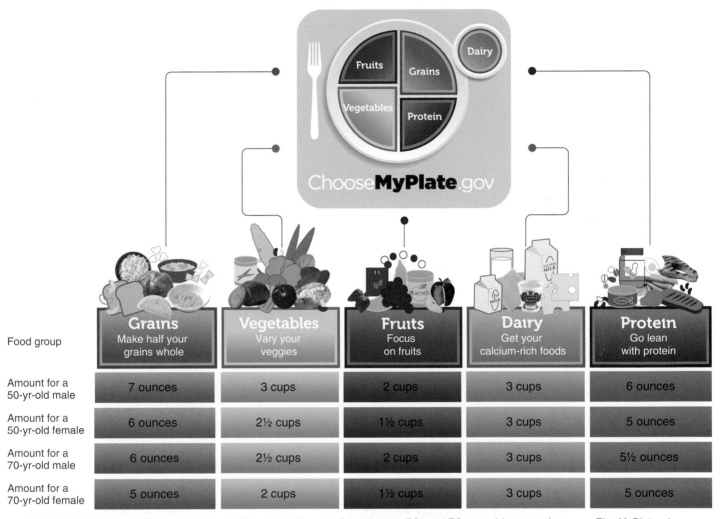

Food group	**Grains** Make half your grains whole	**Vegetables** Vary your veggies	**Fruits** Focus on fruits	**Dairy** Get your calcium-rich foods	**Protein** Go lean with protein
Amount for a 50-yr-old male	7 ounces	3 cups	2 cups	3 cups	6 ounces
Amount for a 50-yr-old female	6 ounces	2½ cups	1½ cups	3 cups	5 ounces
Amount for a 70-yr-old male	6 ounces	2½ cups	2 cups	3 cups	5½ ounces
Amount for a 70-yr-old female	5 ounces	2 cups	1½ cups	3 cups	5 ounces

FIGURE 13-7 The MyPlate recommendations shown here are for sedentary 50- and 70-year-old men and women. The MyPlate plan for sedentary 70-year-old women allows only 120 empty Calories/day. To meet this allowance, food choices must be nutrient dense. High-fiber foods are important because fiber is often low in the diets of older adults. Sufficient beverage consumption is important to ensure that water needs are met.

MyPlate image courtesy of US Department of Agriculture. Reprinted with permission of John Wiley & Sons, Inc.

The grains section of MyPlate continues to emphasize whole grains, and the protein section should focus more on lean proteins such as seafood and chicken. In addition to milk, a variety of liquids such as soup and juice are very important for older adults to decrease their risk of dehydration. Low-sodium products are encouraged to keep blood pressure down, along with the use of herbs and spices. Older adults should also get regular physical activity through daily errands, chores, walks, as well as traditional exercise.

When planning meals for older adults, use these guidelines:

1. Offer moderately sized meals and half portions. Older adults frequently complain when given too much food because they hate to see waste. Restaurants might reduce the size of their entrées by 15 to 25 percent.

2. Emphasize high-fiber foods such as fruits, vegetables, whole grains, and beans. Older people requiring softer diets might have problems chewing some high-fiber foods. High-fiber foods that are soft in texture include cooked beans and peas, bran cereals soaked in milk, canned fruits, and cooked vegetables such as potatoes, corn, green peas, and winter squash.

3. Moderate the use of fat. Many older adults don't like to see an entrée swimming in a pool of butter. Use lean meats, poultry, or fish and sauces prepared with vegetable or fruit purées. Have low-fat dairy products available, such as nonfat or low-fat milk.

4. Dairy products such as milk and yogurt are important sources of calcium, vitamin D, protein, potassium, and vitamin B_{12}.

5. Offer adequate protein but not too much. Use a variety of both animal and vegetable sources. Providing protein on a budget, as in a nursing home, need not be a problem. Lower-cost protein sources include beans and peas, cottage cheese, macaroni and cheese, eggs, chicken, and ground beef.

6. Moderate the use of salt and salty ingredients. Many older adults are on low-sodium diets and recognize a salty soup when they taste it. Avoid highly salted soups, sauces, and other dishes. It is better to let older adults season food to taste.

7. Use herbs and spices to make foods flavorful. Older adults are looking for tasty foods just like anyone else, and they may need them more than ever!

8. Offer a variety of foods, including traditional menu items and cooking from other countries and regions of the United States.

9. Fluid intake is critical, so offer a variety of beverages. Diminished sensitivity to dehydration may cause older adults to drink less fluid than needed. Special attention must be paid to fluids, particularly for those who need assistance with eating and drinking. Beverages such as water, milk, juice, coffee, and tea and foods such as soup contribute to fluid intake.

10. If chewing or swallowing is a problem, softer foods can be chosen to provide a well-balanced diet. Following are some guidelines for soft diets:

    - Use tender meats and, if necessary, chop or grind them. Ground meats can be used in soups, stews, and casseroles. Cooked beans and peas, soft cheeses, and eggs are additional softer protein sources.

    - Cook vegetables thoroughly and dice or chop them by hand if necessary after cooking.

    - Serve mashed potatoes or rice, with a sauce if desired.

    - Serve chopped salads.

    - Soft fruits such as fresh or canned bananas, berries, peaches, pears, and melon, as well as applesauce, are some good choices.

    - Soft breads and rolls can be made even softer by dipping them briefly in milk.

    - Puddings and custard are good dessert choices.

    - Many foods that are not soft can be easily chopped by hand or puréed in a blender or food processor to provide additional variety.

1. The older population is growing rapidly in the United States.
2. As we age, sensitivity to taste and smell decline and the movement of food through the gastrointestinal tract slows down, which can cause heartburn and constipation. Metabolism declines between 8 to 10 percent from age 30 to 70, largely because of loss of muscle tissue and decreased activity, so older adults need to eat fewer kcalories (but still take in lots of nutrients) and get exercise.
3. Table 13-4 gives physical activity recommendations.
4. After age 35, the functional abilities of almost every organ system start to decline.
5. Nutrients of concern include water (due to decreased thirst sensation), calcium, vitamin B_{12} (due mostly to less being absorbed), and vitamin D (less is made in the skin). Fiber and potassium can also be problem nutrients (both are found in fruits, vegetables, beans, and peas).
6. Figure 13-7 shows MyPlate for Older Adults.
7. Menu planning tips are listed.

EATING DISORDERS

An eating disorder is an illness that causes serious disturbances to your everyday diet, such as eating extremely small amounts of food or severely overeating. A person with an eating disorder may have started out just eating smaller or larger amounts of food, but at some point, the urge to eat less or more spiraled out of control. Severe distress or concern about body weight or shape may also characterize an eating disorder (Figure 13-8).

Common eating disorders include anorexia nervosa, bulimia nervosa, and binge-eating disorder. Eating disorders affect both men and women, but are much more common in women. Eating disorders frequently appear during the teen years or young adulthood but may also develop during childhood or later in life.

Some teens have serious eating behaviors that mimic eating disorders, but their symptoms do not meet all the criteria to fit the diagnosis for anorexia or bulimia. These types of eating disorders are called "eating disorders not otherwise specified" (EDNOS). Binge-eating disorder is a type of eating disorder called EDNOS.

Researchers have found that stringent dieting to achieve an "ideal" figure can play a key role in triggering eating disorders. The actual cause of eating disorders is not entirely understood, but many risk factors have been identified, such as a high degree of perfectionism, low self-esteem, genetics, and troubled relationships. Eating disorders are considered a psychological disorder.

Eating disorders are also real medical illnesses. They frequently coexist with other illnesses such as depression, substance abuse, or anxiety disorders. Other symptoms, described in the next section, can become life threatening if a person does not receive treatment. People with anorexia nervosa are 18 times more likely to die early compared with people of similar age in the general population.

ANOREXIA NERVOSA

Anorexia nervosa is characterized by:

- Extreme thinness
- A relentless pursuit of thinness and unwillingness to maintain a normal or healthy weight
- Intense fear of gaining weight
- Distorted body image, a self-esteem that is heavily influenced by perceptions of body weight and shape, or a denial of the seriousness of low body weight
- Lack of menstruation among girls and women
- Extremely restricted eating

Many people with anorexia nervosa see themselves as overweight, even when they are clearly underweight. Eating, food, and weight control become obsessions. People with anorexia

Anorexia nervosa A life-threatening eating disorder that is characterized by self-starvation and excessive weight loss.

FIGURE 13-8 It is common to feel a lot of pressure to be thin.

Courtesy of hkannn/Shutterstock.

nervosa typically weigh themselves repeatedly, portion food carefully, and eat very small quantities of only certain foods. Some people with anorexia nervosa may also engage in binge-eating followed by extreme dieting, excessive exercise, self-induced vomiting, and/or misuse of laxatives, diuretics, or enemas.

Some who have anorexia nervosa recover with treatment after only one episode. Others get well but have relapses. Still others have a more chronic, or long-lasting, form of anorexia nervosa, in which their health declines as they battle the illness.

Other symptoms may develop over time, including:

- Thinning of the bones (osteopenia or osteoporosis)
- Brittle hair and nails
- Drop in internal body temperature, causing a person to feel cold all the time
- Lethargy, sluggishness, or feeling tired all the time
- Growth of fine hair all over the body (lanugo)
- Mild anemia and muscle wasting and weakness
- Damage to the heart, brain, and other organs

BULIMIA NERVOSA

Bulimia nervosa An eating disorder characterized by a destructive pattern of excessive overeating followed by vomiting or other "purging" behaviors to control weight.

Bulimia nervosa is characterized by recurrent and frequent episodes of eating unusually large amounts of food and feeling a lack of control over these episodes. This binge-eating is followed by behavior that compensates for the overeating, such as forced vomiting, excessive use of laxatives or diuretics, fasting, excessive exercise, or a combination of these behaviors.

Unlike anorexia nervosa, people with bulimia nervosa usually maintain what is considered a healthy or normal weight, while some are slightly overweight. But like people with anorexia nervosa, they often fear gaining weight, want desperately to lose weight, and are intensely unhappy with their body size and shape. Usually, bulimic behavior is done secretly because it is often accompanied by feelings of disgust or shame. The binge-eating and purging cycle happens anywhere from several times a week to many times a day.

Other symptoms include:

- Chronically inflamed and sore throat
- Swollen salivary glands in the neck and jaw area
- Worn tooth enamel, and increasingly sensitive and decaying teeth as a result of exposure to stomach acid
- Damage to the gastrointestinal tract
- Severe dehydration from purging of fluids

BINGE-EATING DISORDER

Binge-eating disorder An eating disorder characterized by episodes of uncontrolled eating or bingeing.

With **binge-eating disorder** a person loses control over his or her eating. Unlike bulimia nervosa, periods of binge-eating are not followed by purging, excessive exercise, or fasting. As a result, people with binge-eating disorder often are overweight or obese. People with binge-eating disorder who are obese are at higher risk for developing cardiovascular disease and high blood pressure. They also experience guilt, shame, and distress about their binge eating, which can lead to more binge eating.

Adequate nutrition, reducing excessive exercise, and stopping purging behaviors are the foundations of treatment for eating disorders. Specific forms of psychotherapy, or talk therapy, and medication are effective for many eating disorders. However, in more chronic cases, specific treatments have not yet been identified. Treatment plans often are tailored to individual needs and may include one or more of the following:

- Individual, group, and/or family psychotherapy
- Medical care and monitoring
- Nutritional counseling
- Medications

Some patients may also need to be hospitalized to treat problems caused by malnutrition or to ensure they eat enough if they are very underweight.

SUMMARY

1. Common eating disorders include anorexia nervosa, bulimia nervosa, and binge-eating disorder. Eating disorders affect both men and women, but are much more common in women. Eating disorders frequently appear during the teen years or young adulthood but may also develop during childhood or later in life.
2. Researchers have found that stringent dieting to achieve an "ideal" figure can play a key role in triggering eating disorders. The actual cause of eating disorders is not entirely understood, but many risk factors have been identified, such as a high degree of perfectionism, low self-esteem, genetics, and troubled relationships. Eating disorders are considered a psychological disorder.
3. Treatment plans are tailored to individual needs and may include psychotherapy, medical care, nutritional counseling, and medications.

NUTRITION FOR THE ATHLETE

Many athletes (Figure 13-9) require 3,000 to 6,000 kcalories daily. The amount of energy required by an athlete depends on the type of activity and its duration, frequency, and intensity. In addition, the athlete's basal metabolic rate, body composition, age, and environment must be taken into account.

Carbohydrates (from glycogen and blood glucose) and fat are the primary fuel sources for exercise. The availability of carbohydrate—more specifically, the amount of glycogen stores—heavily influences athletic performance. Glucose is the main source of energy for intense exercise, while fat is the main source of energy during low to moderate exercise.

An appropriate diet for many athletes consists of 6 to 10 grams of carbohydrate per kilogram of body weight and enough protein to provide 1.2 to 1.4 grams per kilogram of body weight for endurance athletes and 1.2 to 1.7 grams for strength athletes. The remaining kcalories come from fat.

Although many athletes take vitamin and mineral supplements, these supplements will not enhance performance unless there is a deficiency. Most athletes get plenty of vitamins and minerals in their regular diets, although young athletes and women need to pay special attention to iron and calcium.

Water is the most crucial nutrient for athletes (Figure 13-10). They need about 1 liter of water for every 1,000 kcalories consumed. At least four hours before exercise, athletes should drink about 2 to 3 ml of fluid for each pound they weigh. So an athlete weighing 180 pounds should drink between 360 to 540 ml of fluids (1.5 to 2.25 cups). During the workout, athletes need about 4 to 8 ounces every 15 minutes. One gulp is about 1 ounce. For persons performing prolonged physical activity (especially in hot weather), sports drinks are recommended.

Each person's fluid needs are different. Your fluid loss will depend on how big you are (larger people sweat more), your fitness level (the more fit you are, the more you sweat), exercise duration, temperature, and opportunities to drink. A good way to determine how much fluid to replace after exercising is to weigh in before and after exercise. For every pound that is lost, you need to at least drink 2 cups, or 16 ounces, of water.

Sports drinks contain a dilute mixture of carbohydrate and electrolytes. Most contain about 50 kcalories per cup, about 3 to 4 teaspoons of carbohydrate per cup, and small amounts of sodium and potassium. Sodium, which is lost during sweating, helps maintain

FIGURE 13-9 Many athletes require higher kcalorie intakes.

Courtesy of AVAVA/Shutterstock.

FIGURE 13-10 Water is the most crucial nutrient for athletes. Sports drinks are recommended when working out at least 60 minutes and especially in hot weather.

Courtesy of pkchai/Shutterstock.

fluid balance in your body. Sports drinks are purposely made to be weak solutions so that they can empty faster from the stomach, and the nutrients they contain are therefore available to the body more quickly. They are primarily designed to be used during exercise lasting 60 minutes or more to help replace water and electrolytes and provide some carbohydrates for energy.

Carbohydrates taken during exercise can help you maintain a normal blood-sugar level and enhance (as well as lengthen) performance. Athletes often consume ½ to 1 cup of sports fluids every 15 to 20 minutes during exercise.

Carbohydrate or glycogen loading is a regimen involving three or more days of decreasing amounts of exercise and increased consumption of carbohydrates (up to 60–70 percent of total kcalories) before an event to increase glycogen stores. The theory is that increasing glycogen stores will enhance performance by providing more energy during lengthy competition. It is most appropriate for endurance athletes.

Here are some menu-planning guidelines for athletes:

1. Include a variety of foods from MyPlate.

2. Good sources of carbohydrates to emphasize on menus include pasta, brown rice, other grain products such as whole-grain breads and cereals, legumes, and fruits and vegetables. On the eve of the New York City Marathon each year, marathon officials typically host a pasta dinner for runners that features spaghetti with marinara sauce and cold pasta primavera. Pasta also provides needed B vitamins, minerals, and fiber (more if whole grain). Here are some ways to include carbohydrates in your menu:

   • At breakfast, offer a variety of pancakes, waffles, cold and hot cereals, breads, and rolls. Make sure many are whole grain.

   • At lunch, make sandwiches with different types of bread, such as pita pockets, raisin bread, various whole-grain breads, and wraps.

   • Serve pasta and rice as a side or main dish with, for example, chicken and vegetables. Cold pasta and rice salads are great, too.

   • Potatoes, whether baked, mashed, or boiled, are an excellent source of carbohydrates.

   • Always have available as many types of fresh fruits and salads as possible.

   • Don't forget to use beans and peas in soups, salads, entrées, and side dishes.

   • Nutritious desserts that emphasize carbohydrates include frozen yogurt with fruit toppings, oatmeal cookies, and fresh fruit.

3. Don't offer too much protein and fat in the belief that athletes need the extra kcalories. They do, but many of those extra calories should come from carbohydrates such as whole grains, fruits, and vegetables. The days of steak-and-egg dinners are over for athletes. The protein and fat present in these meals do nothing to improve performance. Here are some ways to moderate the amount of fat and protein:

   • Use lean, well-trimmed cuts of beef.

   • Offer chicken, turkey, and fish—all lower in fat than beef. Broiling, roasting, and grilling are the preferred cooking methods, with frying being acceptable occasionally.

   • Offer larger serving sizes of meat, poultry, and fish, perhaps 1 to 2 ounces more, but don't overdo it.

   • Offer fried food in moderation.

   • Offer low-fat and fat-free milk.

   • Offer high-fat desserts, such as ice cream and many types of sweets, in moderation. Frozen yogurt and ice milk generally contain less fat than ice cream and can be topped with fruit or crushed oatmeal cookies. Fruit ice and sorbet contain no fat.

4. Offer a variety of fluids, not just soft drinks and other sugared drinks. Good beverage choices include fruit juices, iced tea and iced coffee (preferably freshly brewed decaffeinated), plain and flavored mineral and seltzer water, spritzers (fruit juice and mineral water), and smoothies made with yogurt and fruit. Soft drinks and juice drinks—both are loaded with sugar—should be offered in moderation.

5. Make sure iodized salt is on the table.

6. Be sure to include sources of iron, calcium, and zinc at each meal. Good iron sources include red meats, legumes, and iron-fortified breakfast cereal. Moderate iron sources include raisins, dried fruit, bananas, nuts, and whole-grain and fortified grain products. Be sure to include good vitamin C sources at each meal, as vitamin C assists in iron absorption. Vitamin C sources include citrus fruits and juices, cantaloupe, strawberries, broccoli, and potatoes. Calcium (found in milk and dairy products) and zinc (found in shellfish, meat and poultry, legumes, dairy foods, whole grains, and fortified cereals) are also important.

7. The most important meal is the one closest to the competition, commonly called the **precompetition meal**. The functions of this meal include getting the athlete fueled up both physically and psychologically, helping to settle the stomach, and preventing hunger. The meal should consist of easily digestible carbohydrate-rich foods and should be moderate in protein and low in fat and fiber. High-fat foods take longer to digest and can cause sluggishness. Substantial precompetition meals are usually served three to four hours before the competition to allow enough time for stomach emptying (to avoid cramping and discomfort during the competition). Menus might include a fruit and yogurt smoothie with low-fat granola; oatmeal with brown sugar, almonds, skim milk, and banana; or turkey and Swiss sandwich with fruit and a sports drink. The meal should include 2+ cups of fluid for hydration and typically provide 300 to 1,000 kcalories. Smaller precompetition meals may be served two to three hours before competition. Many athletes have specific comfort foods they enjoy before competition. It is recommended that the stomach be fairly empty during competition so that blood can be sent to the working muscles instead of to the stomach to help in digestion.

Precompetition meal The meal for an athlete closest to the time of a competition or event.

8. After competition and workouts, again emphasize fluids along with complex carbohydrates to restore glycogen. Protein also needs to be replenished, as well as minerals lost in sweating.

SUMMARY

1. Athletes have increased needs for kcalories, carbohydrate, and protein. Glucose is the main source of energy for intense exercise, while fat is the main source of energy during low to moderate exercise.

2. Water is the most crucial nutrient for athletes.

3. Sports drinks contain small amounts of carbohydrate and electrolytes that are designed to be used during exercise lasting 60 minutes or more to enhance performance.

4. Carbohydrates taken during exercise can help you maintain a normal blood-sugar level and enhance (as well as lengthen) performance. Athletes often consume ½ to 1 cup of sports fluids every 15 to 20 minutes during exercise.

5. Carbohydrate or glycogen loading is a regimen involving three or more days of decreasing amounts of exercise and increased consumption of carbohydrates (up to 60–70 percent of total kcalories) before an event to increase glycogen stores. The theory is that increasing glycogen stores will enhance performance by providing more energy during lengthy competition.

6. Menu planning guidelines for athletes, including precompetition meals, are listed.

CHECK-OUT QUIZ

1. Problem nutrients for children include fiber, vitamin A, and vitamin C.
   a. True
   b. False

2. Breakfast is an important meal for dieters, children, and adolescents.
   a. True
   b. False

3. Children and adolescents should get at least one hour of moderate-intensity physical activity most days of the week, and limit television, video games, and computer use to no more than two hours per day.
   a. True
   b. False

4. Preschool children enjoy strong-flavored and highly salted foods.
   a. True
   b. False

5. By the time children are three years old, they can eat amounts that count as regular servings by older family members.
   a. True
   b. False

6. Adolescent males need more kcalories, protein, magnesium, and zinc for muscle and bone development than do females; however, females need increased iron due to the onset of menstruation.
   a. True
   b. False

7. An older adult needs to pack lots of nutrients into fewer kcalories because he/she is less active and his/her basal metabolic rate has gone down.
   a. True
   b. False

8. Older adults don't get in enough vitamin C.
   a. True
   b. False

9. A person with an eating disorder may have started out just eating smaller or larger amounts of food, but at some point, the urge to eat less or more spiraled out of control.
   a. True
   b. False

10. The most crucial nutrient for athletes is carbohydrate, since glycogen stores heavily influence athletic performance.
    a. True
    b. False

11. For an athlete, the precompetition meal should be low in fat and fiber.
    a. True
    b. False

12. List five menu-planning guidelines for preschoolers, five for school-age children, and five for adolescents.

13. List five menu-planning guidelines for older adults.

14. List five menu-planning guidelines for athletes.

NUTRITION WEB EXPLORER

For a complete list of websites for the following activities, please visit the companion book page at www.wiley.com/college/drummond.

CULINARY INSTITUTE OF AMERICA MENU FOR HEALTHY KIDS

At this website, click on "Foodservice Directors." Then click on several of the links under "Helpful Resources for Foodservice Directors" such as "Culinary Techniques." Write up a list of five ways to use some of the resources on this website.

MEAL MAKEOVER MOMS

The Meal Makeover Moms make healthy meals that kids like. Click on "Cooking Videos" and watch how they prepare two recipes. Write up a summary of what particular ingredients or techniques they used to produce a healthy dish that kids like.

COOKING LIGHT AND EATING WELL MAGAZINES

Use both these websites to get ideas for lunches that kids take to school. Write up ideas for three box lunches for an eight-year-old.

THE HEALTHY LUNCHTIME CHALLENGE & KIDS' "STATE DINNER"

Check out the recipes that children developed and cooked for this contest. Which three do you like the most, and why?

MEDLINE PLUS: NUTRITION AND ATHLETIC PERFORMANCE

Read this article on nutrition and athletic performance. Write about how carbohydrate, protein, and water are important for athletes.

IN THE KITCHEN

Kid's Meals and Kid's Snacks

You will be given ingredients to make a healthy version of one or more of the following items.

Chicken strips

Hamburger

Pizza

Taco

Macaroni and cheese

Grilled cheese sandwich

Entrée salad

Snack kebab

Fruit pop/ice cream sandwich

Popcorn snack mix

Decorated apple/fruit wedges

Mini wrap

Veggies and dip

For the item(s) you make, you will do a nutrient analysis using the SuperTracker website. Your instructor will hand out a worksheet that you will use to describe why your menu item is healthier than its traditional counterpart. You will also evaluate each item on appearance and taste.

HOT TOPIC

FIGHTING CHILDHOOD OBESITY

Overweight for children is defined as a BMI between the 85th to 94th percentile for children of the same age and sex using the CDC growth charts (Appendix D). Obesity is defined as a BMI at or above the 95th percentile.

Just take a look at these statistics at how the number of obese children has grown since the 1970s.

	In the 1970s	In 2010
Children ages 2–5	5% were obese	12% were obese
Children ages 6–11	4% were obese	18% were obese
Children 12–19	6% were obese	18% were obese

America's children and youth are less active, consume more fat and sweetened beverages, and eat fewer healthy foods, especially fruits and vegetables, than children of previous generations.

Contributing Factors

At the individual level, childhood overweight is the result of an imbalance between the kcalories a child consumes as food and beverages and the kcalories used to support normal growth and development, metabolism, and physical activity. The imbalance between kcalories consumed and kcalories used can result from the influences and interactions of a number of factors, including genetic, behavioral, and environmental factors. It is the interactions among these factors—rather than any single factor—that is thought to cause overweight.

Studies indicate that certain genetic characteristics may increase an individual's susceptibility to overweight. However, this genetic susceptibility may need to exist in conjunction with contributing environmental factors (such as the ready availability of high-kcalorie foods) and behavioral factors (such as minimal physical activity) to have a significant effect on weight.

Genetic factors alone can play a role in specific cases of overweight. Yet the rapid rise in the rates of overweight and obesity in the general population cannot be attributed solely to genetic factors. The genetic characteristics of the human population have not changed in the last three decades, but the prevalence of overweight has tripled among school-aged children during that time.

American society has become characterized by environments that promote increased consumption of less healthy food and physical inactivity. It can be difficult for children to make healthy food choices and get enough physical activity when they are exposed to environments in their home, child care center, school, or community that are influenced by the following:

- **High-sugar drinks and less healthy foods on school campuses.** More than half of US middle and high

schools still offer sugary drinks and less healthy foods for purchase throughout the day from vending machines and school canteens and at fundraising events, school parties, and sporting events.

- **Advertising of less healthy foods.** More than half of US middle and high schools still offer sugary drinks and less healthy foods for purchase throughout the day from vending machines and school canteens and at fundraising events, school parties, and sporting events.

- **Lack of daily, quality physical activity in all schools.** Most adolescents fall short of the *2008 Physical Activity Guidelines for Americans* recommendation of at least 60 minutes of aerobic physical activity each day, as only 18 percent of students in grades 9 to 12 met this recommendation in 2007. Daily, quality physical education in school can help students meet the recommendations. However, in 2009 only 33 percent attended daily physical education classes.

- **No safe and appealing place, in many communities, to play or be active.** Many communities are built in ways that make it difficult or unsafe to be physically active. For some families, getting to parks and recreation centers may be difficult, and public transportation may not be available. For many children, safe routes for walking or biking to school or play may not exist. Half of the children in the United States do not have a park, community center, or sidewalk in their neighborhood.

- **Limited access to healthy affordable foods.** Some people have less access to stores and supermarkets that sell healthy, affordable food such as fruits and vegetables, especially in rural, minority, and lower-income neighborhoods. Supermarket access is associated with a reduced risk for obesity. Choosing healthy foods is difficult for parents who live in areas with food retailers that tend to sell less healthy food, such as convenience stores and fast-food restaurants.

- **Greater availability of high-energy-dense foods and sugar drinks.** High-energy-dense foods are ones that have a lot of kcalories in each bite. A recent study among children showed that a high-energy-dense diet is associated with a higher risk for excess body fat during childhood. Sugar drinks are the largest source of added sugar and an important contributor of kcalories in the diets of children in the United States. High consumption of sugar drinks, which have few, if any, nutrients, has been associated with obesity. On a typical day, 80 percent of youth drink sugar drinks.

- **Increasing portion sizes.** Portion sizes of less healthy foods and beverages have increased tremendously in restaurants, grocery stores, and vending machines. Research shows that children eat more without realizing it if they are served larger portions. This can mean they are consuming a lot of extra kcalories, especially when eating high-kcalorie foods.

- **Lack of breastfeeding support.** Breastfeeding protects against childhood overweight and obesity. However, in the United States, while 75 percent of mothers start out breastfeeding, only 13 percent of babies are exclusively breastfed at the end of six months.

- **Television and media.** American children spend an average of 7 hours a day using entertainment media, including television, computers, video games, and movies. Of those 7 hours, about 4.5 hours is spent watching television. Television viewing is a contributing factor to childhood obesity because it takes away from the time children spend in physical activities, leads to increased energy intake through snacking and eating meals in front of the television, and influences children to make unhealthy food choices through exposure to food advertisements.

Consequences

Childhood overweight is associated with various health-related consequences:

- High blood pressure and high cholesterol, which are risk factors for cardiovascular disease (CVD)

- Increased risk of impaired glucose tolerance, insulin resistance, and type 2 diabetes

- Breathing problems, such as sleep apnea, and asthma

- Joint problems and musculoskeletal discomfort

- Fatty liver disease, gallstones, and gastro-esophageal reflux (i.e., heartburn)

Obese children and adolescents also have a greater risk of social and psychological problems, such as discrimination and poor self-esteem, which can continue into adulthood.

Type 2 diabetes is increasingly being reported among children and adolescents who are overweight. Although this is normally only seen in overweight adults, type 2 diabetes is now a health concern for children and adolescents. Onset of diabetes in children and adolescents can result in complications such as cardiovascular disease and kidney failure.

Obese children are more likely to become obese adults. As discussed in "How Obesity Affects Your Health" in Chapter 12, adult obesity is associated with a number of serious health conditions such as heart disease and cancer.

If children are overweight, obesity in adulthood is likely to be more severe.

Tips for Parents

To help your child maintain a healthy weight, balance the kcalories your child consumes from foods and beverages with the kcalories your child uses through physical activity and normal growth. Remember that the goal for overweight children and teens is to reduce the rate of weight gain while allowing normal growth and development. Children and teens should *not* be placed on a weight-reduction diet without the consultation of a health-care provider.

One part of balancing kcalories is to eat foods that provide adequate nutrition and an appropriate number of kcalories. You can help children eat better by preparing and eating regular family meals, eating healthfully yourself to provide a good role model, and reducing kcalorie-rich temptations. There's no great secret to healthy eating. To help your children and family develop healthy eating habits:

- Make sure everyone starts the day with a healthy breakfast.

- Provide at least one fruit or vegetable at every meal and snack.

- Choose lean meats, poultry, fish, beans, and peas for protein.

- Serve appropriate portion sizes.

- Serve whole-grain breads and cereals as much as possible.

- Use low-fat or nonfat milk or dairy products. Mix whole or reduced-fat (2%) milk with low-fat (1%) or fat-free milk to help your kids get used to the different texture.

- Encourage your family to drink lots of water.

- Limit (better yet, eliminate!) sugar-sweetened beverages.

- Don't use food as a reward for good behavior.

- Remember that small changes every day can lead to a recipe for success!

- Look for ways to make favorite dishes healthier. The recipes that you may prepare regularly, and that your family enjoys, with just a few changes can be healthier and just as satisfying.

- Involve your children in helping buy foods and prepare meals.

Although everything can be enjoyed in moderation, reducing the kcalorie-rich temptations of high-fat, high-sugar, or salty snacks can also help your children develop healthy eating habits. Instead, only allow your children to eat them sometimes, so that they truly will be treats! Here are examples of easy-to-prepare, low-fat, and low-sugar treats for kids;

- Apple slices with peanut butter
- Banana and string cheese
- Yogurt with fresh fruit
- Sliced carrots, broccoli, or bell peppers with 2 tablespoons hummus
- Half a whole grain bagel with peanut butter
- Low-fat yogurt, fresh fruit and granola parfait
- Low-fat popcorn

Another part of balancing kcalories is to engage in an appropriate amount of physical activity and avoid too much sedentary time (Figure 13-11). In addition to being fun for children and teens, regular physical activity has many health benefits, including helping with weight management and increasing self-esteem. Children and teens should participate in at least 60 minutes of moderate-intensity physical activity most days of the week, prefer-ably daily. Remember that children imitate adults. Start adding physical activity to your own daily routine and encourage your child to join you. Some examples of moderate-intensity physi-cal activity include brisk walking, play-ing tag, jumping rope, playing soccer, swimming, or dancing.

In addition to encouraging physical activity, help children avoid too much sedentary time. Although quiet time for reading and homework is fine, limit the time your children watch television, play video games, or surf the web to no more than two hours per day (Figure 13-12). Instead, encourage your children to find fun activities to do with

FIGURE 13-11 Children should get exercise from activities that are fun.
Courtesy Jim Gathany and Centers for Disease Control and Prevention.

family, friends, or on their own that simply involve more activity.

Tips for Foodservice Operators and Chefs

As foodservice providers, working in schools, camps, day-care centers, or restaurants where a portion of your menu may be dedicated to kid-friendly selections, it is our responsibility to provide appropriate choices. Your food options should be creative and colorful, fun and exciting, relevant to children's needs, and approachable.

One key to a successful nutritional menu plan is for those in charge of the kitchens to have a solid understanding of health and wellness. They must be knowledgeable and motivated to translate

FIGURE 13-12 Limit children to no more than two hours per day of television, video games, or computer time.
Courtesy of Lisa F. Young/Shutterstock.

nutrition concepts into tasty, healthy meals. Children want their favorites, what is comfortable and familiar to them. The easiest way to win over your small clients is to continue to serve the foods they crave, but with a reengineering of the ingredients, methods, and presentations (see Table 13-6). Attractive shapes and colorful combinations, interesting containers such as Bento boxes or cardboard car meal boxes, and easily recognized menu items should be your starting point to accomplish this process in your facility (see Table 13-7). You can make healthy options

TABLE 13-6	Ingredients for Children's Meals
Spaghetti and Meatballs	Turkey meatballs, tomato sauce, whole grain pasta
Quesadillas	Skinless chicken or lean meat, moderate amount of flavorful cheese, fresh salsa, whole-wheat tortillas
Chicken fingers	Coat chicken in nonfat yogurt and spices—then roll in ground corn flakes and panko
Macaroni and cheese	Nonfat milk, moderate amounts of flavorful cheeses, arrowroot thickener, butternut squash puree for color and flavor
Tacos	Whole-wheat tortillas, skinless chicken or lean meat with seasoning, fresh lettuce, diced tomatoes, moderate amount of flavorful cheeses
French fries—white potatoes or sweet potatoes	Oven baked in a hot oven with vegetable spray
Pizza	Whole-wheat dough or tortilla shell pizza, sauce, low-fat mozzarella cheese, slices of green pepper, broccoli
Thai lettuce rolls	Skinless chicken, vegetables, water chestnuts, low-sodium teriyaki sauce
Barbeque pizza	Grilled skinless chicken, barbecue sauce, small amount of cheddar cheese, served with salad
Baked nachos	Crisp, multigrain tortilla triangles with grilled chicken, salsa, light sour cream, moderate amount of shredded cheddar cheese, lettuce
Salads	• Fajita salad—grilled chicken with tomato, olives, and cucumbers • Tortilla and chili salad—baked tortilla chips, turkey chili with beans • Burger salad—cut-up turkey burger, lettuce, tomatoes, onions, cheese strips • Spinach salad—spinach, strawberries, feta cheese, chicken strips, berry vinaigrette dressing • Buffalo chicken salad—buffalo chicken, shredded lettuce, tomatoes, celery, light ranch dressing

TABLE 13-7	Menu Ideas for Children	
BREAKFAST	**LUNCH**	**SNACKS**
House-made granola	**Shake It Up Salads**	Cheddar and grapes
Hot cereals with berries and agave syrup	Chopped vegetables with buffalo chicken strips	Hummus and freshly cut vegetables
Glazed grapefruits	Spinach salad with light ranch dressing	Yogurt-ranch dressing with freshly cut vegetables
Yogurt sauce with fruits and a happy face		
Breakfast burritos	Caesar salad with grilled chicken or shrimp	Cheesy popcorn
Breakfast tacos	**Bento Box Lunch**	Homemade power bars
	Egg salad with lettuce on mini pumpernickel bread, broccoli and cherry tomatoes, yogurt with blueberries and bananas	
Egg in a hole	**Wraps/Sandwiches**	Baked whole-wheat tortillas with dip—pesto or roasted pepper
Crustless quiche	Chicken pesto—olives, roasted peppers, zucchini, and eggplant	Sliced apples and celery with nut butter

TABLE 13-7   **Menu Ideas for Children** (*Continued*)

Egg white and turkey Swiss wrap

Egg and broccoli stuffed in pita bread

Whole-wheat waffles with peanut butter and pure fruit jelly

Egg white frittata

Egg white and chicken sausage on whole wheat

Whole-wheat bagel with nut butter, bananas, and honey

Yogurt parfaits with nonfat yogurt, honey, and fresh berries

Pizza—sauce, cheese, bell pepper slices

Strawberry cream cheese triangles

Pita sandwiches—lettuce, tomato, cucumber, hummus or chicken or falafel

BBQ chicken on whole-wheat roll with lettuce and tomato—serve with julienne carrots and light ranch dressing

Crêpe sandwiches

Hot Entrées

Pasta with tomato sauce and chicken

Chicken quesadillas with vegetables

Tacos—chicken, fish, beef, or tofu with vegetables

Tortilla-baked pizza

Baked cornflake crusted chicken fingers with roasted potato cubes and salad

Soup

Alphabet soup with vegetables and chicken

Chicken noodle or rice with vegetables

Minestrone soup

Vegetable soup with ravioli/tortellini

Creamy tomato soup

Kebabs

Turkey cubes, cheddar, grapes, strawberries, and dried apricots

Sliced turkey wrapped bread sticks with honey mustard

Wheat Chex, pretzels, and almonds tossed with melted bittersweet chocolate

Applesauce and raisins

Cheese and fruit kebabs

Salsa and baked chips

Zucchini or carrot bread

Frozen fruit pops

Homebaked sweet potato chips

Fruit smoothie

Mini whole-wheat pizza bagel

DINNER ENTRÉES

Grilled cheese with tomato or zucchini

Chicken chili with baked chips

Beef noodle bowl

Minestrone soup with pasta and meatballs

Boneless buffalo wings

Lo mein—carrots, nappa cabbage, peppers, broccoli, snow peas, chicken, pork or beef

Fajitas—chicken, beef, or fish

Oven-baked fried chicken

Rigatoni and meatballs

Penne with olives, lemon, garlic, chicken, pesto, or herbs

Chicken, beef, or fish tacos with shredded lettuce, tomato, cheese, and avocado slices

Turkey burritos

Sloppy Joes on whole-wheat buns

Hamburger noodle casserole

Baked fish sticks

Baked chicken cutlets with BBQ sauce

DINNER SIDES

Roasted broccoli

Baked onion rings

Roasted baby potatoes

Grilled vegetables

Sweet potato fries

Oven-fried potatoes

Pasta with tomato sauce

Mashed potatoes

Macaroni and cheese

DESSERT

Cinnamon oranges

Oatmeal white

Yogurt parfaits

Baked spiced apples

Tortilla fruit pizza

Chocolate strawberries

Gingersnap bread pudding

Yogurt fruit pops with pretzel stick

Homemade apricot fruit rolls

that taste good and are fun for kids; let the kids be the judge.

Following are tips for creating a children-friendly food program:

1. Be passionate! Use an energetic approach with a focus on taste, presentation, moderation, and quality. You are a trusted facility for families believing that you are serving the best for their children. You have a very captive audience that will consume the foods you prepare. This, coupled with good, solid nutritional information, will be instrumental in children's eating habits for years to come.

2. Serve foods they like! Send out surveys asking kids their opinions of nourishing foods: what they want, what they like, what are their favorite fruits, vegetables, snacks, lunch, and dinner items. Advertise your menus as nourishing not healthy, familiar not different, typical and traditional and home-style, not weird and intimidating. Make health exciting and engaging, not dictating and boring.

3. Try to "Launch a Campaign." Have the kids draw posters, rename their eating facility, then promote ideas created by the kids with adult supervision. Cool names, fun presentations, lingo, and buzzwords are how you get your customers psyched.

4. Educate your small customers! Give them short quizzes, and have prizes for simple nutritious facts. Put up posters of fruits, vegetables, and tasty meals. Have a poster contest!

5. Create a picture! Children love consistency, repetition, stability. Every Thursday the pasta is spaghetti and meatballs, and snacks always include a balanced dip and colorful cut vegetables. Desserts variety includes sugar-free puddings, oatmeal cookies, and fresh fruits. Rename vegetables with exciting adjectives (sugary beets, candied carrots, long broccoli, sweet peas and corn) Relocate colorful fruits to a better location to increase impulse buying of those items instead of the fatty chips.

6. Promote basic staff knowledge! The kitchen/service staff must have a basic understanding of nutrition knowledge with cooking methods and ingredients. Be aware of the effects of adding additional salt, oil, butter, or sugar when preparing food to guidelines and proper balance of foods with the key ingredients. A very simple manual can prepare your entire staff to be more health savvy, providing better service and quality.

7. Use appropriate ingredients and procedures! Good ingredients, fresh foods such as fruits and vegetables, lean meats and poultry, fresh herbs, stocks, and seasonings are all important. The basic procedures described in this book can reduce fat, supply the proper carbohydrates, limit simple sugars, and reduce kcalories while supplying proper nutritional balance. Following standardized recipes is important.

Of course, many children, especially younger children, are picky about what they eat, and you will need some additional help getting children to pick up new foods. Here are some ideas:

1. It's not just the taste of the vegetables that kids don't enjoy; often, it's the texture of the food. Many kids would rather munch on raw carrots than cooked carrots, simply because of the difference in mouthfeel. For vegetables and dip, blanch and shock vegetables such as green beans, broccoli, and asparagus for 30 seconds. This keeps the vegetables crunchy while adding a vibrant color. Add these together with crunchy celery, radishes, and cucumbers to make an exciting and colorful presentation.

2. Some vegetables like Brussels sprouts, cauliflower, zucchini, summer squash, and green beans need to be blanched and grilled or roasted before eating. Try seasoning with a little pizza or taco seasoning, or BBQ rub, a drizzle of olive oil, and a squirt of lime or lemon juice for a different flavor additive. Get creative and try

different seasoning blends and spices until you find the ones that are most popular.

3. For breakfast, prepare a delicious fruit bowl. Additionally, make some oatmeal with honey and ground cinnamon. Whole-wheat pancakes or waffles with chicken sausage or egg white omelets with vegetables and turkey work well, too.

4. To start kids on whole-wheat breads, disguise it in a grilled cheese sandwich, French toast, or a bagel with nut butter and banana. To introduce brown rice and grains, make it similar to fried rice, with colorful vegetables adding an Asian flavor profile.

5. Include the children in "what to cook for breakfast, lunch, and dinner." Believe it or not, children love to be included in adult things, and cooking is very much an adult responsibility. Get them involved and excited about learning about cooking healthy food while they are still young. Most children will become more selective and responsible in their choices when healthy options are consistently offered. The biggest obstacle will be their perception that wholesome, nutritious, and balanced means the food lacks flavor and satisfaction. But over time, they will realize that healthy foods taste better than processed and refined foods.

6. Reduce the amount of refined-food snacks. Offering healthier alternatives will take some due diligence and persistence, but the end result is worth the effort. Whole fruits, homemade fruit rolls made with pureed dried apricots and apple juice; nut butter and celery; and turkey, cheddar, grape and strawberry kebabs are just some tasty options that kids like. Make your own frozen fruit sticks or pops with fresh juices, nonfat yogurt studded with whole chunks of fruits and berries, then dipped in a little chocolate for an extra treat. Smoothies are another great snack that can be made with nonfat yogurt, fat-free milk, and ice blended into a smooth shake—and

sweetened with a little honey and ripe, crushed fruit.

7. Make food fun. Cut fruits, and other foods such as sandwiches, into interesting and fun shapes such as stars, half moon circles, squares, and so on. Put cut foods on a stick—fruits, vegetables, meats, cheese or a combination. Decorate a bowl of whole-grain cereal with a smiley face: blueberry eyes, raisin nose, peach or apple slices for the mouth.

8. Use familiar terms, such as dips and chips, mac and cheese, spaghetti and meatballs.

9. Add fresh and dried fruits, as well as vegetables, to whole-grain baked goods such as blueberry, banana, or strawberry pancakes or muffins with zucchini, carrots, bananas, or sweet potatoes.

As you incorporate these food choices into your main food production in your facility, educate your staff about the importance of following recipes and portion control.

Resources

There are many ways to get new menu ideas and promote good eating habits in your foodservice with a variety of resources available from the US Department of Agriculture and many other agencies.

- Chefs Move to Schools. This new website is for chefs and schools, allowing them to sign up for the program; access training, recipes and resources; and learn how chefs and schools across the country are partnering to create healthier school environments.

- The Lunchbox: Healthy Tools to Help All Schools. The Lunch Box is an online toolkit with Healthy Tools for All Schools, packed with solutions at your fingertips. Use any of these free tools to transform your school food into healthy and delicious food for all children.

- Food Family Farming Foundation. Founded by Chef Ann Cooper, the Food Family Farming Foundation was created to empower schools to serve nutritious whole food to all students.

- The Culinary Institute of America—Menu for Healthy Kids. This website aims to generate positive change in children's health by creating smarter school foodservice operations and improving access to healthier and more flavorful school meals.

- Team Nutrition. Team Nutrition is an initiative of the USDA Food and Nutrition Service to support the Child Nutrition Programs through training and technical assistance for foodservice, nutrition education for children and their caregivers, and school and community support for healthy eating and physical activity.

- *Cooking Light* magazine. This website has many recipe ideas for children's meals and snacks.

SERVING SIZES FOR MYPLATE FOOD GROUPS AND HOW TO COUNT OILS AND EMPTY KCALORIES

WHAT COUNTS AS A CUP OF FRUIT?

In general, 1 cup of fruit, 100 percent fruit juice, or ½ cup of dried fruit can be considered as 1 cup from the Fruit Group. The following specific amounts count as 1 cup of fruit (in some cases, equivalents for ½ cup are also shown) toward your daily recommended intake.

	AMOUNT THAT COUNTS AS 1 CUP OF FRUIT	AMOUNT THAT COUNTS AS ½ CUP OF FRUIT
Apple	½ large (3.25" diameter)	
	1 small (2.5" diameter)	
	1 cup sliced or chopped, raw or cooked	½ cup sliced or chopped, raw or cooked
Applesauce	1 cup	1 snack container (4 oz.)
Banana	1 cup sliced	1 small (less than 6" long)
	1 large (8" to 9" long)	
Cantaloupe	1 cup diced or melon balls	1 medium wedge (1/8 of a med. melon)
Grapes	1 cup whole or cut-up	
	32 seedless grapes	16 seedless grapes
Grapefruit	1 medium (4" diameter)	½ medium (4" diameter)
	1 cup sections	
Mixed fruit (fruit cocktail)	1 cup diced or sliced, raw or canned, drained	1 snack container (4 oz.) drained = $3/8$ cup
Orange	1 large ($3^1/_{16}$" diameter)	1 small ($2^3/_8$" diameter)
	1 cup sections	
Orange, mandarin	1 cup canned, drained	
Peach	1 large (2¾" diameter)	1 small (2" diameter)
	1 cup sliced or diced, raw, cooked, or canned, drained	1 snack container (4 oz.) drained = $3/8$ cup
	2 halves, canned	
Pear	1 medium pear (2.5 per lb)	1 snack container (4 oz.) drained = $3/8$ cup
	1 cup sliced or diced, raw, cooked, or canned, drained	
Pineapple	1 cup chunks, sliced or crushed, raw, cooked or canned, drained	1 snack container (4 oz.) drained = $3/8$ cup
Plum	1 cup sliced raw or cooked	
	3 medium or 2 large plums	1 large plum

Strawberries	About 8 large berries	
	1 cup whole, halved, or sliced, fresh or frozen	½ cup whole, halved, or sliced
Watermelon	1 small wedge (1" thick)	6 melon balls
	1 cup diced or balls	
Dried fruit (raisins, prunes, apricots, etc.)	½ cup dried fruit is equivalent to 1 cup fruit:	¼ cup dried fruit is equivalent to ½ cup fruit;
	½ cup raisins	
	½ cup prunes	1 small box raisins (1.5 oz.)
	½ cup dried apricots	
100% fruit juice (orange, apple, grape, grapefruit, etc.)	1 cup	½ cup

WHAT COUNTS AS A CUP OF VEGETABLES?

In general, 1 cup of raw or cooked vegetables or vegetable juice, or 2 cups of raw leafy greens can be considered as 1 cup from the Vegetable Group. The chart lists specific amounts that count as 1 cup of vegetables (in some cases' equivalents for ½ cup are also shown) towards your recommended intake.

	AMOUNT THAT COUNTS AS 1 CUP OF VEGETABLES	AMOUNT THAT COUNTS AS ½ CUP OF VEGETABLES
Dark Green Vegetables		
Broccoli	1 cup chopped or florets	
	3 spears 5" long, raw or cooked	
Greens (collards, mustard greens, turnip greens, kale)	1 cup cooked	
Spinach	1 cup, cooked	
	2 cups raw = 1 cup of vegetables	1 cup raw = ½ cup of vegetables
Raw leafy greens: Spinach, romaine, watercress, dark green leafy lettuce, endive, escarole	2 cups raw = 1 cup of vegetables	1 cup raw = ½ cup of vegetables
Red and Orange Vegetables		
Carrots	1 cup, strips, slices, or chopped, raw or cooked	
	2 medium	1 medium carrot
	1 cup baby carrots (about 12)	About 6 baby carrots
Pumpkin	1 cup mashed, cooked	
Red peppers	1 cup chopped, raw, or cooked	1 small pepper
	1 large pepper (3" diameter, 3¾" long)	
Tomatoes	1 large raw whole (3")	1 small raw whole (2¼" diameter)
	1 cup chopped or sliced, raw, canned, or cooked	1 medium canned
Tomato juice	1 cup	½ cup
Sweet potato	1 large baked (2¼" or more diameter)	
	1 cup sliced or mashed, cooked	
Winter squash (acorn, butternut, hubbard)	1 cup cubed, cooked	½ acorn squash, baked = ¾cup

Beans and Peas

Dry beans and peas (such as black, garbanzo, kidney, pinto, or soy beans, or black eyed peas or split peas)	1 cup whole or mashed, cooked

Starchy Vegetables

Corn, yellow or white	1 cup	
	1 large ear (8" to 9" long)	1 small ear (about 6" long)
Green peas	1 cup	
White potatoes	1 cup diced, mashed	
	1 medium boiled or baked potato (2½" to 3" diameter)	
	French fried: 20 medium to long strips (2½" to 4" long) (Contains added calories from solid fats)	

Other Vegetables

Bean sprouts	1 cup cooked	
Cabbage, green	1 cup, chopped or shredded, raw or cooked	
Cauliflower	1 cup pieces or florets, raw or cooked	
Celery	1 cup, diced or sliced, raw, or cooked	
	2 large stalks (11" to 12" long)	1 large stalk (11" to 12" long)
Cucumbers	1 cup raw, sliced, or chopped	
Green or wax beans	1 cup cooked	
Green peppers	1 cup chopped, raw, or cooked	
	1 large pepper (3" diameter, 3¾" long)	1 small pepper
Lettuce, iceberg or head	2 cups raw, shredded, or chopped = 1 cup of vegetables	1 cup raw, shredded, or chopped = ½ cup of vegetables
Mushrooms	1 cup raw or cooked	
Onions	1 cup chopped, raw, or cooked	
Summer squash or zucchini	1 cup cooked, sliced, or diced	

WHAT COUNTS AS AN OUNCE EQUIVALENT OF GRAINS?

In general, 1 slice of bread, 1 cup of ready-to-eat cereal, or ½ cup of cooked rice, cooked pasta, or cooked cereal can be considered as 1 ounce equivalent from the Grains Group.

The chart lists specific amounts that count as 1 ounce equivalent of grains toward your daily recommended intake. In some cases, the number of ounce-equivalents for common portions is also shown.

		AMOUNT THAT COUNTS AS 1 OUNCE EQUIVALENT OF GRAINS	COMMON PORTIONS AND OUNCE EQUIVALENTS
Bagels	WG*: whole wheat	1 "mini" bagel	1 large bagel = 4 oz. equivalents
	RG*: plain, egg		
Biscuits	(baking powder/ buttermilk—RG*)	1 small (2" diameter)	1 large (3" diameter) = 2 oz. equivalents
Breads	WG*: 100% Whole wheat	1 regular slice	2 regular slices = 2 oz. equivalents
	RG*: white, wheat, French, sourdough	1 small slice French	
		4 snack-size slices rye bread	
Bulgur	cracked wheat (WG*)	½ cup cooked	
Cornbread	(RG*)	1 small piece (2½" × 1¼" × 1 ¼")	1 medium piece (2½" × 2½" × 1¼") = 2 oz. equivalents
Crackers	WG*: 100% whole wheat, rye	5 whole-wheat crackers	
		2 rye crispbreads	
	RG*: saltines, snack crackers	7 square or round crackers	
English muffins	WG*: whole wheat	½ muffin	1 muffin = 2 oz. equivalents
	RG*: plain, raisin		
Muffins	WG*: whole wheat	1 small (2½" diameter)	1 large (3½" diameter) = 3 oz. equivalents
	RG*: bran, corn, plain		
Oatmeal	(WG)	½ cup cooked	
		1 packet instant	
		1 oz. (1/3 cup) dry (regular or quick)	
Pancakes	WG*: Whole-wheat, buck-wheat	1 pancake (4½" diameter)	3 pancakes (4½" diameter) = 3 oz. equivalents
		2 small pancakes (3" diameter)	
	RG*: buttermilk, plain		
Popcorn	(WG*)	3 cups, popped	1 mini microwave bag or 100-calorie bag, popped is 2 oz. equivalents
Ready-to-eat breakfast cereal	WG*: toasted oat, whole-wheat flakes	1 cup flakes or rounds	
		1¼ cup puffed	
	RG*: corn flakes, puffed rice		
Rice	WG*: brown, wild	½ cup cooked	1 cup cooked = 2 oz. equivalents
	RG*: enriched, white, polished	1 oz. dry	
Pasta–spaghetti, macaroni, noodles	WG*: whole wheat	½ cup cooked	1 cup cooked = 2 oz. equivalents
	RG*: enriched, durum	1 oz. dry	
Tortillas	WG*: whole wheat, whole-grain corn	1 small flour tortilla (6" diameter)	1 large tortilla (12" diameter) = 4 oz. equivalents
	RG*: Flour, corn	1 corn tortilla (6" diameter)	

*WG = whole grains, RG = refined grains. This is shown when products are available both in whole grain and refined grain forms.

WHAT COUNTS AS AN OUNCE EQUIVALENT IN THE PROTEIN FOODS GROUP?

In general, 1 ounce of meat, poultry or fish, ¼ cup cooked beans, 1 egg, 1 tablespoon of peanut butter, or ½ ounce of nuts or seeds can be considered as 1 ounce equivalent from the Protein Foods Group. The chart lists specific amounts that count as 1 ounce equivalent in the Protein Foods Group toward your daily recommended intake.

	AMOUNT THAT COUNTS AS 1 OUNCE EQUIVALENT IN THE PROTEIN FOODS GROUP	COMMON PORTIONS AND OUNCE EQUIVALENTS
Meats	1 oz. cooked lean beef	1 small steak (eye of round, filet) = 3½ to 4 oz. equivalents
	1 oz. cooked lean pork or ham	1 small lean hamburger = 2 to 3 oz. equivalents
Poultry	1 oz. cooked chicken or turkey, without skin	1 small chicken breast half = 3 oz. equivalents
	1 sandwich slice of turkey (4½ × 2 ½ × 1/8")	½ Cornish game hen = 4 oz. equivalents
		1 can of tuna, drained = 3 to 4 oz. equivalents
Seafood	1 oz. cooked fish or shellfish	1 salmon steak = 4 to 6 oz. equivalents
		1 small trout = 3 oz. equivalents
Eggs	1 egg	3 egg whites = 2 oz. equivalents
		3 egg yolks = 1 oz. equivalent
Nuts and seeds	½ oz. nuts (12 almonds, 24 pistachios, 7 walnut halves)	
	½ oz. seeds (pumpkin, sunflower, or squash seeds, hulled, roasted)	1 oz. nuts or seeds = 2 oz. equivalents
	1 Tbsp. peanut butter or almond butter	
Beans and peas	¼ cup of cooked beans (such as black, kidney, pinto, or white beans)	1 cup split pea soup = 2 oz. equivalents
	¼ cup of cooked peas (such as chickpeas, cowpeas, lentils, or split peas)	1 cup lentil soup = 2 oz. equivalents
	¼ cup of baked beans, refried beans	1 cup bean soup = 2 oz. equivalents
	¼ cup (about 2 oz.) of tofu	
	1 oz. tempeh, cooked	
	¼ cup roasted soybeans	1 soy or bean burger patty = 2 oz. equivalents
	1 falafel patty (2¼", 4 oz.),	
	2 Tbsp. hummus	

WHAT COUNTS AS A CUP IN THE DAIRY GROUP?

In general, 1 cup of milk, yogurt, or soymilk (soy beverage), 1½ ounces of natural cheese, or 2 ounces of processed cheese can be considered as 1 cup from the Dairy Group. The chart lists specific amounts that count as 1 cup in the Dairy Group toward your daily recommended intake.

	AMOUNT THAT COUNTS AS A CUP IN THE DAIRY GROUP	COMMON PORTIONS AND CUP EQUIVALENTS
Milk (choose fat-free or low-fat milk)	1 cup milk 1 half-pint container milk ½ cup evaporated milk	
Yogurt (choose fat-free or low-fat yogurt)	1 regular container (8 fl oz.) 1 cup yogurt	1 small container (6 oz.) = ¾ cup 1 snack size container (4 oz.) = ½ cup
Cheese (choose reduced-fat or low-fat cheeses)	1½ oz. hard cheese (cheddar, mozzarella, Swiss, Parmesan) ½ cup shredded cheese 2 oz. processed cheese (American) ½ cup ricotta cheese 2 cups cottage cheese	1 slice of hard cheese = ½ cup milk 1 slice of processed cheese = ½ cup milk ½ cup cottage cheese = ¼ cup milk
Milk-based desserts (choose fat-free or low-fat types)	1 cup pudding made with milk 1 cup frozen yogurt 1½ cups ice cream	1 scoop ice cream = ½ cup milk
Soymilk (soy beverage)	1 cup calcium-fortified soymilk 1 half-pint container calcium-fortified soymilk	

HOW DO I COUNT THE OILS I EAT?

The chart gives a quick guide to the amount of oils in some common foods.

	AMOUNT OF FOOD	AMOUNT OF OIL Teaspoons/grams	CALORIES FROM OIL Approximate Calories	TOTAL CALORIES Approximate Calories
Oils:				
Vegetable oils (such as canola, corn, cottonseed, olive, peanut, safflower, soybean, and sunflower)	1 Tbsp.	3 tsp./14 g	120	120
Foods rich in oils:				
Margarine, soft (trans fat–free)	1 Tbsp.	2½ tsp./11 g	100	100
Mayonnaise	1 Tbsp.	2½ tsp./11 g	100	100
Mayonnaise-type salad dressing	1 Tbsp.	1 tsp./5 g	45	55
Italian dressing	2 Tbsp.	2 tsp./8 g	75	85
Thousand Island dressing	2 Tbsp.	2½ tsp./11 g	100	120
Olives*, ripe, canned	4 large	½ tsp./ 2 g	15	20
Avocado*	½ med	3 tsp./15 g	130	160
Peanut butter*	2 Tbsp.	4 tsp./ 16 g	140	190
Peanuts, dry roasted*	1 oz.	3 tsp./14 g	120	165

Mixed nuts, dry roasted*	1 oz.	3 tsp./15 g	130	170
Cashews, dry roasted*	1 oz.	3 tsp./13 g	115	165
Almonds, dry roasted*	1 oz.	3 tsp./15 g	130	170
Hazelnuts*	1 oz.	4 tsp./18 g	160	185
Sunflower seeds*	1 oz.	3 tsp./14 g	120	165

*Avocados and olives are part of the Vegetable Group; nuts and seeds are part of the Protein Foods Group. These foods are also high in oils. Soft margarine, mayonnaise, and salad dressings are mainly oil and are not considered to be part of any food group.

HOW DO I COUNT THE EMPTY CALORIES I EAT?

The chart provides a quick guide to the number of empty calories in some common foods. It is very easy to exceed your empty calorie allowance, even when making careful food choices. Fats are concentrated sources of calories. Even small amounts of foods high in solid fats will use up the empty calorie allowance quickly.

FOOD	AMOUNT	ESTIMATED TOTAL CALORIES	ESTIMATED EMPTY CALORIES (CALORIES FROM SOLID FATS AND ADDED SUGARS)
DAIRY GROUP			
Fat-free milk (skim)	1 cup	83	0
1% milk (low fat)	1 cup	102	18
2% milk (reduced fat)	1 cup	122	37
Whole milk	1 cup	149	63
Low-fat chocolate milk	1 cup	158	64
Cheddar cheese	1½ oz.	172	113
Nonfat mozzarella cheese	1½ oz.	59	0
Whole milk mozzarella cheese	1½ oz.	128	76
Fruit flavored low-fat yogurt	1 cup (8 fl oz.)	250	152
Frozen yogurt	1 cup	224	119
Ice cream, vanilla	1 cup	275	210
Cheese sauce	¼ cup	120	64
PROTEIN FOODS GROUP			
Extra lean ground beef, 95% lean	3 oz., cooked	146	0
Regular ground beef, 80% lean	3 oz., cooked	229	64
Turkey roll, light meat	3 slices (1 oz. each)	165	0
Roasted chicken breast (skinless)	3 oz., cooked	138	0
Roasted chicken thigh with skin	3 oz., cooked	209	47
Fried chicken with skin & batter	3 medium wings	478	382
Beef sausage, precooked	3 oz., cooked	345	172
Pork sausage	2 patties (2 oz.)	204	96
Beef bologna	3 slices (1 oz. each)	261	150

FOOD	AMOUNT	ESTIMATED TOTAL CALORIES	ESTIMATED EMPTY CALORIES (CALORIES FROM SOLID FATS AND ADDED SUGARS)
GRAINS GROUP			
Whole wheat bread	1 slice (1 oz.)	69	0
White bread	1 slice (1 oz.)	69	0
English muffin	1 muffin	132	0
Blueberry muffin	1 small muffin (2 oz.)	259	69
Croissant	1 medium (2 oz.)	231	111
Biscuit, plain	1 medium (2½" diameter)	186	71
Cornbread	1 piece (2½" × 2½" × 1¼")	167	52
Corn flakes cereal	1 cup	90	8
Frosted corn flakes cereal	1 cup	147	56
Graham crackers	2 large pieces	118	54
Whole wheat crackers	5 crackers	85	25
Round snack crackers	7 crackers	106	42
Chocolate chip cookies	2 large	161	109
Chocolate cake	1 slice of two-layer cake	408	315
Glazed doughnut, yeast type	1 medium, 3¾" diameter	255	170
Cinnamon sweet roll	1 medium roll	223	137
VEGETABLE GROUP			
Baked potato	1 medium	159	
French fries	1 medium order	431	185
Onion rings	1 order (8 to 9 rings)	275	160
FRUIT GROUP			
Unsweetened applesauce	1 cup	105	
Sweetened applesauce	1 cup	173	68
OTHER			
Pepperoni pizza	1 slice of a 14" pizza, regular crust	340	
Regular soda	1 can (12 fl oz.)	136	136
Regular soda	1 bottle (19.9 fl oz.)	192	192
Fruit-flavored drink	1 cup	128	128
Butter	1 tsp.	36	33
Stick margarine	1 tsp.	36	32
Cream cheese	1 Tbsp.	41	36
Heavy (whipping) cream	1 Tbsp.	51	45
Frozen whipped topping (nondairy)	¼ cup	60	55
Table wine	1 glass (5 fl oz.)	121	121*
Beer (regular)	1 can (12 fl oz.)	155	155*
Beer (light)	1 can (12 fl oz.)	104	104*
Distilled spirits (80 proof)	1 standard drink (1½ fl oz.)	96	96*

Calories from alcohol are not from solid fats or added sugars, but they count against your limit for empty calories—calories from solid fats and added sugars. The calories per serving are listed on the Nutrition Facts label on food packages. Be sure to compare the stated serving size to the amount actually eaten. If you eat twice the stated serving size, you will have twice the calories.

Source: US Department of Agriculture. ChooseMyPlate.gov. Washington, DC.

BODY MASS INDEX CHART

Locate the height of interest in the left-most column and read across the row for that height to the weight of interest. Follow the column of the weight up to the top row that lists the BMI. BMI of 19–24 is the healthy weight range, BMI of 25–29 is the overweight range, and BMI of 30 and above is in the obese range.

BMI	19	20	21	22	23	24	25	26	27	28	29	30	31	32	33	34	35
HEIGHT							WEIGHT IN POUNDS										
4'10"	91	96	100	105	110	115	119	124	129	134	138	143	148	153	158	162	167
4'11"	94	99	104	109	114	119	124	128	133	138	143	148	153	158	163	168	173
5'	97	102	107	112	118	123	128	133	138	143	148	153	158	163	168	174	179
5'1"	100	106	111	116	122	127	132	137	143	148	153	158	164	169	174	180	185
5'2"	104	109	115	120	126	131	136	142	147	153	158	164	169	175	180	186	191
5'3"	107	113	118	124	130	135	141	146	152	158	163	169	175	180	186	191	197
5'4"	110	116	122	128	134	140	145	151	157	163	169	174	180	186	192	197	204
5'5"	114	120	126	132	138	144	150	156	162	168	174	180	186	192	198	204	210
5'6"	118	124	130	136	142	148	155	161	167	173	179	186	192	198	204	210	216
5'7"	121	127	134	140	146	153	159	166	172	178	185	191	198	204	211	217	223
5'8"	125	131	138	144	151	158	164	171	177	184	190	197	203	210	216	223	230
5'9"	128	135	142	149	155	162	169	176	182	189	196	203	209	216	223	230	236
5'10"	132	139	146	153	160	167	174	181	188	195	202	209	216	222	229	236	243
5'11"	136	143	150	157	165	172	179	186	193	200	208	215	222	229	236	243	250
6'	140	147	154	162	169	177	184	191	199	206	213	221	228	235	242	250	258
6'1"	144	151	159	166	174	182	189	197	204	212	219	227	235	242	250	257	265
6'2"	148	155	163	171	179	186	194	202	210	218	225	253	241	249	256	264	272
6'3"	152	160	168	176	184	192	200	208	216	224	232	240	248	256	264	272	279
	HEALTHY WEIGHT						OVERWEIGHT					OBESE					

Source: Evidence Report of Clinical Guidelines on the Identification, Evaluation and Treatment of Overweight and Obesity in Adults, 1998. NIH/National Heart, Lung and Blood Institute (NHLBI).

ANSWERS TO CHECK-OUT QUIZZES

CHAPTER 1

1. True
2. False
3. True
4. False
5. False
6. False
7. False
8. True
9. False
10. True
11. False
12. False
13. True
14. True
15. False
16. Many factors influence what you eat: flavor, food cost, convenience, nutrition, demographics, culture, religion, health concerns, social and emotional influences, marketing and the media, and environmental concerns.
17. You may think that taste and flavor are the same thing, but taste is actually a component of flavor. Flavor is an attribute of a food that includes its taste, smell, feel in the mouth or texture, temperature, and even the sounds made when it is chewed. Flavor is a combination of all five senses: taste, smell, touch, sight, and sound. The taste buds in your mouth and the smell receptors in your nose work together to deliver signals to the brain that are translated into the flavor of food.
18. Factors that increase basal metabolism are gender (men have higher BMR), growth, height (tall people have higher BMR), temperature extremes, fever, stress, exercise, smoking, and caffeine. Factors that decrease basal metabolism are aging and sleep.
19. Carbohydrate and protein yield 4 kcal/gram. Fats provide 9 kcal/gram.
20. • Carbohydrates are a large class of nutrients, including sugars, starches, and fibers, that function as the body's primary source of energy. Fiber can't be digested in the body and is found in legumes, fruits, vegetables, whole grains, nuts, and seeds.

• Fats include triglycerides and cholesterol, that are soluble in fat, not water. Familiar lipids include fats and oils as well as the fat in meat, chicken, and dairy products.

• Protein is the main structural component of all the body's cells. It is made of units called amino acids. Protein is present in significant amounts in animal foods such as meat, poultry, fish, eggs, milk, and cheese.

• Vitamins and minerals are noncaloric nutrients found in a wide variety of foods that are essential in small quantities to regulate body processes, maintain the body, and allow growth and reproduction.

• Water is very important as it plays a vital role in all bodily processes and makes up just over half the body's weight.

21. Fresh fruits and vegetables are all whole foods—we get processed foods from them such as applesauce and frozen broccoli. Whole wheat, a whole grain, is a whole food that is processed to make flours, cereals, and breads. Milk is a whole food that we process to make cheese and ice cream.

22. Mouth

• Teeth and tongue help in chewing.

• Saliva lubricate food and has digestive enzymes to start digestion.

Esophagus

• Through peristalsis, food is moved from pharynx to stomach.

Stomach

• The stomach produces hydrochloric acid and an enzyme to break down protein.

• Hydrochloric acid also destroys harmful bacteria and increases the ability of calcium and iron to be absorbed.

• Food is churned up more and further digested.

Small intestine

• The small intestine furthers digestion with the help of bile and digestive enzymes.

• Many nutrients are absorbed into the blood or lymph systems.

Large intestine
- The large intestine receives waste production of digestion and passes them on to rectum.
- It absorbs water, some minerals and vitamins made by bacteria.

CHAPTER 2

1. False
2. True
3. False
4. True
5. False
6. True
7. False
8. False
9. False
10. True
11. False
12. True
13. False
14. True
15. True
16. • 6 ounces or the equivalent of grains
    - 2.5 cups of vegetables (five servings)
    - 2 cups of fruit (or four servings)
    - 3 cups of milk or the equivalent
    - 5.5 ounces or the equivalent of lean meat and beans
    - 6 teaspoons of oil
    - 258 empty kcalories
17. Saturated fats are found in butter, shortening, lard, beef and chicken fat, full-fat cheese, whole milk, cream, ice cream, fatty meats, poultry skin, and many baked goods. Diets high in saturated fats raise levels of LDL cholesterol (called bad cholesterol) in the blood that then increases your risk for cardiovascular disease.
18. According to the *Dietary Guidelines for Americans, 2010,* you should eat more vegetables, fruits, whole grains, milk and milk products, and seafood. You should eat less sodium, solid fats, added sugars, and refined grains.
19. To lose weight, you can increase intake of fruits, vegetables, and whole grains; reduce intake of sugar-sweetened beverages; focus on the total number of kcalories consumed; monitor food and alcoholic beverage intake; prepare, serve, and eat smaller portions of foods and beverages; eat a nutrient-dense breakfast; eat smaller portions or lower kcalorie options when eating out; limit screen time; and be physically active.
20. Because consumption of vegetables, fruits, whole grains, milk and milk products, and seafood is lower than recommended, intake by Americans of some nutrients is low enough to be of concern. These are potassium, dietary fiber, calcium, and vitamin D.
21. For most foods, all ingredients must be listed on the label and identified by their common names so that consumers can identify the presence of any of eight major food allergens (milk, eggs, fish, shellfish, tree nuts, peanuts, wheat, soybeans). Food manufacturers must use plain English words such as milk or wheat rather than less familiar words such as casein or semolina so consumers can identify the mos common food allergens on ingredients lists.
22. Nutrient content claims, such as "good source of calcium" and "fat-free," can appear on food packages only if they follow legal definitions. For example, a food that is a good source of calcium must provide 10 to 19 percent of the Daily Value for calcium in one serving. Health claims state that certain foods or components of foods (such as calcium) may affect or reduce the risk of a disease or health-related condition.

CHAPTER 3

1. True
2. True
3. False
4. False
5. True
6. False
7. False
8. True
9. False
10. True
11. True
12. False
13. False
14. True
15. True
16. 32 grams divided by 4 grams/teaspoon = 8 teaspoons sugar
17. When starches gelatinize, the granules absorb water and swell, making the liquid thicken. Around the boiling point, having absorbed a lot of water, the granules burst and starch pours out into the liquid. When this occurs, the liquid quickly becomes still thicker.
18. Plant foods rich in soluble fiber include oats, barley, beans, many fruits such as apples, and many vegetables such as carrots. Plant foods rich in insoluble fiber include wheat bran, whole grains, beans, peas, and lentils, and many vegetables and fruits.
19. All grains have a large center area high in starch known as the endosperm. The endosperm, about 85 percent of the kernel, also contains protein and B vitamins. At

one end of the endosperm is the germ, the area of the kernel that germinates to grow into a new plant. The germ is rich in vitamins and minerals and contains some oil. The bran, containing much fiber and other nutrients, covers and protects the endosperm and germ. The seed contains everything needed to reproduce the plant.

20. Carbohydrates are the primary source of the body's energy and glucose is the body's number one source of energy. Under most circumstances, the brain and other nerve cells will only use glucose for energy. Carbohydrates are also important to help the body use fat efficiently. When fat is burned for energy without any carbohydrates present, the process can harm the body. Carbohydrates are part of various materials in the body such as connective tissues, some hormones and enzymes, and genetic material.

21. You need 14 grams of fiber for every 1000 kcalories you eat. The American Heart Association (2009) recommends limiting your intake of added sugars to half of your empty kcalorie allowance found in MyPlate. For most American women, that is no more than 100 kcalories/day (about 6 teaspoons of sugar) and for men, no more than 140 kcalories/day (about 9 teaspoons of sugar) from added sugars.

22. Read the ingredient list on the food label. For many whole-grain products, "whole" or "whole grain" will appear before the grain ingredient's name. The whole grain should be the first ingredient listed. Look for whole wheat, whole oats, whole rye, whole grain corn, brown rice, or whole grain barley. If the first ingredient is wheat flour, enriched flour, or degerminated cornmeal, the product is not a good source of whole grains.

CHAPTER 4

1. False
2. True
3. True
4. False
5. True
6. False
7. True
8. False
9. False
10. True
11. False
12. True
13. True
14. True
15. False
16. True
17. True
18. False
19. The food groups that contribute the most fat are: meat, poultry, and fish; dairy foods, and fats, oils, and condiments.
20. The biggest sources of saturated fat in the American diet are animal foods: fat and eggs used to make grains-based desserts (such as cookies), cheese, beef, pork (sausage, franks, and bacon), French fries, dairy desserts such as ice cream, and whole milk. Saturated fat is also found in eggs and poultry skin.

   Trans fats can be found in many foods—such as many fried foods like French fries and doughnuts, baked goods including pastries, piecrusts, biscuits, pizza dough, cookies, crackers, and some stick margarines and shortenings.
21. To increase your intake of omega-3 fatty acids, you should eat about 8 ounces per week of a variety of seafood, which provides an average of 250 mg per day of EPA and DHA.
22. Fish to avoid because of mercury include shrimp, canned light tuna, salmon, Pollock, and catfish.
23. Fat stores under the skin provide insulation and a cushion around critical organs. Fat is an important part of cell membranes and provides energy. Fat also transports the fat-soluble vitamins throughout the body. Dietary fat also provides the essential fatty acids.

   The essential fatty acids are needed for normal growth and development in infants and children. They are used to maintain the structural parts of cell membranes, and they play a role in the immune system. From the essential fatty acids, the body makes hormone-like substances that are important for heart health.

   The body uses cholesterol to make bile acids, maintain cell membranes, make many hormones, and make vitamin D.
24. Chylomicrons transport mostly triglycerides and some cholesterol from the intestine through the blood to the body's cells. Low-density lipoprotein (LDL) carries cholesterol to the body's tissues. High-density lipoprotein (HDL) picks up cholesterol from the cells and takes it to the liver for removal.
25. To prevent heart disease: eat less than 10 percent of kcalories from saturated fatty acids and less than 1 percent from trans fat, increase intake of soluble fiber as found in oats, and consume foods fortified with sterols or stanols. Also, lose weight if you are overweight and be physically active.

   To prevent cancer, maintain a healthy weight, be physically active, eat a healthy diet with an emphasis on plant foods, and drink alcohol in moderation. Eat five or more servings of a variety of vegetables and fruits each day, choose whole grains, and limit intake of processed meats and red meats.

CHAPTER 5

1. True
2. True
3. False
4. True
5. False
6. False
7. False
8. True
9. False
10. False
11. False
12. True
13. False
14. True
15. True
16. In the body, the instructions to make proteins reside in the DNA in the nucleus of each cell. Each DNA molecule in the nucleus is housed in a chromosome. You have 23 pairs of chromosomes in each of your cells. Segments of each DNA molecule are called genes.
17. One ounce of meat, poultry, or fish is equal to one egg, one tablespoon peanut butter, ½ ounce of nuts or seeds, or ½ cup of cooked bean, peas, or tofu.
18. Plant foods are low in one or more essential amino acids. When certain plant foods, such as peanut butter and whole-wheat bread, are eaten over the course of a day, the limiting amino acid in each of these proteins is supplied by the other food. Such combinations are called complementary proteins.

Grains, Nuts, or Seeds + Legumes = A Complete Protein

19. Protein is part of most body structures; builds and maintains the body; is a part of many enzymes, hormones, and antibodies; transports substances around the body; maintains fluid and acid-base balance; can provide energy for the body; and helps in blood clotting.
20. Multiply your weight in pounds 20 0.36 to get your daily protein requirements.
21. Protein needs increase during growth, pregnancy, infections, fevers, and surgery.
22. **Beef:** bottom round steak or roast, flank steak, eye round roast, top sirloin steak, tenderloin filet, top round roast or steak, 90/10 or 95/5 ground beef

   Lamb or Veal: loin or rib chop, top round

   Pork: pork tenderloin, top loin chops, whole loin

   Poultry: breast or thigh (skinless or skin removed after cooking)

   Fish: all fish and shellfish

CHAPTER 6

1. False
2. False
3. True
4. True
5. False
6. True
7. True
8. False
9. True
10. True
11. a. vitamin B_{12}
    b. thiamin
    c. vitamin C
    d. vitamin B_{12}
    e. vitamin D
    f. niacin
    g. vitamin D
    h. vitamin C
    I. vitamin A
    j. Vitamin D
    k. vitamins C and E (also beta carotene)
    l. vitamins C & D
    m. vitamin A
    n. vitamin D

CHAPTER 7

1. False
2. False
3. True
4. False
5. True
6. False
7. True
8. False
9. False
10. a. Calcium, phosphorus, magnesium
    b. Calcium
    c. Sodium, potassium, chloride
    d. Sodium
    e. Potassium and magnesium
    f. Sodium, chloride, and iodine (if iodized)
    g. Iron

h. Calcium, magnesium, potassium

i. Sodium

j. Iron, zinc

k. Zinc

l. Iodine

CHAPTER 8

1. False
2. True
3. False
4. True
5. True
6. False
7. True
8. False
9. True
10. False
11. Answers will vary.
12. **Reduction** means boiling or simmering a liquid down to a smaller volume. In reducing, the simmering or boiling action causes some of the liquid to evaporate. The purpose may be to thicken the product, to concentrate the flavor, or both. A soup or sauce is often simmered for one or both reasons. The use of reduction eliminates thickeners and intensifies and increases the flavor so that you can serve a smaller portion.

    Searing means exposing the surfaces of a piece of meat to high heat in a hot pan with little or no oil or in a hot oven. Searing is done to give color and to produce a distinctive flavor. Dry searing can be done over high heat in a nonstick pan, using vegetable oil spray, or brushing with olive or canola oil.

    Deglazing means adding liquid to a hot pan with a fond—browned bits and caramelized drippings of meat and vegetables that are stuck to the bottom of a pan after sautéing or roasting (Figure 8-5). Any browned bits of food sticking to the pan are scraped up, adding flavor and color to the sauce.

    Sweating means cooking slowly in a small amount of fat over low or moderate heat without browning. You can sweat vegetables and other foods without fat. Instead, sweat in stock, juice, or wine.

    Pureeing of vegetables or starchy foods is commonly done as a way to thicken soups, stews, sauces, and other foods without using any fat, and also add flavor. While the food processor is useful for this process because you can slowly pulse the mixture, emulsion blenders and high-speed table blenders can make a very smooth puree with a silky texture.

    Rubs and marinades both add flavor to proteins, and marinades help tenderize. A marinade usually contains an acidic ingredient such as wine, beer, vinegar, citrus or tomato juice to break down tough meat. Other ingredients (stock or herbs) add flavor.

13. When sautéing, use a shallow pan to let moisture escape and allow space between the food items in the pan. Use a well-seasoned or nonstick pan and add about half a teaspoon or two sprays of oil per serving after preheating the pan.

    To dry sauté, heat a nonstick pan, spray with vegetable-oil cooking spray, then wipe out the excess with a paper towel. Heat the pan again, then add the food. If browning is not important, you can simmer the ingredient in a small amount of liquid such as wine, vermouth, flavored vinegar, juices, or defatted stock to bring out the flavor. Vegetables naturally high in water content, such as tomatoes and mushrooms, can be cooked with little or no added fluid at a very high heat.

    When sautéing or dry sautéing, add shallots, garlic, or other seasonings, then deglaze the pan with stock, wine, or another liquid and reduce to a sauce or add a previously reduced sauce to accompany your dish.

    To stir-fry, coat the cooking surface with a thin layer of oil. Peanut oil works well because it has a strong flavor (so that you can use a small amount) and a high smoking point. You can also use vegetable cooking spray and wipe away any excess. Have all your ingredients ready and next to you because this process is fast.

    Partially blanch the thick vegetables first (such as carrots or broccoli) so that they will cook completely without excessive browning. Preheat the equipment to a high temperature. Foods that require the longest cooking times—usually meat and poultry—should be the first ingredients you start to cook. Stir the food rapidly during cooking and don't overfill the pan. Use garlic, scallion, ginger, rice wine vinegar, low-sodium soy sauce, and chicken/vegetable stock for flavor.

    When roasting, always place meats and poultry on a rack so that the drippings fall to the bottom of the pan and the meat therefore doesn't cook in its own juices. Also, cooking on a rack allows for air circulation and more even cooking. Roasting adds rich flavors to meat, poultry, and seafood. Remove fat from the drippings and add stock to make a jus. Use rubs and marinades as appropriate and cook with vegetables.

    To flavor foods that will be broiled or grilled, consider marinades, rubs, herbs, and spices. Lean fish can be marinated, sprinkled with Japanese breadcrumbs, and glazed in a broiler. If grilling, consider placing soaked hardwood directly on the coals to add a smoked essence to the food. Serve immediately.

    Steam foods carefully so they are not overcooked. Add flavor to steamed and poached foods by adding herbs, spices, citrus juices, and other flavorful ingredients to the water.

    Foods to be braised are also often marinated before searing to develop flavor and tenderize the meat. When browning meat for braising, sear it in as little fat as possible without scorching and then place it in a covered braising pan to simmer in a small amount of liquid. To add flavor, place roasted vegetables, herbs, spices, and other flavoring ingredients in the bottom of the braising pan before adding the liquid.

CHAPTER 9

1. Change/add healthy preparation techniques.

   Change/add healthy cooking methods.

   Change an ingredient by reducing it, eliminating it, or replacing it.

   Add a new ingredient(s),

2. 1. Nutritional analysis

   2. Flavor

   3. Ingredient functions

   4. Cooking techniques

   5. Acceptability in a taste testing

3. The balanced pizza has a lot more vegetables, uses turkey sausage, which has less fat and saturated fat than regular sausage or pepperoni, has crushed red pepper and garlic for more flavor, has sauce with less sodium, and has a moderate amount of cheese.

4. To decrease fat, don't use a traditional crust—instead, omit a bottom crust and make a crumble topping using oatmeal, etc. as in a cobbler. Most fruit pie recipes use too much sugar. To use less sugar consider using more spices, such as apple pie spice, or adding dried fruit— such as dried cherries—to the pie. You can also make a glaze of a tablespoon of honey with water to add sweetness instead of 1 cup of sugar.

CHAPTER 10

1. • 800 kcalories or less

   • 20 to 35 percent or fewer kcalories from fat, emphasizing oils high in monounsaturated and polyunsaturated fats; 10 percent or less of total kcalories from saturated fat; no trans fat, 100 milligrams or less of cholesterol

   • 10 grams or more of fiber

   • 10 percent or fewer kcalories from added sugars

   • 15 to 25 percent kcalories from protein

   • 800 milligrams or less of sodium (about $1/3$ teaspoon of salt)

2. To develop balanced menu items, you can modify existing items to make them more nutritious or create new selections.

3. Any of the following questions.

   • Does the menu item possess a dynamic flavor profile? Taste is the key to customer acceptance and the successful marketing of these items.

   • Can the cooking staff execute this menu item efficiently and consistently?

   • Can the cooking staff execute this menu item efficiently and consistently?

   • Does the menu item blend well and complement the rest of the menu selections?

   • Does the menu item meet the food style and preferences of the guests?

   • Does the price of the item create a value for the customer?

   • How much preparation time is required for each item?

   • Is there a balance of color that complements the overall components?

   • Are there a variety of textures, such as crisp, crunchy, smooth, tender, chewy, creamy, and hard?

   • Is there a range of shapes, etc. adding an interesting visual appearance?

   • What is the flavor profile? Global, regional, classical, local?

   • Are the food combinations innovative, sensible, and easily acceptable?

   • Are the cooking techniques and methods of preparations varied and interesting?

4. Answers will vary.

5. Answers will vary.

6. Principles of presentation: height gives a plate interest and importance, color and shape are important, match the layout of the menu items with the shape of the plate, be sure your combination of foods make sense, use garnishes that enhance the plate but are not the focus.

   Five simple garnishes: a fruit sauce on angel food cake, red and golden beet chips, dried potato galette, dried potato lattice, fruit chips such as apples, etc.

CHAPTER 11

1. True
2. False
3. False
4. True
5. True
6. False
7. True
8. True
9. False
10. False
11. True
12. True

13. Answers will vary.

14. Fatty meats, bacon, sausage, poultry skin, butter, cheese, whole milk, ice cream, eggs, fried foods, doughnuts, Danish, croissants, biscuits, cakes, pies, cookies.

15. Regular soda, sweetened fruit drinks and iced tea, cakes, cookies, pies, ice cream, candy, gelatin salad, pancake syrup.

16. Milk, eggs, peanuts, tree nuts, fish, shellfish, soy, wheat

17. Determine who will answer guest's questions, who will keep recipe information up-to-date as well as check ingredients used in menu items, what steps everyone needs to take to avoid cross-contact, and how staff should handle an allergic reaction.

18. Cross-contact occurs when an allergen-containing food, such as peanuts, comes in contact with a safe food and contaminates it with the allergen. You should store, prepare, and serve safe foods separately from foods with allergens. Make sure all work surfaces are clean. Always use dedicated utensils, and don't contaminate safe foods with allergens that could be in frying oil or something in a container of jelly that is used by many other cooks. Avoid cross-contact in toasters and the microwave.

19. Milk, cream, ice cream, sherbet, yogurt, some cheeses, butter. Sometimes in breads and baked goods, cereals, baking mixes, instant potatoes, lunchmeats, frozen dinners, margarine.

20. Use a variety of plant protein sources and vegetables, choose low-fat and fat-free varieties of dairy products, moderate the use of eggs, offer dishes made with soybean-based products, use recipe books and websites, and uses cuisines of other countries.

CHAPTER 12

1. True
2. False
3. True
4. False
5. False
6. True
7. True
8. False
9. 10 tips to lose weight: eat foods high in water and fiber, eat smaller portions, eat breakfast, avoid fried foods, moderate use of alcohol, don't overly restrict kcalories, get 250-300 minutes per week of moderate-intensity activity, do strength training, set realistic goals, self monitor, use cue control, reward yourself, plan meals, get support from others, lose weight with a group.
10. No one diet is better than any other.

CHAPTER 13

1. False
2. True
3. True
4. False
5. False
6. True
7. True
8. False
9. True
10. False
11. True
12. *Menu planning for preschoolers:* Offer simply prepared foods; avoid casseroles or any foods that are mixed together; present new foods repeatedly if the child rejects it; offer colorful foods; serve vegetables raw and cut up as finger foods; use brightly colored and mild-flavored vegetables; provide soft and crispy/chewy foods; avoid strong-flavored and highly salted foods; offer whole grain carbohydrates; make sure foods such as mashed potatoes have no lumps; cut foods up well for children under 4; serve foods warm—not hot; serve good sources of fiber, iron, and calcium; and minimize choking hazards for children under 4 by avoiding large chunks of any food.

    Menu planning for school-age children: Serve a wide variety of foods; pick nutritious snacks such as popcorn; balance menu items that are higher in fat with those containing less fat; pay attention to serving sizes; offer foods high in iron, fiber, and calcium; offer a breakfast with lots of protein and whole-grain carbohydrates; get kids involved in planning and making meals.

    Menu planning for adolescents: Emphasize high-fiber carbohydrates such as whole-grain breads and cereals and fruits; offer well-trimmed lean beef, poultry, and fish; offer nonfat and low-fat milk along with other dairy foods for calcium and vitamin D; have nutritious snacks available, emphasize quick and nutritious breakfast; serve good sources of iron and potassium.

13. *Menu planning for older adults:* Offer moderate portions and half portions; *emphasize high fiber foods; moderate the use of fat; offer dairy products as they contain nutrients of concern (calcium, vitamin D, protein, vitamin B_{12}, potassium); offer adequate protein using both animal and vegetable sources; moderate the use of salt and salty ingredients; use herbs and spices; offer a variety of foods; offer a variety of beverages; and offer softer foods for people with swallowing/chewing issues.*

14. *Menu planning for athletes: Include a variety of foods from MyPlate; offer good sources of carbohydrates emphasizing whole grains, etc.; offer lean proteins; offer a variety of fluids; make sure iodized salt is on the table; include sources of iron, calcium, and zinc at each meal; at the precompetition meal offer lots of easily digestible carbohydrate-rich foods, some protein, and don't overdo the fat or fiber; emphasize fluids and balanced meals after competitions and workouts.*

Absorption The passage of digested nutrients through the walls of the intestines or stomach into the body's tissues. Nutrients are then transported through the body in the blood to the cells.

Acceptable Macronutrient Distribution Ranges (AMDR) The percent of total kilocalories coming from carbohydrate, fat, or protein that is associated with a reduced risk of chronic disease while providing adequate intake.

Acid–base balance The process by which the body buffers the acids and bases normally produced in the body so that the blood is neither too acidic nor too basic.

Added sugars Sugars, syrups, and other kcaloric sweeteners that are added to foods during processing, preparation, or consumed separately; do not include naturally occurring sugars such as those in fruit or milk.

Adequate diet A diet that provides enough kcalories, essential nutrients, and fiber to maintain health.

Adequate Intake (AI) The dietary intake that is used when there is not enough scientific research to support an RDA.

Adipose cell A type of cell in the body that readily takes up and stores triglycerides; also called a fat cell.

Agar-agar A seaweed derivative that is flavorless and becomes gelatinous when dissolved in water, heated, and then cooled. Used in cooking as a thickener.

Alpha-linolenic acid (ALA) An omega-3 fatty acid found in canola, flaxseed, soybean, and walnut oils; ground flaxseed, walnuts, and soy products.

Alpha-tocopherol The active vitamin E compound; powerful anti-oxidant.

Amino acid pool The overall amount of amino acids distributed in the blood, organs, and body cells.

Amino acids The building blocks of protein.

Anorexia nervosa A life-threatening eating disorder that is characterized by self-starvation and excessive weight loss.

Antibodies Proteins in the blood that bind with foreign bodies or invaders.

Antioxidant A compound that combines with oxygen to prevent oxygen from oxidizing or destroying importance substances; antioxidants prevent the oxidation of unsaturated fatty acids in the cell membrane, DNA, and other cell parts that substances called free radicals try to destroy.

Anus The opening of the digestive tract through which feces travel out of the body.

Atherosclerosis A disease in which arteries have plaque buildup along the arterial walls.

Attention deficit hyperactivity disorder (ADHD) In children, a problem with inattentiveness, overactivity, impulsivity, or a combination.

Baba ganoush Middle Eastern dish made primarily of roasted eggplant and tahini, a paste made from sesame seeds.

Balanced diet A diet in which foods are chosen to provide kcalories, essential nutrients, and fiber in the right proportions.

Basal metabolism The minimum energy needed by the body for vital functions when at rest and awake.

Behavior modification Use of demonstrated behavior change techniques to increase or decrease the frequency of behaviors, such as restricting eating to the kitchen or dining room table to eat fewer kcalories.

Beta-carotene A precursor of vitamin A that functions as an antioxidant in the body; the most abundant carotenoid.

Bile A substance that is released when fat enters the small intestine because it emulsifies, or breaks up, fat.

Binge-eating disorder An eating disorder characterized by episodes of uncontrolled eating or bingeing.

Bioavailability The degree to which a nutrient is absorbed and available to be used in the body.

Blood glucose level The amount of glucose found in the blood; glucose is vital to the proper functioning of the body, as it provides most of your energy.

Body mass index (BMI) A measure of weight in kilograms relative to height in meters squared. BMI is considered a reasonably reliable indicator of total body fat, which is related to the risk of disease and death. BMI categories include underweight, healthy weight, overweight, and obese.

Bottled water Water that is intended for human consumption that is tested for safety and sealed in bottles with no added ingredients except that it may optionally contain safe and suitable antimicrobial agents. May be used as an ingredient in beverages such as flavored bottled waters.

Braising Searing or browning food (usually meat) in a small amount of oil or its own fat and then adding liquid and simmering until done.

Bran In grains, the part that cover the grain and contains much fiber and other nutrients.

Broiling Cooking with radiant heat from above.

Bulimia nervosa An eating disorder characterized by a destructive pattern of excessive overeating followed by vomiting or other "purging" behaviors to control weight.

Butterhead A type of lettuce with generally small, loose and tender leaves with a delicate sweet flavor.

Caffeine A naturally occurring stimulant found in the leaves, seeds, or fruits of over

60 plants around the world—such as the coffee bean, tea leaf, coco bean, and kola nut. Is mildly addicting and has side effects at high doses.

Carbohydrates   A large class of nutrients, including sugars, starch, and fibers, that function as the body's primary source of energy.

Carotenoids   A class of pigments that contribute a red, orange, or yellow color to fruits and vegetables; some can be converted to vitamin A in the body.

Carpaccio   Very thin slices of meat, fish, or vegetables (massaged or pounded), served raw.

Celiac disease   An inherited autoimmune disease. When a person with celiac disease consumes any food, beverage, or medication containing gluten (found in wheat, barley, and rye—or oats that are contaminated by contact with these other grains), his or her immune system is "triggered" and responds by damaging the lining of the intestinal tract. Symptoms include abdominal pain, diarrhea, and/or a severe skin rash.

Cholesterol   The most abundant sterol (a class of lipids); a soft, waxy substance found only in foods of animal origin; it is present in every cell of the body.

Chromosome   Rod-like structures in the nucleus that house DNA.

Chutney   A strongly spiced sauce from India, either hot or sweet, that is made with fruits, vegetables, and herbs.

Chylomicron   The lipoprotein responsible for carrying mostly triglycerides (and some cholesterol) from the intestines through the lymph and blood to the body's cells.

Coenzyme   A molecule that combines with an enzyme to make the enzyme active.

Cofactor   A substance that is necessary for the activity of an enzyme.

Collagen   The most abundant protein in the body; a fibrous protein that is a component of skin, bone, teeth, ligaments, tendons, and other connective structures.

Complementary proteins   The ability of two protein foods to make up for the lack of certain amino acids in each other when eaten over the course of a day.

Complete proteins   Food proteins, generally from animals, that provide all the essential amino acids in the proportions needed by the body.

Complex carbohydrates or polysaccharides   Carbohydrates made of chains of sugars, including starches and fibers.

Compote   A preparation of fruit, fresh or dried, cooked in syrup flavored with spices such as vanilla, citrus peel, toasted spices, vanilla bean, or liqueur.

Confit   Classically, a food cooked in its own fat, such as duck or pork. Reinvented, a food saturated with an ingredient, such as vinegar in vegetables or sugar or alcohol in fruits.

Constipation   Infrequent passage of feces.

Cos or romaine   A type of lettuce that forms an upright, elongated head with tall leaves.

Coulis   A sauce made of a puree of vegetables or fruits.

Court bouillon   Water containing herbs, seasonings, and an acidic product used to cook fish.

Crisphead   .A type of lettuce, such as iceberg, that contain curved, overlapping leaves that form crispy, firm round heads.

Cross-contact   This occurs when an allergen-containing food, such as peanuts, comes in contact with a safe food and contaminates it with the allergen.

Crystalline vitamin B$_{12}$ The type of vitamin B$_{12}$ used in supplements and fortified foods; much more easily absorbed than natural vitamin B$_{12}$ in foods.

Culture   The behaviors and beliefs of a certain social, ethnic, or age group.

Daily Value (DV)   A set of nutrient-intake values developed by the Food and Drug Administration that are used as a reference for expressing nutrition content on nutrition labels.

DASH diet (Dietary Approaches to Stop Hypertension)   A diet that helps reduce blood pressure and emphasizes vegetables, fruits, and low-fat milk and milk products; includes whole grains, poultry, seafood, and nuts; and is lower in sodium, red and processed meats, sweets, and sugar-containing beverages than the typical American diet.

Deglazing   Adding liquid to the hot pan used in making sauces and meat dishes; any browned bits of food sticking to the pan are scraped up and added to the liquid, adding flavor and color.

Denaturation   The process in which protein structure changes due to high temperature, addition of acid, or whipping,

resulting in the protein becoming firm, shrinking in size, and losing moisture.

Deoxyribonucleic acid (DNA)   Molecules in the nucleus of cells that carry your genetic information.

Diabetes   A disorder of carbohydrate metabolism characterized by high blood sugar levels and inadequate or ineffective insulin.

Diet   The food and beverages you normally eat and drink.

Dietary fiber   Polysaccharides (and lignin) found in plants that are not digested by human digestive enzymes.

Dietary Guidelines for Americans, 2010   A set of dietary recommendations for Americans (ages 2 and older) that is updated every five years and designed to promote dietary changes to improve health, reduce the risk of chronic diseases, and reduce the number of people who are overweight or obese.

Dietary recommendations   Guidelines that discuss food groups, foods, and nutrients to eat for optimal health.

Dietary Reference Intakes (DRIs)   Nutrient standards for healthy Americans and Canadians that estimate how much you need daily of various nutrients, as well as when you might be taking in excessive amounts.

Digestion   The process by which food is broken down into its components in the mouth, stomach, and small intestine with the help of digestive enzymes and fluids.

Disaccharides   Pairs of monosaccharides linked together, including sucrose, maltose, and lactose.

Docosahexaenoic acid (DHA) and eicosapentaenoic acid (EPA)   Two omega-3 fatty acids found mostly in fish that are especially important for heart health, growth, and proper brain and eye development during pregnancy and infancy.

Dry sautéing   Sautéing with a minimal amount of vegetable oil cooking spray or a small amount of liquid such as wine, flavored vinegar, or defatted stock.

Eating pattern   The combination of foods and beverages that constitute an individual's complete dietary intake over time.

Electrolytes   Minerals in your body that carry an electric charge when dissolved in water. Examples include sodium, potassium, and chloride.

Empty-kcalorie foods Foods with added solid fats or sugar that provide few nutrients for the number of kcalories they contain.

Emulsifiers Substances that keep fats and water from separating in a mixture such as oil and vinegar.

Endosperm In grains, a large center area high in starch.

Energy drinks Drinks that contain varying amounts of caffeine and/or other plant-based stimulants such as ginseng.

Energy-yielding nutrients Nutrients that can be burned as fuel to provide energy for the body, including carbohydrates, fats, and proteins.

Enhanced drinks Drinks such as fruit drinks, teas, dairy drinks, and waters that contain added vitamins, minerals, phytochemicals, and/or herbs.

En papillote Allowing food that is sealed in a folded packet to be steamed so that it cooks in its own flavorful liquids.

Enriched A food to which nutrients are added to replace the same nutrients that were lost in processing.

Enzymes Compounds that speed up the breaking down of food so that nutrients can be absorbed. Also perform other functions in the body.

Esophagus The muscular tube that connects the pharynx to the stomach.

Essential amino acids Amino acids that either cannot be made in the body or cannot be made in the quantities needed by the body; must be obtained in foods.

Essential fatty acids Two polyunsaturated fatty acids (linoleic and alpha-linolenic acids) that the body can't make so they must be consumed in the diet; vital to growth and development, maintenance of cell membranes, and the immune system.

Essential nutrients Nutrients that either cannot be made in the body or cannot be made in the quantities needed by the body; therefore, we must obtain them from food.

Farro A grain that is the ancestor of modern wheat.

Fat A lipid that is solid at room temperature, generally of animal origin, such as butter.

Fat-soluble vitamins A group of vitamins that generally occurs in foods containing fats; include vitamins A, D, E, and K.

Fatty acids Major component of most lipids. Three fatty acids are present in each triglyceride.

Fitness waters Drinks that contain fewer kcalories, carbohydrates, and electrolytes than sports drinks but taste better than plain water due to some sugar and artificial sweeteners.

Flambée To light alcohol in preparation.

Flavor An attribute of a food that includes its taste, smell, feel in the mouth, texture, temperature, and even the sounds made when it is chewed.

Flavorings Substances used in cooking to add a new flavor or modify the original flavor.

Fluid balance The process of maintaining the proper amount of water in each of the three body compartments: inside the cells, outside the cells, and in the blood vessels.

Food allergens Specific components of food that cause food allergies.

Food allergies An abnormal response to a food triggered by your body's immune system. Allergic reactions to food can sometimes cause serious illness and death.

Food guides Guidelines that tell us the kinds and amounts of foods to make a nutritionally adequate diet. They are typically based on current dietary recommendations, the nutrient content of foods, and the eating habits of the targeted population.

Food intolerance An unpleasant reaction to food that, unlike a food allergy, does not involve an immune system response. Symptoms can include gas, bloating, constipation, or dizziness.

Food jags A habit of young children in which they eat mostly one food for a period of time.

Fortified A food to which nutrients are added that were not present originally or nutrients are added that increase the amount already present.

Free radicals An unstable compound resulting from cell metabolism and the functioning of the immune system that reacts quickly with other molecules in the body.

Fructose A monosaccharide found in fruits and honey; the sweetest natural sugar.

Functional beverages Drinks that have been enhanced with added ingredients to provide specific health benefits beyond general nutrition.

Functional fibers Fibers (such as oat fiber or vegetable gums) extracted from plants and then added to a wide variety of foods, such as juices, yogurt, bread, and cereal.

Galactose A monosaccharide found linked to glucose to form lactose (milk sugar).

Gastrique A mixture of caramelized sugar and vinegar used to flavor sauces.

Gastrointestinal tract A hollow tube running down the middle of the body in which digestion of food and absorption of nutrients take place.

Gelatinization A process in which starches, when heated in liquid, absorb water and swell in size.

Genes A tiny section of DNA that has a code to make a specific product—usually a protein.

Germ In grains, the area of the kernel rich in vitamins and minerals that sprouts when allowed to germinate.

Glaze A stock reduced to a thick, gelatinous consistency with flavoring and seasonings.

Glucagon A hormone that stimulates the liver to convert glycogen to glucose and release the glucose into the bloodstream to bring the blood glucose level to normal levels.

Glucose The most significant monosaccharide; the body's primary source of energy.

Gluten sensitivity (also called non-celiac gluten intolerance) A condition that is not a wheat allergy or an autoimmune disease that does not seem to damage the intestines. Common symptoms of gluten sensitivity include abdominal pain, diarrhea, fatigue, and headaches.

Glycemic load An index of the glycemic response that occurs after eating a specific food and which takes into account portion size.

Glycemic response How quickly, how high, and how long your blood sugar rises after eating.

Glycogen The storage form of glucose in the body; stored in the liver and muscles.

Grilling Cooking on an open grid over a heat source.

Growth spurts Periods of rapid growth.

Health claims Claims on food labels that state that certain foods or food substances—as part of an overall healthy diet—may affect the risk of disease or health-related condition.

Heart attack When flow of blood to a section of the heart muscle suddenly becomes blocked.

Heartburn A painful burning feeling in your chest or throat that happens when stomach acid backs up into your esophagus, the tube that carries food from your mouth to your stomach.

Heme iron The main form of iron in animal foods; it is absorbed and used more readily than iron in plant foods.

Hemoglobin A protein in red blood cells that carries oxygen to the body's cells.

Herbs The leafy parts of certain plants that grow in temperate climates. They are used to season and flavor foods.

High-density lipoprotein (HDL) Lipoproteins that carry cholesterol away from body cells and tissues to the liver for excretion from the body. Called "good cholesterol."

High-fructose corn syrup Corn syrup that has been treated with an enzyme that converts part of the glucose it contains to fructose; found in some sweetened foods.

Homeostasis A constant internal environment in the body.

Hormones Chemical messengers secreted into the bloodstream by various organs that travel to a target organ and influence what it does.

Hydrogenation A process in which liquid vegetable oils are converted into solid fats (such as margarine) by the use of heat, hydrogen, and catalysts.

Hypoglycemia Abnormally low blood glucose levels.

Immune response The body's response to a foreign substance, such as a virus, in the body.

Incomplete proteins Food proteins that contain at least one limiting amino acid.

Insoluble fiber A category of fiber that does not swell in water; found in wheat bran, whole grains, legumes, many fruits and vegetables, and seeds.

Insulin A hormone that is necessary for glucose to leave the bloodstream and enter the body's cells.

Intrinsic factor A protein secreted by stomach cells that is necessary for the absorption of vitamin B_{12}.

Ion An atom or group of atoms carrying a positive or negative electrical charge.

Iron deficiency A condition in which iron stores are used up.

Iron-deficiency anemia A condition in which red blood cells are smaller than usual and carry less hemoglobin; may result from severely depleted iron stores or blood loss. Symptoms include fatigue, decreased work and school performance, and decreased immune function.

Iron overload (hemochromatosis) A common genetic disease in which individuals absorb too much iron from their food and supplements, which can damage the liver. It is more common in men than women.

Jus lié Juices from a roast thickened lightly with a starch slurry.

Kamut A grain that is an ancient relative to wheat.

Kilocalorie balance The balance between kcalories consumed in foods and beverages and kcalories expended through physical activity and metabolic processes.

Kilocalories A unit of measure used to express the amount of energy found in different foods.

Lactase An enzyme needed to split lactose into its components in the intestines.

Lactose A disaccharide found in milk and milk products formed by joining glucose with galactose.

Lactose intolerance A condition caused by a deficiency of the enzyme lactase, resulting in symptoms such as flatulence and diarrhea after drinking milk or eating most dairy products.

Large intestine (colon) The part of the gastrointestinal tract between the small intestine and the rectum. Contributes to the build-up of cholesterol in your arteries which can lead to heart disease.

Leaf lettuce A type of lettuce with crisp leaves loosely arranged on a stalk; the leaves are often colorful.

Legumes Plant species that have seed pods that split along both sides when ripe. Some of the more common legumes are beans, peas, and lentils.

Limiting amino acid An essential amino acid in lowest concentration in a protein

that limits the protein's usefulness unless another food in the diet contains it.

Linoleic acid An omega-6 fatty acid found in vegetable oils such as soybean, sunflower, and corn oils.

Lipids A group of fatty substances, including triglycerides, cholesterol, and lecithi, that provide a rich source of energy and structure to cells; not soluble in water.

Lipoprotein Protein-coated packages that carry fat and cholesterol through the bloodstream; the body makes four types classified according to their density.

Low-density lipoprotein (LDL) Lipoproteins that contain most of the cholesterol in the blood and carry cholesterol to the body's tissues.

Macronutrients Nutrients needed by the body in large amounts, including carbohydrates, lipids, and proteins.

Major minerals Minerals needed in relatively large amounts in the diet—over 100 milligrams daily.

Manchego cheese A Spanish hard cheese made from sheep's milk.

Marinade A seasoned liquid used before cooking to flavor and moisten foods; usually based on an acidic ingredient.

Metabolism All the chemical processes by which nutrients are used to support life.

Microgreens The first true leaves that develop after a seed sprouts.

Micronutrients Nutrients needed by the body in small amounts, including vitamins and minerals.

Minerals Noncaloric, inorganic chemical substances found in a wide variety of foods; needed to regulate body processes, maintain the body, and allow for growth and reproduction.

Mirepoix A mixture of rough-cut or diced vegetables, herbs, and spices, used for flavoring stocks, sauces, soups, and other foods.

Moderate diet A diet that avoids excessive amounts of kcalories or any particular food/food group or nutrient.

Mojo A spicy Caribbean and South American sauce; it is a mixture of garlic, citrus juice, oil, and fresh herbs.

Monosaccharides Simple sugars, including glucose, fructose, and galactose, which are the building blocks for other carbohydrates, such as disaccharides and starch.

Monounsaturated fat   A triglyceride made of mostly monounsaturated fatty acids.

Mouthfeel   How the texture of a food is perceived in the mouth.

Myoglobin   A muscle protein that stores and carries oxygen that the muscles will use to contract.

MyPlate   A food guide developed by the US Department of Agriculture as a guide to the amounts of different types of foods needed to provide an adequate diet and comply with current nutrition recommendations.

Natural   Meat or poultry products that contain no artificial ingredient or added color and are only minimally processed; for other foods, *natural* means that there are no added colors, artificial flavors, or synthetic ingredients.

Neural tube defects   Diseases in which the brain and/or spinal cord form improperly in early pregnancy.

Niacin equivalents   The unit for measuring niacin. One niacin equivalent is equal to 1 milligram of niacin or 60 milligrams of tryptophan.

Nonessential amino acids   Amino acids that can be made in the body.

Nonheme iron   A form of iron found in all plant sources of iron and also as some of the iron in animal foods.

Nutrient content claims   Claims on food labels about the nutrient composition of a food, such as low in kcalories. Regulated by the Food and Drug Administration.

Nutrient density   A measure of the nutrients provided in a food per kcalorie of that food.

Nutrients   The nourishing substances in food that provide energy and promote the growth and maintenance of your body.

Nutrition   A science that studies nutrients and other substances in foods and in the body and the way those nutrients relate to health and disease. Nutrition also explores why you choose particular foods and the type of diet you eat.

Nutrition Facts   Mandatory nutrition information that appears on all packaged foods and beverages that can guide you in selecting healthy foods.

Obese   Having a body mass index of 30 or greater.

Oil   A lipid that is usually liquid at room temperature, generally of plant origin.

Omega-3 fatty acids   Fatty acid with double bonds after the third carbon in the chain; tend to be inadequate in many American diets.

Omega-6 fatty acids   Fatty acids with double bonds after the sixth carbon in the chain; Americans get plenty of omega-6 fatty acids in diet.

Organic foods   Food produced without antibiotic or growth hormones, most conventional pesticides, fertilizers made with synthetic ingredients or sewage sludge, bioengineering, or ionizing radiation.

Osteomalacia   Softening of the bones in adults due to a lack of vitamin D.

Osteoporosis   The most common bone disease, characterized by loss of bone density and strength, it is associated with debilitating fractures (such as hip and wrist), especially in people 50 years and older.

Overweight   Having a body mass index of 25 to 29.9.

Panada   A paste made by mixing bread crumbs, flour, and possibly other ingredients with water, milk, or stock. It is used as a binder for making soups or thickening sauces.

Peptic ulcer   A sore on the lining of the stomach or small intestine.

Peristalsis   Involuntary muscular contraction that forces food through the entire digestive system.

Pharynx   A passageway that connects your mouth to the esophagus.

Photosynthesis   A process in which plants use energy from sunlight to convert carbon dioxide and water to carbohydrate.

Phytochemicals   A wide variety of compounds produced by plants that promote health.

Plancha   Flattop grill that originated in Spain.

Plaque   (1) The sticky film of bacteria, protein, and polysaccharides that forms on the teeth and gums. (2) Deposits on arterial walls that contain cholesterol and other biological debris.

Poaching   Cooking a food submerged in liquid at a temperature of 160° to 180°F (71° to 82°C).

Polyunsaturated fat   A triglyceride made of mostly polyunsaturated fatty acids.

Precompetition meal   The meal for an athlete closest to the time of a competition or event.

Precursors   Forms of vitamins that the body converts to active vitamin forms.

Processed foods   Foods that have been prepared using a certain procedure such as canning, cooking, freezing, dehydration, or milling.

Protein   Major structural component of the body's cells that is made of amino acids assembled in chains. It performs many functions; is particularly rich in animal foods.

Pureeing   Mashing or straining a food to a smooth pulp.

Qualified health claims   Health claims for which there is not yet well-established evidence and that must be accompanied by an explanatory statement to ensure that they do not mislead consumers.

Recommended Dietary Allowance (RDA)   The dietary intake value that represents what you need to take in on a daily basis.

Rectum   The last section of the large intestine, in which feces, the waste products of digestion, are stored until elimination.

Reduction   Boiling or simmering a liquid down to a smaller volume.

Refined grains   Grains and grain products missing the bran, germ, and/or endosperm; any grain product that is not a whole grain.

Relish   Cooked, pickled foods typically served as a cold condiment.

Retinol   An animal form of vitamin A; one of the active forms of vitamin A in the body.

Rickets   A childhood disease in which bones do not grow normally, resulting in bowed legs and knock knees; it is generally caused by a vitamin D deficiency.

Roasting   Cooking with heated dry air such as in an oven.

Roux   A thickener of fat and flour in a one-to-one ratio by weight.

Rub   A dry marinade made of herbs and spices (and other seasonings), sometimes moistened with a little oil, vinegar, mustard, or other flavoring liquid, that is rubbed or patted on the surface of meat, poultry, or fish (which is then refrigerated and cooked at a later time).

Saliva A fluid secreted into the mouth from the salivary glands that contains important digestive enzymes and lubricates the food so that it may readily pass down the esophagus.

Salsa Mixtures of vegetables and/or fruits and flavor ingredients.

Satiety The feeling of fullness and satisfaction after eating.

Saturated fat A triglyceride made of mostly saturated fatty acids.

Saturated fatty acids A type of fat found mostly in animal foods such as meat, chicken, eggs, milk, butter, and cheese, as well as some oils that have been hydrogenated to make stick margarine and shortening. Known to raise levels of bad cholesterol in the blood, which is a risk factor for cardiovascular disease. A type of fatty acid that is filled to capacity with hydrogens.

Sautéing Cooking foods in single portions or small pieces very quickly on high heat with a small amount of fat.

Scurvy A vitamin C deficiency disease marked by bleeding gums, weakness, loose teeth, and broken capillaries under the skin.

Searing Exposing the surfaces of a piece of meat to high heat in a hot pan with little or no oil, or in a hot oven, to give the meat color and a distinctive flavor.

Seasonings Substances used in cooking to bring out a flavor that is already present.

Simple carbohydrates or sugars A form of carbohydrate that includes sugars occurring naturally in foods, such as fructose in fruits, as well as sugars added to foods, such as brown sugar in a cookie.

Small intestine The digestive tract organ that extends from the stomach to the opening of the large intestine. The site of most digestion and absorption.

Smoke-roasting Cooking with dry heat in the presence of wood smoke.

Solid fats Fats found in most animal foods but also can be made from vegetable oils through hydrogenation. Some common solid fats include butter, beef fat, stick margarine, and shortening. Solid fats contain more saturated and/or trans fats than vegetable oils.

Soluble fiber A category of fiber that swells in water into a gel-like substance; digested by some bacteria in the colon; found in oats, barley, beans, and many fruits and vegetables.

Spices The roots, bark, seeds, flowers, buds, and fruits of certain tropical plants. They are used to season and flavor foods.

Sports drinks Drinks that contain a dilute mixture of carbohydrate and electrolytes that are designed to be used during exercise lasting 60 minutes or more.

Sprouts Young plants that have just emerged from their seeds before any leaves grow.

Starch A complex carbohydrate made up of a long chain of glucoses linked together; found in grains, legumes, vegetables, and some fruits.

Steaming Cooking by direct contact with steam.

Stem lettuce A type of lettuce, also called Chinese lettuce, grown mostly for its thick stem.

Stir-frying An Asian cooking method involving a quick sauté over high heat, occasionally followed by a brief steam in a flavored sauce.

Stomach J-shaped muscular sac that holds about 4 cups of food when full and helps in digestion. Some alcohol is absorbed through the stomach.

Stroke Damage to brain cells resulting from an interruption of blood flow to the brain.

Sucrose A disaccharide formed by linking glucose and fructose; commonly called white sugar, table sugar, or simply sugar.

Sweating Cooking slowly in a small amount of fat over low or moderate heat without browning. Vegetables and other foods can be sweated in stock, juice, or wine as well.

Taste Sensations perceived by the taste buds on the tongue.

Taste buds Clusters of cells found on the tongue, cheeks, throat, and roof of the mouth. Each taste bud houses 60 to 100 receptor cells that bind food molecules dissolved in saliva and alert the brain to interpret them.

Texture Those physical properties of food that can be felt with the tongue, mouth, teeth, or fingers—such as tender, juicy, or firm.

Tolerable Upper Intake Level (UL) The maximum intake level above which you may feel adverse health effects.

Trace minerals Minerals needed in smaller amounts in the diet—less than 100 milligrams daily.

Trans fats (trans fatty acids) Unsaturated fatty acids found mostly in foods that have been hydrogenated such as margarine. Act like saturated fats in the body to raise blood cholesterol levels. Occur naturally in low levels in meat and dairy.

Triglycerides The major form of lipid in food and in the body; it is made of three fatty acids attached to a glycerol backbone.

Triticale A hybrid of wheat and rye.

Umami A taste often referred to as "savory" that is characteristic of monosodium glutamate and is associated with meats, mushrooms, tomatoes, Parmesan cheese, and other foods. It is a basic taste along with sweet, sour, salty, and bitter.

Varied diet A diet in which you eat a wide selection of foods to get necessary nutrients.

Very-low-kcalorie diet A doctor-supervised diet that usually uses commercially prepared formulas to promote rapid weight loss in obese patients. These formulas, usually liquid shakes or bars, replace all food intake for several weeks or months.

Villi Tiny fingerlike projections in the wall of the small intestines that are involved in absorption.

Vitamins Noncaloric, organic nutrients found in a wide variety of foods that are essential in small quantities to regulate body processes, maintain the body, and allow for growth and reproduction.

Water-soluble vitamins A group of vitamins that are soluble in water and are not stored appreciably in the body; include vitamin C, thiamin, riboflavin, niacin, vitamin B_6, folate, vitamin B_{12}, pantothenic acid, and biotin.

Whole foods Foods as we get them from nature; some may be minimally processed.

Whole grains Grains and grain products made from the entire grain seed, usually called the kernel, which consists of the bran, germ, and endosperm.